POSTIMPLANTATION DEVELOPMENT IN THE MOUSE

The Ciba Foundation is an international scientific and educational charity. It was established in 1947 by the Swiss chemical and pharmaceutical company of CIBA Limited — now CIBA-GEIGY Limited. The Foundation operates independently in London under English trust law.

The Ciba Foundation exists to promote international cooperation in biological, medical and chemical research. It organizes about eight international multidisciplinary symposia each year on topics that seem ready for discussion by a small group of research workers. The papers and discussions are published in the Ciba Foundation symposium series. The Foundation also holds many shorter meetings (not published), organized by the Foundation itself or by outside scientific organizations. The staff always welcome suggestions for future meetings.

The Foundation's house at 41 Portland Place, London W1N 4BN, provides facilities for meetings of all kinds. Its Media Resource Service supplies information to journalists on all scientific and technological topics. The library, open five days a week to any graduate in science or medicine, also provides information on scientific meetings throughout the world and answers general enquiries on biomedical and chemical subjects. Scientists from any part of the world may stay in the house during working visits to London.

Ciba Foundation Symposium 165

POST-IMPLANTATION DEVELOPMENT IN THE MOUSE

A Wiley–Interscience Publication

1992

JOHN WILEY & SONS

Chichester · New York · Brisbane · Toronto · Singapore

©Ciba Foundation 1992

Published in 1992 by John Wiley & Sons Ltd.
Baffins Lane, Chichester
West Sussex PO19 1UD, England

Other Wiley Editorial Offices

John Wiley & Sons, Inc., 605 Third Avenue,
New York, NY 10158-0012, USA

Jacaranda Wiley Ltd, G.P.O. Box 859, Brisbane,
Queensland 4001, Australia

John Wiley & Sons (Canada) Ltd, 22 Worcester Road,
Rexdale, Ontario M9W 1L1, Canada

John Wiley & Sons (SEA) Pte Ltd, 37 Jalan Pemimpin #05-04,
Block B, Union Industrial Building, Singapore 2057

Suggested series entry for library catalogues:
Ciba Foundation Symposia

Ciba Foundation Symposium 165
x + 315 pages, 69 figures, 22 tables

Library of Congress Cataloging-in-Publication Data
Postimplantation development in the mouse.
      p.   cm.—(Ciba Foundation symposium; 165)
      Editors, Derek J. Chadwick and Joan Marsh.
      "A Wiley-Interscience publication.
      Includes bibliographical references and indexes.
      ISBN 0-471-93384-8
      1. Mice—Embryos—Congresses.   2. Mice—Development—Congresses.
3. Developmental cytology—Congresses.   4. Mice—Embryos—
—Physiology—Congresses.   I. Chadwick, Derek.   II. Marsh, Joan.
III. Series.
      [DNLM: 1. Embryo—growth & development—congresses.   2. Gene
Expression—Congresses.   3. Mice—embryology—congresses.
4. Postimplantation Phase—physiology—congresses. W3 C161F v. 165]
QL971.P67 1992
599.32′33—dc20
DNLM/DLC
for Library of Congress   92-113
                                                                                CIP

British Library Cataloguing in Publication Data
A catalogue record for this book is available from the British Library

ISBN 0 471 93384 8

Phototypeset by Dobbie Typesetting Limited, Tavistock, Devon.
Printed and bound in Great Britain by Biddles Ltd., Guildford.

Contents

v

Participants

R. Balling   Department of Developmental Biology, Max-Planck Institute of Immunobiology, Stübeweg 51, 7800 Freiburg, Germany

R. S. P. Beddington   AFRC Centre for Genome Research, The Old ABRO Building, King's Buildings, West Mains Road, Edinburgh EH9 3JQ, UK

A. Bradley   Institute of Molecular Genetics, Baylor College of Medicine, One Baylor Plaza, Houston, TX 77030, USA

N. A. Brown   MRC Experimental Embryology & Teratology Unit, St George's Hospital Medical School, Cranmer Terrace, London SW17 0RE, UK

M. E. Buckingham   Department of Molecular Biology, Institut Pasteur, 28 rue du Dr Roux, F-75724, Paris 15, France

A. J. Copp   ICRF Developmental Biology Unit, Department of Zoology, University of Oxford, South Parks Road, Oxford OX1 3PS, UK

M. Evans   University of Cambridge, Wellcome/CRC Institute, Tennis Court Road, Cambridge CB2 1QR, UK

P. N. Goodfellow   Laboratory of Human Molecular Genetics, Imperial Cancer Research Fund, PO Box 123, Lincoln's Inn Fields, London WC2A 3PX, UK

N. Hastie   MRC Human Genetics Unit, Western General Hospital, Crewe Road, Edinburgh EH4 2XU, UK

B. G. Herrmann   Max-Planck Institut für Entwicklungsbiologie, Spemannstrasse 35, D-7400, Tübingen, Germany

B. L. M. Hogan   Department of Cell Biology, Vanderbilt University Medical School, Nashville, TN 37232, USA

R. Jaenisch   Whitehead Institute for Biomedical Research, 9 Cambridge Center, Cambridge MA 02142, USA

A. L. Joyner   Division of Molecular & Developmental Biology, Samuel Lunenfeld Research Institute, Mount Sinai Hospital, 600 University Avenue, Toronto, Canada M5G 1X5

M. H. Kaufman   Department of Anatomy, University Medical School, Teviot Place, Edinburgh EH8 9AG, UK

R. Krumlauf   Laboratory of Eukaryotic Molecular Genetics, National Institute for Medical Research, The Ridgeway, London NW7 1AA, UK

K. A. Lawson   Hubrecht Laboratory, Netherlands Institute for Developmental Biology, Uppsalalaan 8, 3584 CT Utrecht, The Netherlands

K. K. H. Lee (*Ciba Foundation Bursar*)   Department of Anatomy, The Chinese University of Hong Kong, Shatin NT, Hong Kong

R. Lovell-Badge   Laboratory of Eukaryotic Molecular Genetics, National Institute for Medical Research, The Ridgeway, London NW7 1AA, UK

A. L. McLaren (*Chairwoman*)   MRC Mammalian Development Unit, University College London, Wolfson House, 4 Stephenson Way, London NW1 2HE, UK

A. P. McMahon   Department of Cell & Developmental Biology, Roche Institute of Molecular Biology, 340 Kingsland Street, Nutley, NJ 07110-1199, USA

E. J. Robertson   Department of Genetics & Development, Columbia University, 701 West 168th Street, New York, NY 10032, USA

L. Saxén   Department of Pathology, University of Helsinki, Haartmaninkatu 3, SF-00290, Helsinki, Finland

J. C. Smith   Laboratory of Developmental Biology, National Institute for Medical Research, The Ridgeway, London NW7 1AA, UK

D. Solter   Max-Planck Institute of Immunobiology, Stübeweg 51, PO Box 1169, D-7800 Freiburg, Germany

P. P. L. Tam   Children's Medical Research Foundation, PO Box 61, Camperdown, NSW 2050, Australia

I. Thesleff   Department of Pedodontics & Orthodontics, University of Helsinki, Mannerheimintie 172, SF-00300 Helsinki, Finland

D. G. Wilkinson   Laboratory of Eukaryotic Molecular Genetics, National Institute for Medical Research, The Ridgeway, London NW7 1AA, UK

Introduction

Anne L. McLaren

MRC Mammalian Development Unit, University College London, Wolfson House, 4 Stephenson Way, London NW1 2HE, UK

It is my task to put this meeting into historical perspective. I shall begin in 1965 with the Ciba Foundation symposium on Preimplantation stages of pregnancy. That volume is interesting to look back on, partly because the participants were such a mixed lot: cell biologists, geneticists, reproductive biologists, including for instance Gregory Pincus, the pioneer of the Pill. C.H. Waddington was in the chair and he focused our minds on experimental embryology—at least every now and then. At that meeting the important lines of work for the future of the field lay in the discussions of the *in vitro* culture systems for preimplantation embryos. That has been basic to so many of the later advances. There were also reports of embryo manipulations, embryo splitting and the making of aggregation chimeras. Both A.K. Tarkowski and Beatrice Mintz were there. The most important unsolved problem that was raised and discussed was the origin of polarity and axis formation in the mammalian embryo—and that is still unsolved.

Ten years later was the symposium on Embryogenesis in mammals, which was intended to focus on the period immediately after implantation. At the meeting the hope was expressed that it would prove as timely as its predecessor. Preimplantation stages of pregnancy really was timely and stimulated a lot of work that perhaps would not have happened otherwise. But Embryogenesis in mammals was premature: the time was not yet ripe. There were interesting discussions on developmental potential, determination and commitment, but there was too little in the way of solid results to produce a symposium that really hung together. The importance of relevant techniques was stressed. In 1965 there were the culture techniques, but there was nothing corresponding to that for post-implantation development in 1975. There was no mention anywhere in that whole symposium of recombinant DNA technology and the opportunities that it might bring. Dorothea Bennett, who sadly is no longer with us, said of that 1975 symposium that mammalian embryogenesis was an area where the techniques of molecular biology were as yet of little help. One of the interesting reports that came up in the discussion in 1975 was Richard Gardner's mention that he and Janet Rossant had been working with embryonal carcinoma cells, injecting them into blastocysts and making chimeras. That was a sign of the future.

In the 1975 meeting it was clear from my introduction that I personally was already becoming side-tracked onto the implications for human development. One result was that in 1985 the Ciba Foundation held a meeting entitled 'Human embryo research: yes or no?' Of course that was concerned with social and ethical problems more than with biological problems. I still regard those social and ethical problems as being important, but I am happy to see that we are now in 1991 back on track, thanks to the initiative of Andy McMahon. I think the time is now abundantly ripe for a symposium on Postimplantation development in the mouse. Once again I stress the crucial importance of relevant techniques: gene cloning, the facility to isolate related genes from organisms as far apart as yeast and flies and mice, the whole transgenic technology, embryonic stem cells, homologous recombination, the polymerase chain reaction, *in situ* hybridization. All these combine to make this the most exciting time scientifically that I have ever known.

I will not take any more time, because we should focus on the future, not the past, but in introducing Kirstie Lawson, I would like to quote from the 1975 symposium: 'The period immediately after implantation opens with the embryo as a radially symmetrical blastocyst containing as far as we know just two sorts of cells, and within a few days we see the development of a bilaterally symmetrical animal with a front and a back and a right side and a left side, with three germ layers, with segmentation in the form of somites and with different tissues and organs beginning to form'. So Kirstie, that's the backdrop, can you now introduce us to some of the action?

References

Embryogenesis in mammals 1976 Elsevier, Amsterdam (Ciba Found Symp 40)
Human embryo research: yes or no? 1986 Tavistock Publications, London (a Ciba Foundation study group)
Preimplantation stages of pregnancy. 1965 Churchill, Lond (Ciba Found Symp)

Clonal analysis of cell fate during gastrulation and early neurulation in the mouse

K. A. Lawson* and R. A. Pedersen†°

*Hubrecht Laboratory, Netherlands Institute for Developmental Biology, Uppsalalaan 8, 3584 CT Utrecht, The Netherlands, †Department of Radiobiology and Environmental Health and °Department of Anatomy, University of California, San Francisco, CA 94143, USA

Abstract. The foundation of the germ layers and the extraembryonic mesoderm from the epiblast between 6.5 and 7.5 days post coitum (p.c.) is accompanied by substantial cell proliferation. It is followed during the next 24 hours by the organization of major systems of the embryo such as the central nervous system, somites, heart and vascular system. Injection *in situ* of a short-term lineage label (horse radish peroxidase) into single epiblast cells at 6.7 days p.c. and analysis of the descendant clones in cultured embryos have been used to trace these processes and led to the following conclusions: (1) There is extensive but not indiscriminate cell mixing at the onset of gastrulation; epiblast cells spread towards the primitive streak and descendants are there progressively incorporated into mesoderm. (2) The fate map of the mouse epiblast at the early primitive streak stage is topologically similar to those of other vertebrates. (3) Germ layers and the · extraembryonic mesoderm are not clonally distinct before gastrulation, the region of overlapping boundaries in the fate map being occupied by cells that will have descendants in more than one layer. (4) Cranial neurectoderm is mainly derived from axial epiblast immediately anterior to the primitive streak of the early streak stage embryo, clonal descendants being spread rostrocaudally in the developing neural tube. Contribution to the putative floor plate is made by progenitors some of which also contribute to notochord and mesoderm.

1992 Postimplantation development in the mouse. Wiley, Chichester (Ciba Foundation Symposium 165) p 3–26

Preimplantation development in the mouse, and presumably in other eutherian mammals, is characterized by the setting apart of cell lineages necessary for implantation and for subsequent nutrition of the conceptus. This insight rests largely on observations of chimeras, both aggregation chimeras (Tarkowski 1961, 1963, Mintz 1964) and blastocyst injection chimeras (Gardner 1968). Results from blastocyst injection chimeras led to the now generally accepted view that the inner cell mass at 4.5 days, about the time of implantation, is specified into two cell lineages, the primitive endoderm and the primitive ectoderm or epiblast. The epiblast is the source of all the fetal tissues, both somatic and germline, as well as of the extraembryonic amnion, yolk sac mesoderm and allantois (Gardner & Papaionnou 1975, Gardner & Rossant 1979).

Blastocyst injection chimeras cannot be used to investigate the divergence and establishment of fetal cell lineages because epiblast cells will not develop in a blastocyst environment (Gardner et al 1985); even if they did, the analysis would confound cell potency and cell fate, since the donor cells would be in an alien environment and out of synchrony developmentally.

Mouse embryos of about 6.5–8.5 days gestation will develop in culture for up to two days, which covers the period from just before gastrulation begins until early organogenesis. This system has been exploited to trace the fate of small groups of epiblast cells, labelled with [³H]thymidine or wheat germ agglutinin–gold conjugate and transplanted orthotopically. A partial fate map of the 7.5 day epiblast emerged from these experiments, i.e. a fate map of an embryo in which the primitive streak is fully developed and a substantial amount of mesoderm has already formed (Beddington 1981, 1982, Tam 1989, Tam & Beddington 1987, this volume). Potency at this stage was also tested by transferring groups of labelled cells to heterotopic sites. The grafts were able to contribute to some, but not all, tissue types outside their normal fate and did not form structures inappropriate to their new position, thereby demonstrating considerable potency of the epiblast at mid-gastrulation. Because groups of cells were used in these grafting experiments, it was not possible to distinguish between multipotentiality at the cell level and cell selection from a mixed population of specified cells. Selection would imply that normal regionalization in the epiblast could occur by sorting out of previously specified cell types, a process for which there is recent evidence in the chick (Stern & Canning 1990).

Labelling single cells *in situ* and following their descendants is an accurate method of obtaining information about cell fate. Such an approach applied to the mouse embryo at the onset of gastrulation could provide answers to the following questions: (1) Can a fate map be drawn for the period during which the epiblast is multipotent? (2) Can a description at the cell level be obtained of the morphogenetic transformation of the epiblast into the germ layers and further into organ primordia? (3) Is there clonal restriction before or during germ layer formation? (4) Do previously specified cells sort out during gastrulation?

Experimental approach

Embryos were obtained from non-inbred Swiss mice of the Dub (ICR) strain at 6.7 days gestation: litters from these mice were usually at the onset of gastrulation and were a mixture of pre-streak and early streak stage embryos (i.e. comparable to the 6.0 and 6.5 day embryos shown in this volume: General Discussion I, Fig. 1). The embryos were cultured for 1 or 1.5 days in a 1:1 mixture of rat serum and DMEM (Dulbecco's modification of Eagle's Minimal Essential Medium) in an atmosphere of 6% CO_2 in air (Lawson et al 1986, 1991).

One epiblast cell/embryo was injected iontophoretically with a mixture of HRP (horse radish peroxidase) and RDX (rhodamine dextran) (Lawson et al 1991). Labelling by iontophoresis has the advantage over pressure injection that relatively high concentrations of label can be injected without appreciable change in the volume of the cell, and that the needle used can be finer (<0.1 µm diameter). HRP has proved a suitable short-term lineage label for mouse embryos (Balakier & Pedersen 1982, Lawson et al 1986, 1987). The fluorescent label was included (1) so that embryos in which the endoderm was unintentionally labelled during needle penetration into or withdrawal from the underlying epiblast could be rejected before culture; (2) to identify the exact position of the labelled cell in the proximodistal direction of the epiblast. Because bilateral symmetry, detected as obvious mesoderm formation, was not always evident at the moment of injection, the circumferential position of the labelled cell, i.e. the angular distance from the anteroposterior axis of the embryo, was identified retrospectively from the descendants of a visceral extraembryonic endoderm cell lying in the same plane as the injected epiblast cell. Such visceral extraembryonic endoderm clones remain coherent (Gardner 1984) and do not shift significantly relative to the embryonic axis during gastrulation (Lawson et al 1991).

HRP activity, and therefore the presence of labelled descendants, can be detected in the intact embryo, but accurate localization and counts of labelled cells require histology and reconstruction from the resulting sections (Lawson et al 1986).

Results mainly from initially early streak stage embryos are presented here with additions from pre-streak stage embryos where appropriate. The complete results on embryos cultured for one day are published elsewhere (Lawson et al 1991).

The dynamics of gastrulation

Early streak stage embryos gastrulate during the first day in culture, reaching late streak and neural plate stages. Prominent changes during the second day involve the establishment of heart and foregut, early somite formation and growth of the cranial neural folds which will form fore-, mid- and hindbrain (Fig. 1; see also General Discussion I, Fig. 1). Gastrulation in the mouse, unlike in other vertebrates, is accompanied by rapid growth of the labelled epiblast (Snow 1977) and this is reflected in the size of the epiblast clones (Fig. 1B); more than 75% of labelled epiblast cells go through three to four cell divisions during the first day in vitro (Lawson et al 1991).

Clones originating in the anterior half of the epiblast (Fig. 1B) have two striking characteristics: (1) the component cells do not form a coherent clone, members of the clone being commonly separated by unlabelled cells, and (2) the labelled cells are not randomly scattered through the uninvaginated epiblast, but are distributed towards the primitive streak. Clones originating in the midline anterior to the streak tend to be arranged in an arc spanning the axis, with the sides of the arc directed towards the streak (Fig. 2).

In order to classify the results from the comprehensively sampled epiblast, we considered the initial epiblast as consisting of three tiers (Fig. 3), the proximal

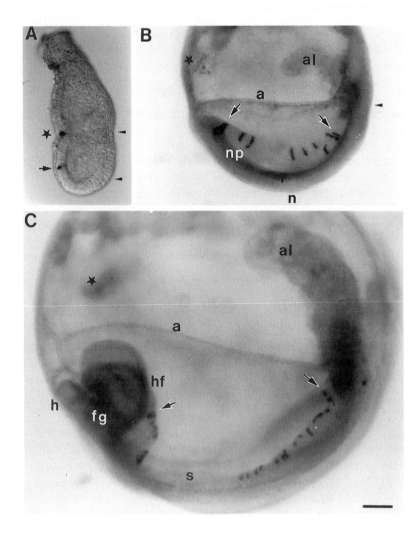

and middle tiers having an anterior axial, lateral, anterolateral, posterolateral and posterior axial zone (zones I–V and VI–X, respectively) and the tier at the distal tip being a single, axial zone (zone XI). For representation, the results from left and right longitudinal halves of the epiblast cup were superimposed, giving a total of 11 arbitrary zones.

A summary of the clonal distribution during the first 24 hours of gastrulation (Fig. 4) shows that the ectoderm at the neural plate stage is derived from the anterior axial and paraxial zones I, II, VI and VII, the distal axial zone XI, the distal, i.e. paraxial, part of zone VIII and the posterior axial zone X. So the ectoderm at the neural plate stage, which is topographically epiblast that has not invaginated through the primitive streak, is formed by expansion towards the streak of epiblast mainly localized in the anterior half of the axis at the early streak stage. The direction of clonal spread is very similar to the directional displacement of marked areas of the rabbit blastodisc (Daniel & Olson 1966) and of the chick blastoderm (Spratt & Haas 1965, Vakaet 1984).

Descendants of initially lateral and posterior epiblast are found in mesoderm, extraembryonic mesoderm and endoderm, and their position indicates an orderly progression of cells through the main body of the growing streak: the most anterior mesoderm is derived from the initially most posterior axial zone V and presumably passes through the streak early, followed by descendants from lateral and anterolateral zones. The bulk of the extraembryonic mesoderm comes from posterolateral and posterior axial zones IV and V and these cells must also be early travellers through the streak. However, cells colonizing the base of the allantois are derived from anterior and anterolateral zones I and II and will arrive in a posterior position after zones IV and V have grown and spread into the developing yolk sac. The head process (notochordal plate) is derived from

---

FIG. 1.   Development of the early streak stage mouse embryo in culture and clonal spread of the descendants of a progenitor cell labelled with horse radish peroxidase (HRP). (A) Lateral view of a 6.7 day early streak stage embryo in which one epiblast cell in the anterior half of the axis has been injected with HRP and stained (arrow). An injected visceral extraembryonic endoderm cell at the embryonic/extraembryonic junction (*) identifies the position relative to the embryonic axis. Arrow heads indicate the anterior and posterior limits of the primitive streak. (B) A similar embryo to that shown in (A) after 21 hours in culture. Descendants of the labelled epiblast cell (arrows) are distributed in ectoderm parallel with the embryonic axis. The position of the cluster of labelled visceral extraembryonic endoderm cells (*) indicates that the position of injection was just to the left of the axial midline. The posterior limit of the primitive streak is shown by an arrow head and the anterior end is defined by the node (n). (C) A similar embryo after 41 hours in culture. The initial site of injection was anterolateral, the visceral extraembryonic endoderm clone (*) is out of focus. The epiblast-derived clone (arrows) is distributed longitudinally in the dorsal part of the prospective hindbrain through the ectoderm to the posterior of the embryo. a, amnion; al, allantois; fg, foregut; h, heart; hf, neural head fold; np, neural plate; s, somite.

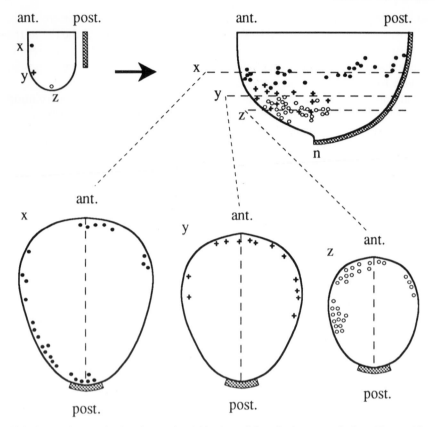

FIG. 2.   Anisotropic clonal growth within the epiblast during gastrulation. The positions of three progenitor cells (x,y,z) in the anterior half of the embryonic axis at the early streak stage are shown in the figure at upper left. The positions of the descendants of x, y and z in the ectoderm after one day culture are shown projected onto a normalized sagittal section (upper right) and onto transverse sections (lower figures). The dashed lines on the sagittal section indicate the levels of the transverse sections. The primitive streak is indicated by a shaded bar. ant, anterior; post, posterior.

material at and near the anterior end of the early streak in zones IX and X. Epiblast derivatives in endoderm come from a similar, but slightly larger region round the anterior end of the streak (zones IX and X).

Thus cells originally in the posterior half of the early streak move into mesoderm and extraembryonic mesoderm and are replaced in the streak by continuously expanding lateral epiblast which, in turn, is replaced by even larger cohorts of cells from the more anterior epiblast.

The anterior part of the streak differs in that cells that were in or near it (zone X) at the early streak stage have descendants at the neural plate stage in the anterior part of the streak and the node, as well as in more anteriorly

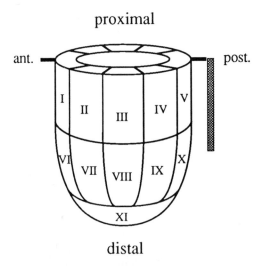

proximal

ant.   post.

distal

FIG. 3.   Schematic representation of the epiblast of the early streak stage embryo. The anterior and posterior limits of the embryonic axis are marked by horizontal bars, the axis running from anterior (ant.) through the distal tip of the embryo to posterior (post.). The extent of the primitive streak is indicated with a hatched bar. The injection zones are indicated (I–XI). For the spatial relationship of the epiblast to other cell layers in the conceptus, see General discussion I (p 55–60).

located mesoderm and endoderm. This suggests that a subpopulation of cells in or near the anterior end of the early streak maintains its position in that part of the streak until late in gastrulation and could form a stem cell population whose progeny will populate successively more posterior regions of the embryo.

The fate map

The consistent behaviour of descendants of progenitors from defined areas of the early streak stage epiblast allows the construction of a fate map for the germ layers (Fig. 5). Similar labelling experiments on pre-streak stage embryos from the same litters (Lawson et al 1991) resulted in a companion map of the epiblast just before the onset of gastrulation (Fig. 5). The fate maps of the two stages do not differ in essentials. Therefore at the onset of gastrulation the mouse epiblast shows determinate development, in the restricted sense that the fate of any cell can be predicted probabilistically (Stent 1985).

Although taking the form of a cup, the rodent epiblast is essentially a pseudostratified columnar epithelium, like the avian blastoderm. If the mouse embryonic axis is straightened and the rest of the epiblast flattened, the presumptive areas are distorted but the boundary relationships remain the same. The fate map of the mouse can then be seen to be remarkably similar to the avian and urodele fate maps at a comparable stage of development (Fig. 6). Topological fate relationships have therefore been conserved in these vertebrates.

Regional restriction or sorting out?

There are considerable overlaps in the areas of presumptive fate at both the pre-streak and early streak stages (Fig. 5), to which biological variation and errors in retrospectively estimating the position of progenitor cells can contribute. However, the boundary regions are interesting in that they can provide information on whether sorting out of committed cells is occurring.

If cells in the boundary regions gave rise to a mixture of pure clones (a pure clone consists of cells in only one of the germ layers), there would be support for the sorting out hypothesis. On the other hand, a boundary region containing progenitors of mixed clones would be evidence against sorting out and would support a hypothesis of gradual regionalization.

Separation of the germ layers is not complete after 24 hours culture, and the anisotropic spread of clonal descendants within the epiblast could be a cause of bias in the clonal classification: some mixed clones with descendants in both mesoderm and ectoderm might later lose their ectoderm component by passage through the streak and become pure mesodermal clones. Early streak stage embryos can be cultured to early somite stages and HRP-labelled descendants identified (Fig. 1C). Extension of the culture period to 36 hours should be sufficient to give an unbiased estimate of clonal fate. The distribution of pure and mixed clones from pre-streak and early streak stage embryos cultured for 36 hours to head fold and early somite stages, respectively, is shown in Table 1 and Fig. 7. Because at least some endoderm is derived from the same lineage as the notochord (Lawson et al 1986), endoderm and mesoderm are considered as a single category, thus giving three categories for defining a pure clone, i.e. descendants in ectoderm, endomesoderm or extraembryonic mesoderm. From Table 1 it is clear that the proportion of pure clones is higher when the clones are derived from progenitors in early streak stage embryos than in pre-streak stage embryos. Fig. 7 shows that there is negligible intermingling on the fate map between progenitors producing pure clones of different categories; in addition, the region producing mixed clones is more extensive at the pre-streak stage than at the early streak stage.

Therefore the clonal analysis provides no support for a process of sorting out of intermixed cells that are specified at or before the onset of gastrulation to form endomesoderm or extraembryonic mesoderm. Rather, it suggests a gradual stabilizing of areas of presumptive fate at the onset of gastrulation. One interpretation of this observation is that the morphogenetic movements of gastrulation are tightly controlled both spatially and temporally; accordingly, each epiblast cell could be a *tabula rasa* until it has reached, or gone through, the streak.

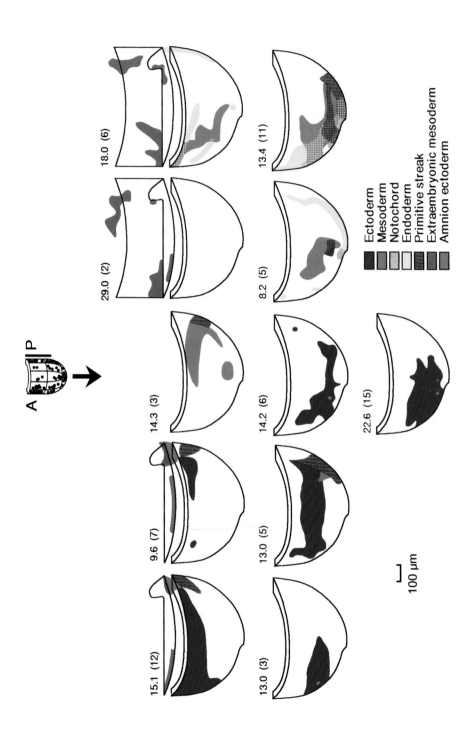

A

P

100 μm

15.1 (12)

9.6 (7)

14.3 (3)

29.0 (2)

18.0 (6)

13.0 (3)

13.0 (5)

14.2 (6)

8.2 (5)

13.4 (11)

22.6 (15)

Ectoderm
Mesoderm
Notochord
Endoderm
Primitive streak
Extraembryonic mesoderm
Amnion ectoderm

Regionalization within the presumptive germ layers

The appearance of structures such as the heart, early somites, notochord, foregut, blood islands, allantois and cranial neural folds during 36 hours in culture (Fig. 1C), opens the possibility of establishing whether there is regionalization within the presumptive germ layers mapped at the early streak stage (Fig. 5). The fate map in Fig. 8 is based on the results from 92 embryos, 36 hours after single cell labelling with HRP at the early streak stage. The positions shown indicate the region where most of the clones contributing to a given structure were initiated. Although some pure clones were found in most structures, at least 60% of the clones contributing to any structure also contributed to one or more of the other structures (Table 2). In other words, none of the structures identifiable at the early somite stage developed from a stem cell population already allocated at the onset of gastrulation.

Clonal distribution during early neurogenesis

The elevated neural folds present at the early somite stage comprise the regions of neurectoderm that will form the fore-, mid- and hindbrain, and the anterior portion of the spinal cord. Clonal analysis provides an opportunity to trace the morphogenetic transformation of the presumptive ectoderm at the early streak stage into the cranial neural folds and to establish whether there is lineage restriction associated with regionalization in the cranial neurectoderm.

The descendants from zones VI and XI are found almost exclusively in neurectoderm (Table 3). This region of axial epiblast anterior to the primitive streak is the region that had been proposed, on indirect evidence, to be the main source of neurectoderm (Snow & Bennett 1978). Substantial contributions to neurectoderm are also made by the adjoining anterior axial zone I and paraxial

---

FIG. 4.   Distribution of descendants of early streak stage epiblast in the different germ layers after culture for one day to late streak/neural plate stage. The injection zones (Fig. 3) are indicated by broken lines in the upper figure, which represents a lateral view of the early streak stage epiblast at the time of injection. Dots mark the position of injected progenitors (one per embryo). The extent of the primitive streak is given by a grey bar. A, anterior; P, posterior. Each of the 11 lower figures shows the area colonized by descendants after injection in a particular zone at the early streak stage. The germ layer location of descendants is colour coded. Each figure represents the longitudinal half of the embryonic part of a cultured embryo of average dimensions. The arrangement of the zonal figures in an upper tier (zones I–V), a lower tier (zones VI–X) and the distal tip (zone XI) follows the position of zones in the initial embryo, e.g. the extreme left figure of the upper tier represents zone I. The amnion and allantois is represented in outline above zones I and II, and the amnion, allantois and yolk sac above zones IV and V. The numbers above each zonal figure are the mean number of labelled cells/embryo; the numbers in parentheses are the number of embryos injected in each zone.

Prestreak   Early streak

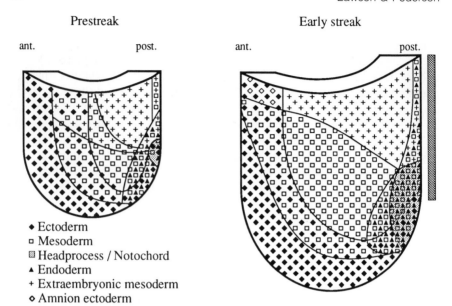

◆ Ectoderm
▫ Mesoderm
▨ Headprocess / Notochord
▲ Endoderm
+ Extraembryonic mesoderm
◇ Amnion ectoderm

FIG. 5.   Fate maps of the epiblast of the pre-streak and early streak stages showing the derivation of germ layers up to the mid/late streak and neural plate stages, respectively. The approximate extent of the primitive streak is indicated by the hatched bar. ant, anterior, post, posterior. From Lawson et al (1991).

zone VII, and quantitatively minor contributions by paraxial zones II and IX. Zone I is the source of most of the surface ectoderm.

The distribution of descendants in neurectoderm, surface ectoderm and postnodal ectoderm from the different zones is summarized in Fig. 9. There are a number of striking features of these zonal contributions: (1) Labelled descendants are dispersed along the length of the embryo, but those from axial zones I, VI and XI also span the ventral midline anteriorly in the prospective forebrain and anterior midbrain. This means that a cell descended from a progenitor in zone I or VI and lying ventrally in the forebrain is more closely related to cells lying dorsally and laterally in the midbrain than to ventral cells of the midbrain. (2) Clones from these axial regions of the early streak stage do not span the midline in the prospective hindbrain; their longitudinal arrangement is such that descendants of cells from originally more anterior, axial ectoderm (zone I) tend to occupy positions dorsal to descendants of more posterior progenitors (zones VI and VII) which, in their turn, lie dorsal to descendants from zone XI. Therefore, an anteroposterior spatial relationship in the epiblast at the early streak stage has been transformed into a dorsoventral one in the neurectoderm of the hindbrain: lineage relationship in the hindbrain is closer in the craniocaudal direction than in the dorsoventral. (3) The ventral

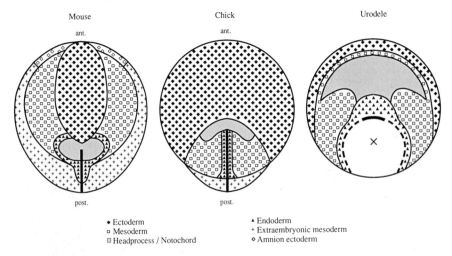

FIG. 6. Fate maps at the early gastrulation stage of mouse, chick and urodele. The mouse epiblast at the early streak stage has been flattened and the overlaps between the presumptive germ layers have been removed. The map of the chick epiblast (area pellucida) at stage 3 is according to Vakaet (1985). The urodele is viewed from the vegetal side (X = vegetal pole); much presumptive ectoderm and some mesoderm is therefore not visible (adapted from Nieuwkoop et al 1985, Nieuwkoop & Sutasurya 1979). The primitive streak of the mouse and chick and the dorsal lip of the blastopore are indicated by a solid bar. From Lawson et al (1991).

midline extending posteriorly from the midbrain to the node is not occupied by descendants of any of the zones considered so far, which account for more than 95% of the labelled cells in the neurectoderm. The only labelled cells found in this region, which will presumably later form the floor plate, were derived from zones IX and X in a position anterolateral to the anterior end of the early streak. Of the six clones contributing to the putative floor plate, only one was a pure clone in neurectoderm; the other five clones had labelled cells in notochord, paraxial mesoderm or primitive streak, or a combination of these. Although the data are few and require both expansion and more detailed analysis, these lineage relationships indicating common progenitors for the floor plate, notochord and paraxial mesoderm at the onset of gastrulation are strikingly similar to those in other vertebrates (reviewed by Jessell et al 1989, Sellek & Stern 1991).

The pattern of clonal distribution in the cranial neurectoderm can best be understood by considering the situation after the first day in culture. The transformation of the epiblast of pre-streak and early streak stage embryos to ectoderm of late streak and neural plate stages by the anisotropic expansion of axial zones towards the streak (Figs 1B, 2, 4) involves more dramatic changes in cell relationships than are required for descendants from these originally

Prestreak   Early streak

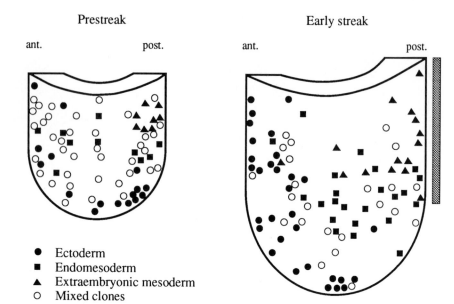

- ● Ectoderm
- ■ Endomesoderm
- ▲ Extraembryonic mesoderm
- ○ Mixed clones

FIG. 7. Distribution of progenitors producing pure and mixed clones 36 hours after labelling at 6.7 days. Pure clones consist exclusively of ectoderm, endomesoderm or extraembryonic mesoderm and are indicated by solid symbols; mixed clones contain labelled cells in two or more of these categories and are indicated by open circles. ant, anterior; post, posterior. The extent of the primitive streak is indicated by the hatched bar.

TABLE 1   Clones of ectoderm, endomesoderm and extraembryonic mesoderm after 36 hours in culture

| | Number of clones | | | |
Initial stage	Total	Pure[a]	Mixed[b]	% Pure
Pre-streak	74	33	41	44.6
Early streak	92	67	25	72.8

[a]A pure clone is a clone with all its members in one category: ectoderm, endomesoderm or extraembryonic mesoderm.
[b]Mixed clones have members in more than one category.

axial zones in their subsequent development as the cranial neural folds (Fig. 1C). The changes in the presumptive ectoderm after 24 hours (Fig. 4) and 36 hours in culture (Fig. 9) are shown schematically in Fig. 10. Formation of the neurectoderm at 36 hours requires continued longitudinal expansion of clones derived from zones I, II, VI, VII and XI after 24 hours, plus a new phenomenon, the insertion of some material from zones IX and X into the ventral midline.

Early streak

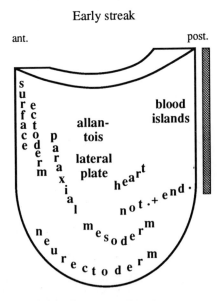

FIG. 8. Fate map of the early streak stage to show the approximate regions from which structures identifiable at the early somite stage are derived. ant, anterior; post, posterior; not + end, notochord + endoderm. The extent of the primitive streak is indicated by the hatched bar.

TABLE 2 Clones originating in the early streak stage embryo and contributing after 36 hours to structures in the early somite stage embryo

	Number of clones			
Structure	*Total*	*Pure*[a]	*Mixed*[b]	*% Pure*
Neurectoderm	35	14	21	40
Surface ectoderm	20	3	17	15
Other ectoderm[c]	8	1	7	13
Notochordal plate	10	0	10	<10
Paraxial mesoderm	14	3	11	21
Lateral plate mesoderm	10	3	7	30
Postnodal mesoderm	7	0	7	<14
Primitive streak	8	2	6	25
Heart	6	1	5	17
Endoderm	12	0	12	<8
Blood islands	10	3	7	30
Allantois	9	2	7	22
Yolk sac and amnion mesoderm	16	4	12	25

[a] A pure clone is a clone with all its members within the structure indicated.
[b] A mixed clone is a clone with members in the structure indicated and in at least one other structure.
[c] Other ectoderm combines amnion ectoderm and non-primitive streak ectoderm posterior to the node.

TABLE 3 Contribution of different epiblast zones of the early streak stage embryo to ectoderm derivatives after 36 hours in culture

	Epiblast zone										
	I	II	III	IV	V	VI	VII	VIII	IX	X	XI
Number of embryos injected	8	14	4	7	6	5	7	12	9	9	10
Labelled cells/embryo[a]											
Total	29.1	9.3	13.8	24.9	9.5	33.4	29.9	18.1	14.9	14.1	29.4
Neurectoderm	13.5 (46)	1.9 (20)	—	—	—	31.0 (93)	12.6 (42)	0.8 (5)	2.6 (17)	0.3 (2)	26.5 (90)
Surface ectoderm (including amnion ectoderm)	11.6 (40)	4.1 (44)	—	—	—	2.4 (7)	0.3 (1)	—	—	—	—
Cranial neural crest	0.6 (2)	—	—	—	—	—	0.7 (2)	—	—	—	—
Postnodal ectoderm	2.8 (10)	0.9 (10)	—	—	—	—	4.4 (15)	0.8 (5)	—	—	1.3 (4)

[a]Numbers represent the average number of labelled cells per embryo in each category. Numbers in parentheses are the percentage of total zonal descendants in ectoderm derivatives. Other descendants were in endoderm or extraembryonic mesoderm (see Table 2).

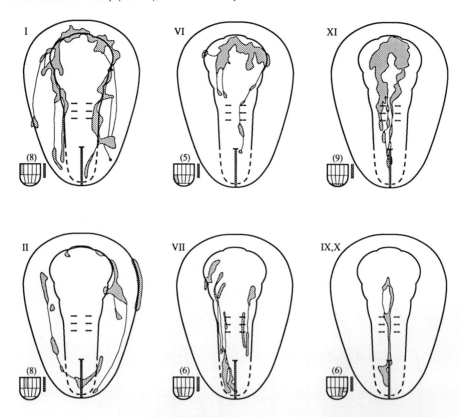

FIG. 9. Distribution of zonal descendants in ectoderm 36 hours after single cell labelling with horse radish peroxidase at the early streak stage. The six large figures show the ectoderm of the early somite stage flattened out as an oval and viewed from the dorsal side. The bulges in the curved outline of the similarly flattened neurectoderm represent the forebrain, midbrain, and anterior and posterior segments of the hindbrain. The level of the posterior limits of the first three somites is indicated by three pairs of dashed lines in the posterior hindbrain region. The primitive streak is indicated by the hatched bar. The longitudinal axis of symmetry in each diagram therefore represents the ventral midline of the neurectoderm and the depicted lateral outline of the neurectoderm represents the dorsal edge of the neural folds; anteriorly, the outline runs from dorsal at the junction of fore- and midbrain to ventral where it crosses the midline. The approximate area occupied by labelled descendants from each early streak zone with descendants in ectoderm is shaded (the injected zone is indicated to the upper left of each large figure). Thin lines joining shaded areas indicate clonal continuity through a region containing no labelled cells. The small figure to bottom left of each large figure represents the flattened longitudinal half of an early streak stage embryo (anterior to the left) and shows the relevant injected zone (shaded) before culture. For clarity, the scale of this figure relative to the main figure has been doubled. The numbers in parentheses are the numbers of injected embryos with zonal descendants in ectoderm.

A

ant.   post.

B

ant.   post.

C

ant.

post.

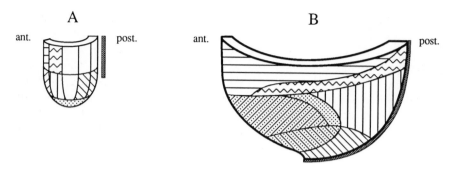

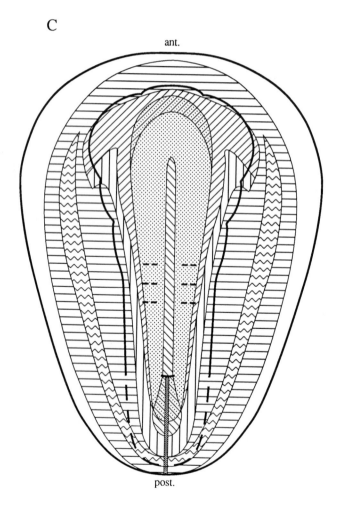

The patterns of clonal distribution in the neurectoderm at eight days are not reflected in morphological patterning in the craniocaudal direction: sculpting into fore-, mid- and hindbrain cuts across the longitudinally aligned clones. The patterns of clonal relationship were initiated several cell divisions earlier, and it is conceivable that subpopulations, particularly for the fore- and midbrain, are established at the late streak stage, when the prospective craniocaudal divisions of the brain can be mapped on the ectoderm (Tam 1989). Additional information is clearly needed on the fate of clones initiated within early neurectoderm and on the degree of cell mixing occurring at this time, and later, before we can see whether there are lineage restrictions within the neurectoderm relevant to the expression patterns of genes such as *En* (Davis et al 1991) and *Wnt*-1 (Wilkinson et al 1987).

Conclusions and perspectives

The dynamic transformation of the cup-shaped epiblast during gastrulation into the three germ layers of the neural plate stage mouse embryo is accomplished by the coordinated, directional spread of expanding epiblast clones to and through the primitive streak. In spite of extensive cell mixing, the predictability and directionality of epiblast behaviour has allowed the construction of a fate map at the onset of gastrulation that resembles those of other vertebrates in its topological fate relationships. Such a fate map implies nothing about the state of commitment of the epiblast cells, and indeed the clonal analysis data exclude lineage restriction for any somatic structure prospectively mapped at the early streak stage.

While several genes have been reported as being expressed in, or near, the primitive streak and in early mesoderm at the late streak stage, and many advances have been made in analysing the establishment of craniocaudal patterns later in development (Nieto et al, this volume), plausible candidates for organizing and maintaining the primitive streak, and thus for setting up the craniocaudal axis in the mouse, are conspicuously lacking, with the possible exception of *T* (Wilkinson et al 1990, Herrmann, this volume). A primary requirement for such a gene would be a high level of expression in a resident population of cells at the anterior end of the streak, evident from or before the onset of streak formation.

FIG. 10. Schematic representation of the transformation of presumptive ectoderm between early streak and early somite stages. (A) Longitudinal half of early streak stage embryo with the relevant injection zones identified by shading. (B) Longitudinal half of neural plate stage embryo (24 hours culture). (C) Dorsal view of flattened early somite stage embryo (36 hours culture). ant, anterior; post, posterior. The extent of the primitive streak is indicated by heavy shading.

Acknowledgements

This work was made possible by a NATO grant for collaborative research and was also supported by US DOE/OHER no. DEAC 03-76-SFO 1012. We are grateful to S. Kerkvliet and M. Flannery for histology and to W. Hage for producing the figures by computer graphics.

References

Balakier H, Pedersen RA 1982 Allocation of cells to inner cell mass and trophectoderm lineage in preimplantation mouse embryos. Dev Biol 90:352–362

Beddington RSP 1981 An autoradiographic analysis of the potency of embryonic ectoderm in the 8th day postimplantation mouse embryo. J Embryol Exp Morphol 64:87–104

Beddington RSP 1982 An autoradiographic analysis of tissue potency in different regions of the embryonic ectoderm during gastrulation in the mouse. J Embryol Exp Morphol 69:265–285

Daniel JC Jr, Olson JD 1966 Cell movement, proliferation and death in the formation of the embryonic axis of the rabbit. Anat Rec 156:123–127

Davis CA, Holmyard DP, Millen KJ, Joyner AL 1991 Examining pattern formation in mouse, chicken and frog embryos with an *En*-specific antiserum. Development 111:287–298

Gardner RL 1968 Mouse chimaeras obtained by the injection of cells into the blastocyst. Nature (Lond) 220:596–597

Gardner RL 1984 An *in situ* cell marker for clonal analysis of development of the extraembryonic endoderm in the mouse. J Embryol Exp Morphol 80:251–288

Gardner RL, Papaionnou VE 1975 Differentiation in the trophectoderm and inner cell mass. In: Balls M, Wild AE (eds) The early development of mammals. Cambridge University Press, Cambridge p 107–132

Gardner RL, Rossant J 1979 Investigation of the fate of 4.5 day *post coitum* mouse inner cell mass cells by blastocyst injection. J Embryol Exp Morphol 52:141–152

Gardner RL, Lyon MF, Evans EP, Burtenshaw MD 1985 Clonal analysis of X-chromosome inactivation and the origin of the germ line in the mouse embryo. J Embryol Exp Morphol 88:349–363

Herrmann BG 1992 Action of the *Brachyury* gene in mouse embryogenesis. In: Postimplantation development in the mouse. Wiley, Chichester (Ciba Found Symp 165) p 78–91

Jessell TM, Bovolenta P, Placzek M, Tessier-Lavigne M, Dodd J 1989 Polarity and patterning in the neural tube: the origin and function of the floor plate. In: Cellular basis of morphogenesis. Wiley, Chichester (Ciba Found Symp 144) p 255–280

Lawson KA, Meneses JJ, Pedersen RA 1986 Cell fate and cell lineage in the endoderm of the presomite mouse embryo, studied with an intracellular tracer. Dev Biol 115:325–339

Lawson KA, Pedersen RA, van de Geer S 1987 Cell fate, morphogenetic movement and population kinetics of embryonic endoderm at the time of germ layer formation in the mouse. Development 101:627–652

Lawson KA, Meneses JJ, Pedersen RA 1991 Clonal analysis of epiblast fate during germ layer formation in the mouse embryo. Development 113:891–911

Mintz B 1964 Formation of genetically mosaic mouse embryos, and early development of "lethal (t^{12}/t^{12})—normal" mosaics. J Exp Zool 157:273–292

Nieto MA, Bradley LC, Hunt P, Das Gupta, Krumlauf R, Wilkinson DG 1992 Molecular mechanisms of pattern formation in the vertebrate hindbrain. In: Postimplantation development in the mouse. Wiley, Chichester (Ciba Found Symp 165) p 92–107

Nieuwkoop PD, Sutasurya LA 1979 Primordial germ cells in the chordates. Cambridge University Press, Cambridge

Nieuwkoop PD, Johnen AG, Albers B 1985 The epigenetic nature of early chordate development; inductive interaction and competence. Cambridge University Press, Cambridge

Sellek MAJ, Stern CD 1991 Fate mapping and cell lineage analysis of Hensen's node in the chick embryo. Development 112:615–626

Snow MHL 1977 Gastrulation in the mouse: growth and regionalization of the epiblast. J Embryol Exp Morphol 42:293–303

Snow MHL, Bennett D 1978 Gastrulation in the mouse: assessment of cell populations in the epiblast of t^{w18}/t^{w18} embryos. J Embryol Exp Morphol 47:39–52

Spratt NT Jr, Haas H 1965 Germ layer formation and the role of the primitive streak in the chick. I. Basic architecture and morphogenetic tissue movements. J Exp Zool 158:9–38

Stent GS 1985 The role of cell lineage in development. Phil Trans R Soc Lond B Biol Sci 312:3–19

Stern CD, Canning DR 1990 Origin of cells giving rise to mesoderm and endoderm in chick embryo. Nature (Lond) 343:273–275

Tam PPL 1989 Regionalization of the mouse embryonic ectoderm: allocation of prospective ectodermal tissues during gastrulation. Development 107:55–67

Tam PPL, Beddington RSP 1987 The formation of mesodermal tissues in the mouse embryo during gastrulation and early organogenesis. Development 99:109–126

Tam PPL, Beddington RSP 1992 Establishment and organization of germ layers in the gastrulating mouse embryo. In: Postimplantation development in the mouse. Wiley, Chichester (Ciba Found Symp 165) p 27–49

Tarkowski AK 1961 Mouse chimaeras developed from fused eggs. Nature (Lond) 190:857–860

Tarkowski AK 1963 Studies on mouse chimaeras developed from fused eggs *in vitro*. Nat Cancer Inst Monogr 11:51–71

Vakaet L 1984 Early development of birds. In: LeDouarin N, McLaren A (eds) Chimaeras in developmental biology. Academic Press, New York p 71–88

Vakaet L 1985 Morphogenetic movements and fate maps in the avian blastoderm. In: Edelman GM (ed) Molecular determinants of animal form. Alan R Liss, New York p 99–109

Wilkinson DG, Bailes JA, McMahon AP 1987 Expression of the proto-oncogene *int*-1 is restricted to specific neural cells in the developing mouse embryo. Cell 50:79–88

Wilkinson DG, Bhatt S, Herrmann BG 1990 Expression pattern of the mouse *T* gene and its role in mesoderm formation. Nature (Lond) 343:657–658

DISCUSSION

McLaren: Kirstie, your beautiful diagrams seemed to me deficient in only one respect and that is that they showed no primordial germ cells. Some of your injected single cells must have landed up in that population. Have you included the primordial germ cells in extraembryonic mesoderm? Did you ever identify them specifically?

Lawson: We found cells that had moved from the anterior region to the posterior region, including the base of the allantois, at the stage when primordial germ cells are found posteriorly. So it seemed plausible that primordial germ cell precursors were arising in the more anterior part of the epiblast. Using a single cell injection technique, we should be able to find out whether the precursors are coming from the anterior region, and to distinguish whether primordial germ cells are already committed when they are in the anterior part of the epiblast or not until they reach the posterior part. If you inject a cell in the anterior region and look for its descendants later, there are three possibilities: (1) none of the descendants are primordial germ cells, which is not informative; (2) all its descendants are primordial germ cells, which could mean that the progenitor cell was already committed at the time of injection or before the first division after injection; or (3) some of the cells in the clone are primordial germ cells, which would indicate that they were committed after the division after injection.

We were unsuccessful in developing a double fluorescent labelling technique for lineage labelling and identification of primordial germ cells that could be used with the confocal laser scanning microscope. We had to settle for a combination of fixable rhodamine dextran as lineage marker and conventional histochemical staining of alkaline phosphatase for primordial germ cell identification at head fold and later stages.

Willem Hage has developed a method whereby he can plot the position of fluorescent cells in sections, then stain for alkaline phosphatase and project the stained sections back onto the fluorescent image objectively and precisely, so that we have two sets of images which match exactly. Unfortunately, in the few embryos that have been analysed completely and in which lineage-labelled cells from the anterior region were in the same region as the primordial germ cells, none of the lineage-labelled cells were primordial germ cells.

Another approach is being used by Karen Downs, who has been working with Rosa Beddington and has spent three months in my lab. She is using a technique originally used by Rosa for 7.5 day embryos of grafting groups of [^{3}H]thymidine-labelled cells into 6.5 day embryos. By injecting a small population of cells orthotopically, one might find out where the progenitors of primordial germ cells come from. These experiments are in progress and there are no results yet.

Jaenisch: At the developmental stage before the cylinder stage, the existence of about 10 epiblast cells has been demonstrated using a very different technique, i.e. retroviral marking (Soriano & Jaenisch 1986). Those experiments suggested that these 10 cells gave rise to the whole somatic lineage. In addition, three cells at this stage were already allocated to the germ cell lineage. How do you envisage that these cells are distributed in the egg cylinder?

Lawson: If there are only three, I am not likely ever to pick them up, simply on the basis of probability. Bearing in mind the possibility that there is a very small population of precursor primordial germ cells and the possibility that I have more chance of hitting a primordial germ cell precursor in the anterior epiblast the younger the embryos I use, I have tried injecting into 6 day embryos.

These can be cultured until the neural fold stage and the primordial germ cells identified, but I have analysed very few of them.

Smith: Kirstie, when do you first see cells whose fate is restricted exclusively to ectoderm or to mesoderm? Claudio Stern's results in the chick embryo would predict that some cells become restricted in their fate prior to primitive streak formation: the monoclonal antibody HNK-1 recognizes an apparently random population of cells in the epiblast, and these cells go on to form the mesoderm and endoderm. The rest of the epiblast cannot form mesoderm, if this population of cells is removed (Stern & Canning 1990).

Lawson: The patterns of cell fate make Stern's hypothesis unlikely for the mouse, but don't tell us anything about the actual times of restriction.

Smith: I was just asking when the cells became restricted in their normal fate.

Lawson: There are areas even in the pre-streak stage which give pure clones, however, the majority (58%) of clones at this stage are mixed clones. By analogy with the chick, this is the stage when a subpopulation of specified 'streak-organizing' and mesoderm-forming cells would be expected. But only 23% of the clones giving rise to endomesoderm are purely endomesoderm.

I haven't looked later at mid-streak and late streak stages at 7.5 days to see whether there is 100% restriction then.

Tam: In your fate map, is there any overlap of the domains occupied by the progenitor cells of the surface ectoderm and neuroectoderm? Where are the neural crest cells and the epidermal placode located?

Lawson: At 6.5 days there is overlap in the anterior region between domains for surface ectoderm and neurectoderm. Clones, particularly from zones I and II, will be producing both; the anterior members of the clone will be in the prosencephalon and will then spread backwards into surface ectoderm.

There were four embryos cultured for 36 hours to early somite stages that had labelled cranial neural crest, i.e. cranial mesoderm in a clone with labelled neuroectoderm or adjacent surface ectoderm. These clones originated just lateral to the midline in the anterior half of the epiblast in zone I, junction I/II, junction I/VI and zone VII (Fig.3), and would have expanded towards the streak by the late streak stage. The descendants that contributed to the neural crest could therefore have been in the region at the late streak stage that you found contributed to the neural crest (Tam 1989). I don't know where the presumptive epidermal placode is located.

Beddington: In the colonization of the notochord, do you get coherent patches of clonal descendants or do you get interspersed cells?

Lawson: They tend to be interspersed. I have one or two embryos where I find periodicity, as has been found in the chick segmental plate (Stern et al 1988). My data are sparse in comparison to the chick data; I don't have many embryos in which notochord is labelled.

Joyner: So you are saying that the cells which will give rise to the notochord are sitting and dividing at the anterior edge of the primitive streak, whereas the cells that will give rise to the rest of the mesoderm-derived tissues in the embryo sweep backwards through the length of the primitive streak from the more anterior region of the embryo.

Lawson: On the fate map at 36 hours, notochord and endoderm were coming not only from the anterior end of the streak but also from cells presumably moving into that region (Fig. 7). If one looks at the distribution of cells from zones VIII, IX and X (Fig. 3) in the notochord, descendants of cells from zone X can be found fairly extensively in the notochordal plate. Cells from zone IX are found only in a shorter part of the notochordal plate near the node at the early somite stage. So cells from zone X go in first and become distributed anteriorly but also leave descendants more posteriorly, whereas those from zone IX begin to contribute later and so are restricted to a more posterior position. In an occasional early somite stage embryo there is a very small contribution posteriorly to the notochordal plate by cells from zone VIII.

Joyner: Do you think there is a group of stem cells that remains at the anterior edge of the streak?

Lawson: Yes. There is a population of cells in or near the node that, when they divide, will throw off cells into the notochord, into the paraxial mesoderm and also into lateral plate mesoderm. A sister cell sometimes stays behind and repeats the process, which presumably is why you get clonal descendants distributed through the whole of the notochord.

Kaufman: Were you surprised by the rapidity of colonization? You are labelling some of the ectoderm cells in the pre-head fold region and envisaging that they migrate all the way to the primitive streak, dive into the primitive streak and then migrate rostrally again to produce mesoderm. How do you account for this? There are only so many cell divisions in 24 hours.

Lawson: A cell from zone V which is delaminated through the streak at the early streak stage would be situated about 200 μm from the anterior end of the embryo. Since the direction of epiblast growth tends to thrust the posterior end of the streak further away from the anterior end of the embryo, rather than to expand the epiblast cup uniformly, the mesoderm cell has to divide and move only as fast as the anterior 200 μm of the egg cylinder expands (or somewhat more slowly because it is internal) in order to maintain its position with respect to the anterior end of the embryonic axis. So to get all the way to the anterior end after 24 hours, mesoderm cells from zone V would have to move only an additional 200 μm. This is easily accommodated by the demonstrated ability of mesoderm cells *in situ* to move 46 μm h^{-1} (Nakatsuji et al 1986).

Similarly, descendants from progenitors in zone III in the pre-streak stage, lateral in the epiblast, are found in the mesoderm of the posterior half of the embryo at the mid-streak stage (data in Lawson et al 1991). Cells from zone III probably began going through the streak after about eight hours when the width of the epiblast cup had increased by about 75%. The first cells through the streak would then only have to maintain their position with respect to the anterior end of the embryo until the mid-streak stage.

McMahon: Several genes have been described as having interesting distributions in the primitive streak, yet it has not always been clear to me whether they really are in the primitive streak. Can you give us some topological limits on what you call the primitive streak?

Lawson: The early streak in the mouse is not a well defined structure: it doesn't have a pronounced groove as in the chick. The anterior end at the late streak stage is clearly defined by the node. Perhaps, at all stages, it can best be defined in transverse section, on the basis of its lateral limits where mesoderm is inserting between epiblast and endoderm, and the basal lamina between epiblast and endoderm is disrupted. I know of no published work in which these limits over the length of the streak and at different stages have been recorded. A spot check on glycol methacrylate sections of a cultured embryo at the late streak stage gave a value of 35 μm (four or five cell diameters) for the width of the streak half way down its length.

Beddington: Is there a permanent resident population of cells in the primitive streak that when you label in the streak region will always retain descendants there? Is there a kind of latent structure to the primitive streak?

Lawson: That applies only to the anterior end of the streak. Some cells or their descendants from the anterior end of the early streak will still be in the anterior end of the streak at the neural plate stage. Cells from the posterior end of the early streak are not found in the streak at the neural plate stage, they are in the extra-embryonic mesoderm and in the anterior mesoderm. One assumes that those posterior cells have gone through the streak, but you find them in extraembryonic mesoderm. The cells from more laterally and from even more anteriorly are found in the posterior end of the streak at the late streak and neural plate stages.

Beddington: Is the relative length of streak that is populated by that resident population a constant size or does it increase?

Lawson: The relative size is much the same, between 0.25 and 0.33 length from the anterior end of the streak; so in absolute terms there is some proximodistal spread.

Beddington: I am interested in whether there is a population that maintains the activity of the streak by staying there or whether there has to be a signal which keeps saying 'Go through the streak'.

Lawson: It could be both: the signal could be coming from the resident population in the anterior part of the streak.

McLaren: You have probably looked at more pre-streak embryos than most people. They are usually assumed to be radially symmetrical, but one suspects that there may be interesting asymmetries. Your single cell injection technique would be ideal for checking that.

Lawson: It is a rather unsatisfactory story. L. J. Smith (1985) proposed that at the time of implantation asymmetry has already developed within the blastocyst and that this is maintained so you can predict where the primitive streak will form. The theory is that asymmetry is produced by the position of the inner cell mass in the trophectoderm and the shape of the trophoblast relative to the site of implantation, which is later reflected in the tilt of the extraembryonic ectoderm and ectoplacental cone.

When you dissect embryos out at the pre-streak stage and at the early streak stage but leave Reichert's membrane so they still have some of the constraints present in the decidua, you can't really see the shape of the embryo. When you remove Reichert's membrane you quite often see a definite asymmetry. On the basis of Smith's criteria, I thought it ought to be possible to predict the future end of the embryonic axis from the tilt of the ectoplacental cone and the asymmetry in the attachment of Reichert's membrane. A visceral extraembryonic endoderm cell was labelled with horse radish peroxidase and its position relative to the putative embryonic axis predicted, according to the above criteria. The frequency of correct predictions, judged from the position of the descendant clone after 24 hours in culture, was only about 40%. Possibly, when one removes Reichert's membrane the normal form of the conceptus is distorted and the asymmetries may have little to do with the asymmetries *in utero*. On the basis of the shape of the embryonic part, even at the late pre-streak stage, my judgement is much better (70% correctly predicted) than when I used what seemed to be quite clear criteria for the extraembryonic part.

Richard Gardner has also been examining Smith's theory and confirms the presence of asymmetry in the blastocyst (unpublished observations, cited in Gardner 1990). His marking experiments (unpublished) have not so far ruled out a correlation between the axes of the blastocyst and the early fetus.

References

Gardner RL 1990 Location and orientation of implantation. In: Edwards RG (ed) Establishment of a successful human pregnancy. (Serono Symposia Publications, vol 66) Raven Press, New York p 225–238

Lawson KA, Meneses JJ, Pedersen RA 1991 Clonal analysis of epiblast fate during germ layer formation in the mouse embryo. Development 113:891–911

Nakatsuji N, Snow MHL, Wylie CC 1986 Cinemicrographic study of the cell movement in the primitive-streak-stage mouse embryo. J Embryol Exp Morphol 96:99–109

Smith LJ 1985 Embryonic orientation in the mouse and its correlation with blastocyst relationships to the uterus. II. Relationships from 4¼ to 9½ days. J Embryol Exp Morphol 89:5–35

Soriano P, Jaenisch R 1986 Retroviruses as probes for mammalian development: allocation of cells to the somatic and germ cell lineages. Cell 46:19–29

Stern CD, Canning DR 1990 Origin of cells giving rise to mesoderm and endoderm in chick embryo. Nature (Lond) 343:273–275

Stern CD, Fraser SE, Keynes RJ, Primmett DRN 1988 A cell lineage analysis of segmentation in the chick embryo. Development (suppl) 104:231–244

Tam PPL 1989 Regionalisation of the mouse embryonic ectoderm: allocation of prospective ectodermal tissue during gastrulation. Development 107:55–67

Establishment and organization of germ layers in the gastrulating mouse embryo

P. P. L. Tam* and R. S. P. Beddington†

*The Children's Medical Research Foundation, PO Box 61, Camperdown, New South Wales 2050, Australia and †AFRC Centre for Genome Research, King's Building, West Mains Road, Edinburgh EH9 3JQ, UK

Abstract. By following the distribution of wheat germ agglutinin–gold-labelled cells in primitive streak stage embryos, we obtained direct evidence for a continuous recruitment of the embryonic ectoderm cells to the definitive endoderm and to the embryonic and extraembryonic mesoderm during gastrulation. The majority of the definitive endodermal cells ingressed through the anterior end of the primitive streak and were incorporated initially into the midline endoderm at the archenteron, but a small population of endodermal cells may be recruited by direct delamination from the embryonic ectoderm. The pre-existing visceral embryonic endoderm was progressively replaced, but not totally, by the newly recruited population which colonized the embryonic foregut and the notochord. The developmental fate of the recruited endoderm and that of cells in the embryonic ectoderm and the mesoderm of late primitive streak stage embryos indicate that concomitant with the establishment of the germ layers, an orderly allocation of prospective fetal tissues to specific parts of the body occurs simultaneously in all three germ layers.

1992 Postimplantation development in the mouse. Wiley, Chichester (Ciba Foundation Symposium 165) p 27–49

The formation of germ layers is a crucial event in the establishment of a basic body plan during the immediate postimplantation development of the mouse embryo. It begins with the differentiation of the inner cell mass of the blastocyst into a discrete epithelium of primitive endoderm and a compact cellular mass (which subsequently becomes a pseudostratified epithelium) of primitive (embryonic) ectoderm (Snell & Stevens 1966, Enders et al 1978). The onset of gastrulation is heralded by the appearance of the primitive streak in the primitive ectoderm (Batten & Haar 1979, Hashimoto & Nakatsuji 1989) and the emergence of a nascent population of embryonic and extraembryonic mesoderm (Poelmann 1981a, Tam & Meier 1982). In contrast to the formation of the mesoderm, there is no overt organization of a new layer of endoderm, suggesting that the definitive endoderm may be established by entirely different mechanisms.

Recent clonal analyses and fate mapping studies have provided evidence for a gradual transition from the primitive endoderm to definitive endoderm during gastrulation. The primitive endoderm of the implanting blastocyst contributes primarily to the endoderm of the extraembryonic membranes but not to any endodermal tissues of the embryo proper (Beddington 1983, Gardner & Beddington 1988). Clones derived from primitive endoderm of embryos as advanced as 5.5 days are still restricted to the extraembryonic lineages (Cockroft & Gardner 1987). However, by 6.5 days, the visceral embryonic endoderm of the egg cylinder is apparently composed of a heterogenous population of cells: most of the axial endodermal cells retain an extraembryonic fate but some descendants of cells located in the distal midline region of the egg cylinder colonize the embryonic gut endoderm (Lawson & Pedersen 1987). As the embryo develops, a progressively larger population of midline endoderm contributes to the embryonic gut (Lawson et al 1986). The immediate question that arose from these studies concerned the origin of this new endodermal population. Fate mapping studies by orthotopic grafting and *in situ* intracellular labelling of ectodermal cells have demonstrated unequivocally that the progenitor population of the gut endoderm is localized in the embryonic ectoderm near the anterior end of the primitive streak (Beddington 1981, Tam 1989, K. A. Lawson, personal communication). During germ layer formation, this progenitor population would have to be relocated to the endodermal layer; this may be accomplished by the exodus of cells from the primitive ectoderm and the integration of these cells into the visceral embryonic endoderm. Several morphological studies have revealed a contiguity of tissues at the anterior end of the primitive streak with midline structures, such as the head process mesoderm and the notochordal plate in the gastrulating embryo. Both the head process and the notochordal plate are closely associated with the midline endoderm, and at the archenteron these structures are exposed to the yolk sac cavity and become a part of the endodermal layer (Poelmann 1981b, Lamers et al 1987). The intimate anatomical relationship between the primitive streak and the archenteron suggests that this may be the site of incorporation of the definitive gut endoderm. The present work was therefore carried out to examine the process of cellular recruitment from the embryonic ectoderm during the establishment of the definitive endoderm. Previous fate mapping studies showed that the formation of the ectoderm and the mesoderm is closely linked to the establishment of an orderly distribution and sequence of differentiation of the definitive fetal tissues (Tam & Beddington 1987, Lawson & Pedersen 1987, Tam 1989). In the present study, special attention was given to the deployment and the developmental fate of ectoderm-derived cells to see if a temporal and spatial allocation of these cells to embryonic structures also occurs during the formation of the definitive endoderm.

The experimental approach

Two different experimental strategies were used to examine the origin and the distribution of the endodermal cells during gastrulation and early organogenesis.

The first approach was to label all the cells in the ectoderm of the primitive streak (PS) stage embryo, then to trace the location of the labelled cells in the endodermal layer during subsequent *in vitro* development. The second approach was to label the pre-existing endoderm, then to examine the relative contributions made by the native cells and by any new population added to the endoderm after labelling. Gastrulating mouse embryos at the pre- and early-PS, mid-PS and late-PS stages were obtained from outbred albino PO and ARC/S strain mice at 6.5, 7.0 and 7.5 days post coitum (p.c.), respectively. The embryonic ectoderm was labelled by microinjecting 10–15 nl of wheat germ agglutinin (WGA)–gold label (30 µg WGA/µl and 8 µg gold/µl, Tam & Beddington 1987) into the proamniotic or amniotic cavity (= ectoderm-labelled embryo). The visceral embryonic endoderm was labelled by incubating the whole embryo in a WGA–gold preparation diluted with M2 medium (= endoderm-labelled embryo). Endodermal cells that were recruited from the ectoderm therefore appeared as WGA–gold labelled endodermal cells in ectoderm-labelled embryo and as unlabelled endodermal cells in the endoderm-labelled embryos.

Embryonic development and analysis

When 6.5 day embryos were examined shortly after microinjection, heavy labelling by WGA–gold conjugate was observed in nearly all cells in the ectoderm (Fig. 1a). In contrast, incubation of the embryo with WGA–gold solution resulted in intense labelling of only the endoderm. Ultrastructural examination revealed the presence of numerous gold-laden endocytotic vesicles in the cytoplasm of the ectoderm and endoderm. Labelled embryos were cultured in Dulbecco's modified Eagle's medium supplemented with 25% human cord serum and 50% rat serum. 66–80% of PS stage embryos developed normally in culture and there was no difference between the development of the WGA–gold labelled embryos and that of the unlabelled control embryos (Table 1A). The total number of endodermal cells in 6.5 day embryos cultured for 48 hours was similar to that in 7.5 day embryos cultured for 24 hours (Table 1B), suggesting that the culture system was capable of supporting normal endodermal development. Embryos were harvested after 4–48 hours of culture and processed for wax histology. Serial sections were cut at 8 µm and the gold particles in labelled cells were visualized by the silver enhancement method (Tam & Beddington 1987).

The number and location of the labelled cells in the endoderm were recorded in every 3rd or 5th section, depending on the size of the specimen. For mapping the position of recruited cells, the endoderm layer of the PS stage embryos was subdivided into four major regions along a vertical plane midway between the anterior and posterior (primitive streak) aspect of the egg cylinder and a horizontal plane halfway from the insertion of the amnion (or amniotic folds) to the tip of the egg cylinder. The four regions (anterior–proximal, anterior–distal, posterior–distal and posterior–proximal) thus represented four tissue

TABLE 1 *In vitro development and the endodermal population of primitive streak stage embryos*

A *In vitro development*

Age of embryo	6.5 d + 24 h		6.5 d + 48 h or 7.5 d + 24 h			
	n	*Late-PS stage embryo*	*n*	*Head fold stage*	*Early somite (1–5) stage*	*Total*
6.5 D Control	10	10 (100%)	10	3	5	8 (80%)
WGA–gold labelled	119	101 (85%)	41	17	10	27 (66%)
7.5 D Control			11	2	6	8 (73%)
WGA–gold labelled			36	15	10	25 (70%)

B *Endodermal cell population*

Age of embryo	Total population		Recruited population	
	No. of cells (n)	*Increase/24 h*	*% Total (n)*	*Increase/24 h*
6.5 days + 4 h *in vitro*	398 ± 44*	—	5.3 ± 0.9 (8)	—
6.5 days + 24 h *in vitro*	880 ± 140 (5)	2.7-fold	16.3 ± 2.1 (21)	3.7-fold
6.5 days + 48 h *in vitro*	2485 ± 560 (5)	2.8-fold	60.4 ± 19 (5)	3.7-fold
7.5 days + 24 h *in vitro*	2735 ± 501 (5)	3.1-fold		

*Estimate based on data of Snow (1976). *n* = number of embryos; PS, primitive streak; WGA, wheat germ agglutinin.

compartments along the craniocaudal axis of the embryo. For embryos that were more advanced than the mid-PS stage, the boundary between the anterior and posterior compartments was demarcated by the archenteron at the anterior end of the primitive streak. There was further subdivision of the endoderm into midline and lateral compartments in the late-PS stage embryo: endoderm lying in the median one-third of the embryo constituted the midline endoderm and the rest belonged to the lateral ectoderm. For the pre-somite to early somite stage embryo, the midline and lateral endoderm corresponded to the endoderm immediately beneath the paraxial and the lateral mesoderm, respectively.

Recruitment of cells to the endoderm

The total endodermal population in the PS stage embryo increased uniformly 2.7–2.8-fold for every 24 hours of *in vitro* development but the recruited population increased at a faster rate of 3.7 times per 24 hours. This differential increase argues strongly in favour of a continuous recruitment of new cells from outside the endoderm, although other factors such as a higher proliferative rate of the recruited population or preferential loss of the native population by cell death or tissue displacement may also contribute to the expansion of the recruited population (Lawson et al 1986, Lawson & Pedersen 1987).

In the 6.5 day ectoderm-labelled embryos examined four hours after labelling, an average of 18.2 ± 4.2 cells in the endoderm (representing $5.3 \pm 0.9\%$ of total endoderm) were labelled with WGA–gold. These newly recruited cells were distributed mostly to the distal endoderm of the egg cylinder (Table 2A, Fig. 2b). Histological examination revealed the presence of a primitive streak in about 80% of the embryos: the streak spanned about half the proximal-distal length of the posterior region of the egg cylinder. Labelled cells located in the posterior endoderm are therefore likely to have been recruited through the primitive streak. However, as many as 43% of the labelled cells were located in the anterior endoderm (Fig. 2a) which was at a considerable distance (100–150 μm) from the primitive streak. In the chick, the formation of the primary hypoblast is accomplished by local delamination of blastoderm cells, mainly in the anterior regions of the embryo (Stern 1990). The presence of labelled endoderm in the anterior region of the early-PS stage embryo suggests that delamination of the ectoderm may be involved in the initial stage of endoderm formation in the mouse embryo as well. The contribution of such non-streak-derived endoderm is likely to be small because those cells constituted only 1–10% of the total anterior endoderm of the early-PS stage embryo. An average of 53 ± 2 labelled cells were found in the endoderm of the mid-PS stage embryo examined four hours after ectoderm labelling. In contrast to the early-PS stage, most of the labelled cells were located in the posterior midline endoderm underneath the elongating primitive streak, particularly in the archenteron (Table 3). These observations indicate that by the mid-PS stage endodermal recruitment is most active at the primitive streak, especially at its anterior end (Fig. 2c). The newly recruited cells were initially concentrated in the midline endoderm. About 9.3% of the cells in the anterior endoderm were labelled; these may have been recruited by local delamination of the ectoderm (Table 3). Regional variation in WGA-binding activity was observed in the ectoderm of late-PS stage embryos examined

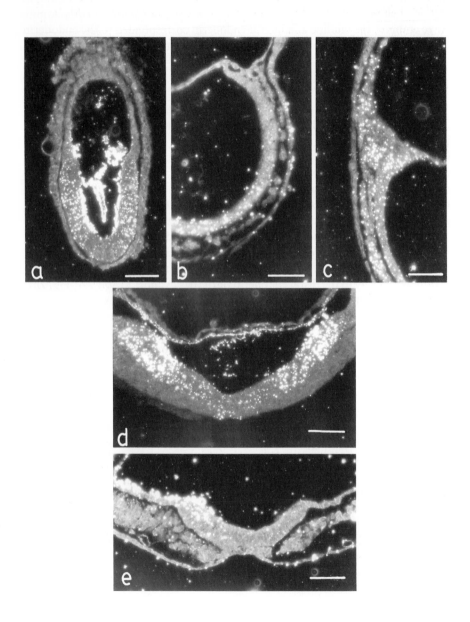

five hours after labelling. Cells in the primitive streak were labelled preferentially compared with cells in other parts of the ectoderm. At the anterior end of the primitive streak, labelled cells were found in the cuboidal epithelium of the archenteron (Fig. 1d) and in the head process immediately anterior to the

TABLE 2   The distribution of recruited cells in the endoderm of gastrulating embryos labelled at the early primitive streak stage with WGA–gold conjugate and cultured *in vitro*

A *Regional localization of the recruited cells four hours after labelling*
% Total recruited cells (n = 8)

Location	Anterior	Posterior	Overall
Proximal	10.5 ± 2.9	18.9 ± 7.4	35.4 ± 8.0
Distal	34.5 ± 7.4	37.1 ± 5.5	65.5 ± 6.9
Overall:	43.2 ± 8.8	56.8 ± 8.5	

B *Extent of cellular recruitment in specific regions of the endoderm*

Time in culture	22–24 h		48 h
Labelled germ layer	Ectoderm (n)	Endoderm (n)	Endoderm (n)
% of total cell population in:			
Anterior–proximal	16.8 ± 2.6 (9)	14.5 ± 4.7 (8)	64.6 ± 7.2 (5)
Anterior–distal	22.3 ± 2.5 (9)	17.9 ± 3.2 (8)	
Posterior–distal	17.2 ± 2.9 (9)	21.0 ± 2.3 (8)	58.5 ± 5.6 (5)
Posterior–proximal	5.6 ± 1.2 (9)	6.2 ± 1.2 (8)	
Anterior–midline	30.5 ± 3.3 (16)	25.3 ± 6.3 (11)	ND
Posterior–midline	13.8 ± 1.4 (16)	21.8 ± 4.0 (11)	ND
Anterior–lateral	20.3 ± 2.0 (16)	16.9 ± 3.0 (11)	ND
Posterior–lateral	11.3 ± 1.5 (16)	13.3 ± 1.7 (11)	ND
Midline	20.4 ± 4.6 (16)	18.0 ± 3.9 (8)	38.2 ± 6.2 (5)
Lateral	16.2 ± 1.6 (16)	13.5 ± 2.1 (8)	72.7 ± 6.8 (5)
Archenteron	72.6 ± 4.2 (10)	ND	ND

n, number of embryos; ND, not determined; WGA, wheat germ agglutinin.

FIG. 1.   Early-PS stage embryos with ectoderm labelled with WGA–gold conjugate and cultured for (a) four hours and (b,c) 22 hours. (a) Heavy labelling of the ectoderm and several labelled cells in the distal endoderm of the egg cylinder. (b) Both labelled and unlabelled cells in the anterior endoderm. (c) Sparsely labelled posterior endoderm and heavily labelled primitive streak, embryonic and extraembryonic mesoderm. (d) A late-PS stage embryo with ectoderm labelled with WGA–gold conjugate and cultured for five hours, showing the incorporation of labelled cells in the midline endoderm at the anterior end of the primitive streak. (e) An early somite stage embryo with ectoderm labelled with WGA–gold conjugate and cultured for 22 hours, showing labelled cells in the midline and lateral endoderm. Silver-enhanced and fast-green counterstained. Bar = 50 µm for all figures.

archenteron. No labelled cells were found in other regions of the endoderm. Results of the present study therefore provide the most compelling evidence for recruitment of ectoderm cells to the endoderm during gastrulation and also point to the role of the archenteron as a gateway for the ectoderm-derived cells to colonize the pre-existing visceral embryonic endoderm.

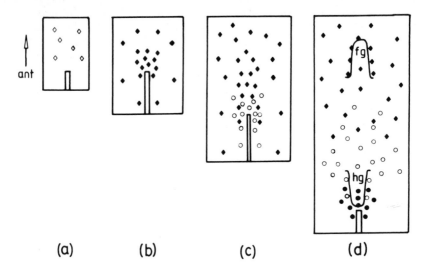

(a)   (b)   (c)   (d)

FIG. 2.   The recruitment and developmental fate of endodermal cells at various stages of embryogenesis. (a) Extra-primitive streak recruitment of cells (◇) in the anterior and distal compartment and (b) from the anterior end of the primitive streak (◆) of the early-PS stage embryo. (c) Continuous recruitment of new cells (○) from the ectoderm at the archenteron and the dispersion of previously recruited cells (◆) to the anterior and lateral endoderm at the mid- to late-PS stage. (d) The distribution of the previously recruited cells (◆,○) to the anterior intestinal portal (fg) and to the prospective foregut. Newly recruited cells (●) contribute to the more caudal gut endoderm (hg) and the notochord.

TABLE 3   **Distribution of newly recruited cells in the endoderm of three mid primitive streak stage embryos four hours after labelling**

	Recruited cells as % total endoderm		
Location	Anterior	Posterior	Overall
Proximal	7.0 ± 3.2	13.3 ± 4.3	
Distal	9.6 ± 3.0	15.0 ± 5.4	
Overall:	9.3 ± 2.7	15.3 ± 4.3	
Midline	11.4 ± 2.6	20.0 ± 3.6	15.8 ± 2.9
Lateral	7.5 ± 2.4	8.3 ± 2.6	8.1 ± 2.6
Archenteron		30.5 ± 5.5	

Deployment of the recruited population

In the embryo that was ectoderm-labelled at 6.5 days and cultured for 22–24 hours, the recruited population made up about 16–22% of the endodermal population in the anterior regions of the late-PS embryo, with more cells in the midline compartment (Fig. 1b, Table 2B). In the posterior region of the embryo, recruited cells were found mostly in the midline endoderm anterior to the primitive streak and particularly in the archenteron. Fewer cells were found in the lateral region and in the endoderm associated with the primitive streak (Fig. 1c, Table 2B). There was a distinct anteroposterior gradient in the distribution of recruited cells in both the midline and lateral endoderm (Fig. 3). A similar distribution of recruited cells was observed in the endoderm-labelled embryos except for a slightly larger recruited population in the posterior midline endoderm near the anterior end of the primitive streak (Table 2).

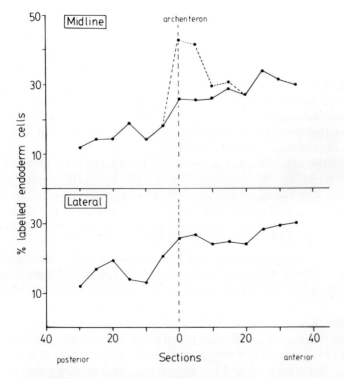

FIG. 3. The distribution of ectoderm-derived labelled cells along the anteroposterior axis in the midline and lateral endoderm of seven embryos labelled at early-PS stage by microinjection of WGA–gold conjugate and cultured for 24 hours. Cells were counted in every 5th section and the embryos were lined up with reference to the archenteron (dashed line). The broken line in the graph for midline endoderm represents the profile of the cell distribution if the archenteron is included with other midline endoderm.

The ultimate distribution of the recruited population in the endoderm is determined by the initial localization and the proliferative activity of the recruited cells, and the degree of intermixing of the recruited and the native populations. Results of the present study indicate that during the formation of the definitive endoderm, prospective endodermal cells are first localized in the archenteron and subsequently distributed to the midline endoderm as the primitive streak regresses. An analysis of the growth of endodermal clones in the early-PS stage embryo has revealed that the ectoderm-derived population may proliferate faster than the native visceral embryonic endoderm (Lawson & Pedersen 1987). A relatively small founder population can therefore make a substantial contribution to the endoderm during development. Eventually, descendants of the recruited cells were distributed to the entire embryonic axis in both the medial and lateral compartments (Fig. 2c,d). Such territorial expansion may be brought about by a change in epithelial cell shape: transformation of the cuboidal cells in archenteron to the squamous epithelium typical of the endodermal layer can result in a fourfold increase in the surface area covered by the population. It is tempting to postulate that the recruited cell population is initially organized as a cohesive cell sheet which, as it expands, will displace the pre-existing visceral embryonic endoderm onto the extraembryonic membranes (Fontaine & Le Douarin 1977, Lawson & Pedersen 1987). However, a complete replacement of the primitive endoderm by the recruited population was never observed, even in the most advanced embryo examined in the present study (Tables 2B, 4), suggesting that there may be instead an extensive intermingling of the ectoderm-derived cells and the native cells in the endoderm. In the avian embryo, a similar mixing of the hypoblast and definitive endoblast has been observed; this is brought about by the fragmentation of the hypoblast epithelium when confronted with the more invasive definitive endoblast (Sanders et al 1978).

Fate of the recruited cells in the endoderm

By the early somite stage, ectodermal cells recruited since the start of gastrulation (unlabelled cells in endoderm-labelled embryos, Tables 2, 4) were found in all parts of the embryonic endoderm and constituted about 60% of the population. Significantly more recruited cells were found in the anterior endoderm (Friedman 2-way ANOVA at $P<0.001$ and paired t-test $P<0.05$). About 65% of the endoderm in the anterior intestinal portal and 52% of 'trunk endoderm', which is a part of the prospective foregut endoderm, was made up of recruited cells (Fig. 1e). The notochord and the archenteron were densely populated by the recruited cells. The endodermal cells recruited during early-PS to late-PS stages are therefore allocated to the presumptive foregut endoderm and the notochord of the early-somite stage embryo (Fig. 2d). Fate mapping studies of the midline visceral embryonic endoderm of the gastrulating embryo have shown that the majority of the cells in the pre- and early-PS embryo are destined for

TABLE 4   The distribution of recruited cells in the endoderm of nine early somite stage embryos labelled at the late primitive streak stage by incubation with WGA–gold conjugate and cultured for 24 hours

Location	Recruited cells as % total
Anterior–midline	51.9 ± 4.9
Anterior–intestinal portal	66.1 ± 6.4
Anterior–lateral	46.5 ± 7.9
Posterior–midline	9.6 ± 1.8
Posterior–lateral	27.7 ± 7.3
Notochord	75.2 ± 2.5
Archenteron	88.8 ± 1.5

WGA, wheat germ agglutinin.

extraembryonic endoderm. Only those descendants of the endoderm adjacent to the anterior end of the primitive streak, which were found to be the newly recruited population in the present study, will contribute to the anterior intestinal portal (Lawson & Pedersen 1987). By the late-PS stage, cells in the head process and the archenteron contribute primarily to the notochord and to more posterior segments of the embryonic foregut. Endoderm posterior to the archenteron received only a minor contribution from the recruited cell population (this study), and it contributes to the post-somite endoderm of the developing embryo (Lawson et al 1986). Results of the present study are therefore entirely consistent with those of the fate mapping study and have, in addition, demonstrated that a substantial fraction of the embryonic foregut endoderm is also derived from the visceral embryonic endoderm of the early egg cylinder. The absence of any significant contribution by the visceral embryonic endoderm to the gut tissues of chimeras produced by blastocyst injection (Gardner & Beddington 1988) suggested that the primitive endoderm may be selected against or diluted to insignificant levels by the more rapid proliferation of the recruited endoderm cells during later stages of gut development. Recruitment of new endodermal cells from the primitive streak continues during early organogenesis: the recruited cells are allocated to the notochord of the trunk and the embryonic midgut and hindgut (Fig. 2d, Tam 1984, Tam & Beddington 1987). Endodermal cells designated for specific segments of the gut and notochord are therefore recruited in a craniocaudal sequence during the establishment of the definitive endoderm.

Results of the present study further showed that ectoderm-derived cells also contribute to the entire mesodermal population of the conceptus (Fig. 1b, c, e). Mesoderm cells are recruited primarily from the embryonic ectoderm by the ingression of cells at the primitive streak (Hashimoto & Nakatsuji 1989, Tam & Beddington 1987). The embryonic mesoderm expands as a result of the cell

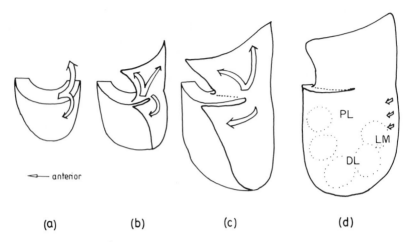

(a)   (b)   (c)   (d)

FIG. 4.   Morphogenetic tissue movement in the formation of the extraembryonic and embryonic mesoderm of gastrulating embryos at (a) pre-PS, (b) early-PS, (c) mid-PS and (d) late-PS stages. Arrows in (a)–(c) indicate the directions of tissue sheet expansion which is proximal–anteriorly for extraembryonic mesoderm and distal–anteriorly for embryonic mesoderm. Arrows in (d) indicate the continuous addition of cells to the embryonic mesoderm from the primitive streak and the circles represent somitomeres in the paraxial mesoderm. DL, distal–lateral; LM, lateral to mid-region of PS; PL, proximal–lateral.

migration at the leading edge of the mesodermal wing (Nakatsuji et al 1986) and the continuous addition of cells from the primitive streak (Poelmann 1981a, Fig. 4a–c). Mesodermal cells destined for the paraxial structures are organized into somitomeres (Fig. 4d) which are laid down in a strict craniocaudal sequence (Tam & Meier 1982). Preliminary fate mapping by orthotopic grafting showed that the mesoderm of the late-PS stage embryo contains the prospective tissues for craniofacial structures. Mesodermal cells in the proximal–lateral regions (Fig. 4d, site PL) contribute to the heart (cardiac mesenchyme, aortic trunk and pericardium), the cranial mesenchyme of the upper hindbrain (4th–6th somitomere, Meier & Tam 1982) and the mesenchyme associated with the foregut and the first two branchial arches. Cells in the distal–lateral mesoderm (site DL) contribute to paraxial mesoderm of the lower hindbrain and the mesenchyme of the branchial arch and foregut. Mesoderm cells located lateral to the mid-region of the primitive streak (site LM) colonize somites 2–8 and the adjacent lateral mesoderm. Comparison of the tissue fate of the mesoderm and that of the overlying ectoderm (Fig. 5, Tam 1989) and the underlying endoderm (this study, Lawson et al 1986) reveals a close topographical association of the prospective tissues allocated to specific body parts. A basic body pattern is therefore organized globally in all three germ layers right from their inception at the start of gastrulation. The allocation of cells to prospective

Ectodermal derivatives

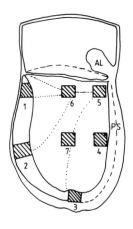

	FB MB HB OT————— SP————————— SE—————————			
1 Ant–Mar	▰ ░			
2 Mid–Ant	▰▰			
3 Arch			▰▰▰	
4 ML–PS			▰▰	▰
5 PL–PS		▰ ░		
6 P–Lat		▰ ░		
7 D–Lat		▰ ▰ ░		

FIG. 5. The developmental fate of cells from different regions of the embryonic ectoderm of the late-PS stage embryo: segmental distribution of cells in different parts of the neural tube (dark bar) and the surface ectoderm (light bar). Ant–Mar, anterior marginal, Mid-Ant, mid-anterior, Arch, archenteron; ML-PS, lateral ectoderm next to the mid region of the primitive streak; PL-PS, lateral region next to the posterior region of the primitive streak; P-Lat, proximal lateral ectoderm; D-Lat, distal lateral ectoderm; AL, allantois; PS, primitive streak; FB, forebrain; MB, midbrain; HB, hindbrain; OT, otic placode; SP, spinal cord; SE, surface ectoderm.

tissues along the craniocaudal axis is likely to be related to the sequence in which the cells are recruited to the mesoderm and endoderm. The ectodermal germ layer that is formed *in situ* after the departure of the mesodermal and endodermal cells is probably deprived of the organizational signal sent by the act of cellular ingression at the primitve streak. The regionalization of tissue fate in the ectoderm is therefore likely to be determined by the positional or inductive cue emanating from the mesoderm and the endoderm (Frohman et al 1990, Stern 1990).

Concluding remarks

The present study provides the most direct evidence that the definitive endoderm is established by the recruitment of ectodermal cells into the existing endoderm layer via cellular insertion at the archenteron. Analysis of the developmental fate of cells in the three germ layers revealed that the prospective tissues for specific body parts are organized in the correct craniocaudal order within the germ layer and also in the appropriate topographical association with their counterparts in other germ layers well before the onset of morphogenesis. The primitive streak thus acts as an organizer for body pattern formation. This is achieved presumably by a concerted ingression and the assortment of related groups of progenitor cells during their transit to different germ layers (Bellairs 1986, Stern & Canning 1990). Such cellular activities are probably related to changes in the expression of adhesion molecules (Takeichi 1988), lectin binding activity (Kimber 1986) and the expression of genes for pattern formation and cell signalling (Sundin et al 1990, Wilkinson et al 1988, 1990). Recent studies on avian embryos suggest that the growth factor-like proteins instrumental in germ layer induction in the amphibian are also involved in the organization of the body pattern (Mitrani et al 1990, Cooke & Wong 1991). Because the process of germ layer formation is remarkably similar in the bird and the mouse, it is possible that these inducing factors play a similar role in mammalian embryogenesis. If the primitive endoderm is the source of the inductive signals, then the persistence of a minor population of visceral embryonic endoderm underneath the primitive streak during gastrulation may be important in the perpetuation of the induction process during axis formation.

Acknowledgements

This work is partly supported by a grant from the NH&MRC of Australia. We thank Professor Peter Rowe for his comments on the manuscript.

References

Batten BE, Haar JL 1979 Fine structural differentiation of germ layers in the mouse at the time of mesoderm formation. Anat Rec 194:125–142
Beddington RSP 1981 An autoradiographic analysis of the potency of embryonic ectoderm in the 8th day postimplantation mouse embryo. J Embryol Exp Morphol 64:87–104
Beddington RSP 1983 The origin of foetal tissues during gastrulation in the rodent. In: Johnson MH (ed) Development in mammals. Elsevier, Amsterdam p 1–32
Bellairs R 1986 The primitive streak. Anat Embryol 174:1–14
Cockroft DL, Gardner RL 1987 Clonal analysis of the developmental potential of 6th and 7th day visceral endoderm in the mouse. Development 101:143–155
Cooke J, Wong A 1991 Growth factor-related proteins that are inducers in early amphibian development may mediate similar steps in amniote (bird) embryogenesis. Development 111:197–212
Enders AC, Given RL, Schlafke S 1978 Differentiation and migration of endoderm in the rat and mouse at implantation. Anat Rec 190:65–78
Fontaine J, Le Douarin NM 1977 Analysis of endoderm formation in the chick embryo at the primitive streak stage as determined by radioautography mapping. Dev Biol 26:323–335
Frohman MA, Boyle M, Martin GR 1990 Isolation of the mouse *Hox-2.9* gene; analysis of embryonic expression suggests that positional information along the anterior-posterior axis is specified by mesoderm. Development 110:589–608
Gardner RL, Beddington RSP 1988 Multi-lineage 'stem' cells in the mammalian embryo. J Cell Sci Suppl 10:11–27
Hashimoto K, Nakatsuji N 1989 Formation of the primitive streak and mesoderm cells in mouse embryos—detailed scanning electron microscopical study. Dev Growth & Differ 31:209–218

Kimber SJ 1986 Distribution of lectin receptors in postimplantation mouse embryo at 6–8 days gestation. Am J Anat 177:203–219

Lamers WH, Spliet WGM, Langemeyer RATM 1987 The lining of the gut in the developing rat embryo. Its relation to the hypoblast (primary endoderm) and the notochord. Anat Embryol 176:259–265

Lawson KA, Pedersen RA 1987 Cell fate, morphogenetic movement and population kinetics of embryonic endoderm at the time of germ layer formation in the mouse. Development 101:627–652

Lawson KA, Meneses JJ, Pedersen RA 1986 Cell fate and cell lineage in the endoderm of the presomite mouse embryo, studied with an intracellular tracer. Dev Biol 115:325–339

Meier S, Tam PPL 1982 Metameric pattern development in the embryonic axis of the mouse. I. Differentiation of the cranial segments. Differentiation 21:95–108

Mitrani E, Ziv T, Thomsen G, Shimoni Y, Melton DA, Bril A 1990 Activin can induce the formation of axial structures and is expressed in the hypoblast of the chick. Cell 63:495–501

Nakatsuji N, Snow MHL, Wylie CC 1986 Cinematographic study of the cell movement in the primitive-streak stage mouse embryo. J Embryol Exp Morphol 96:99–100

Poelmann RE 1981a The formation of the embryonic mesoderm in the early postimplantation mouse embryo. Anat Embryol 162:29–40

Poelmann RE 1981b The head process and the formation of the definitive endoderm in the mouse embryo. Anat Embryol 162:41–49

Sanders EJ, Bellairs R, Portch PA 1978 In vivo and in vitro studies on the hypoblast and definitive endoblast of avian embryos. J Embryol Exp Morphol 46:187–205

Snell GD, Stevens LC 1966 Early embryology. In: Green L (ed) Biology of the laboratory mouse. Dover, New York p 205–245

Snow MHL 1976 Embryo growth during the immediate postimplantation period. In: Embryogenesis in mammals. Elsevier, Amsterdam (Ciba Found Symp 40) p 53–70

Stern CD 1990 The marginal zone and its contribution to the hypoblast and primitive streak of the chick embryo. Development 109:667–682

Stern CD, Canning DR 1990 Origin of cells giving rise to mesoderm and endoderm in chick embryo. Nature (Lond) 343:273–275

Sundin OH, Busse HG, Rogers MB, Gudas LJ, Eichele G 1990 Region-specific expression in early chick and mouse embryos of Ghox-lab and Hox 1.6, vertebrate homeobox-containing genes related to *Drosophila labial*. Development 108:47–58

Takeichi M 1988 The cadherins: cell–cell adhesion molecules controlling animal morphogenesis. Development 102:639–655

Tam PPL 1984 The histogenetic capacity of tissues in the caudal end of the embryonic axis of the mouse. J Embryol Exp Morphol 82:253–266

Tam PPL 1989 Regionalisation of the mouse embryonic ectoderm: allocation of prospective ectodermal tissues during gastrulation. Development 101:55–67

Tam PPL, Beddington RSP 1987 The formation of mesodermal tissues in the mouse embryo during gastrulation and early organogenesis. Development 99:109–126

Tam PPL, Meier S 1982 The establishment of a somitomeric pattern in the mesoderm of the gastrulating mouse embryo. Am J Anat 164:209–225

Wilkinson DG, Bhatt S, Herrmann BG 1990 Expression pattern of the mouse *T* gene and its role in mesoderm formation. Nature (Lond) 343:657–659

Wilkinson DG, Peters G, Dickson C, McMahon AP 1988 Expression of the FGF-related proto-oncogene *int-2* during gastrulation and neurulation in the mouse. EMBO (Eur Mol Biol Organ) J 7:691–695

DISCUSSION

Kaufman: In the initial experiments where you put intact egg cylinders into labelled medium, the cells were harvested after four hours. Exactly how long were the cells in labelled medium? If you labelled for a shorter period, would you get differential labelling?

Tam: The embryos were labelled by incubation in the colloidal gold medium for only 5–10 minutes. They were then washed extensively to remove the unbound label and cultured for four hours before the first sampling for analysis of cell number.

Solter: How often do you think this layer jumping of cells from ectoderm into endoderm, without passing through the primitive streak, occurs? Would such an occurrence in some way affect the interpretation of the fate map that Kirstie Lawson has described? How does it affect your hypothesis that cells become induced (programmed) as they pass through the primitive streak?

Tam: I think that between the blastocyst stage and the onset of gastrulation the primitive endoderm continuously receives new cells from the primitive ectoderm by direct delamination. What we see in the 6.5 day embryo may therefore be the last phase of this recruitment. I would like to believe that the endodermal cells that come from direct delamination are allocated to the extraembryonic lineage. When the primitive streak starts to function, there is a second source of endoderm which contributes to the definitive gut endoderm.

Beddington: Wouldn't Richard Gardner's experiments have then shown that injection of a single smooth epiblast cell into the blastocyst would occasionally lead to colonization in visceral endoderm (Gardner & Rossant 1979)? I don't think he has seen this.

Tam: In the embryos we have studied, at most about 70% of the cells in the endoderm are derived from the recruited population. This suggests that nearly 30% of the cells come from the pre-exisiting primitive endoderm. We also know that both populations colonize the embryonic foregut. The absence of any contribution by primitive endoderm to the chimera produced by blastocyst injection may therefore be explained by selection against the primitive endoderm during development of the gut.

Another possibility is that tissue samples for analysis of glucose phosphate isomerase were taken from the intestine of the chimeras and not from the foregut, where descendants of the primitive endoderm are initially found.

Copp: You mentioned that in the 8.5 day embryo there was intercalation of head process cells into the endoderm on the outside of the embryo. During intercalation, are some of those cells actually contributing to yolk sac? We have cultured embryos around that stage in the presence of iodinated transferrin and looked at receptor-mediated endocytosis (A.J. Copp, J. P. Estibeiro, unpublished). We find a very sharp, single cell cut-off between cells laterally that take up the transferrin, which are probably yolk sac, and cells medially that do not. Our boundary would lie within your region of colonization, so some of your cells may be going into the region that functionally appears to be yolk sac.

Tam: It is possible that cells recruited through the archenteron at the late primitive streak stage also end up in the yolk sac as they disperse laterally. The section (Fig. 1e) through the 'open gut' region of an early somite stage embryo does show the presence of labelled cells in the endoderm further away from the paraxial region.

Copp: Then the single cell injections of smooth cells should have shown cells going through that route into the yolk sac?

Tam: Yes, I agree.

Lawson: Can you be sure that there is no transfer of label as opposed to transfer of cells?

Tam: I cannot be sure of that. However, I have repeated the experiments using DiI to label the cells and found the same pattern of distribution. To test the possibility of indiscriminate transfer of gold label, I have done a grafting experiment using primitive ectoderm cells double-labelled with [³H]thymidine and colloidal gold and found no major discrepancies in any distribution pattern of graft-derived cells.

Lawson: Was that done at these early stages, at 6.5 days?

Tam: Yes. The DiI experiment was done on 6.5 day embryos and the grafting experiment on 7.5 day embryos.

McLaren: You mentioned gene expression in the primitive streak and the idea that the cells might be different from one another because this was the time when they were being specified. Would you care to say anything more about that idea?

Tam: This came from a discussion with Andy McMahon on the expression of genes in the primitive streak. It is no longer appropriate to regard the primitive streak as a single entity. Various genes are expressed by subsets of cells within the primitive streak, some in the ectodermal component, others in the newly immigrated mesodermal cells, and some are expressed throughout the streak. This may depend on the action of the genes, whether they specify pattern or determine fate. If there is a gene responsible for organizing the entire streak, I would expect it to be expressed in all the cells. If a gene were involved in mesodermal patterning, it may be expressed primarily in the mesodermal population.

Herrmann: Concerning the morphology of the ingressing cells, is there a difference between the anterior part of the streak and the posterior part?

Tam: I saw no obvious differences in the morphology of cells in different parts of the primitive streak that could be related to their fate. The anterior end of the streak seems to be contiguous with a group of very tightly packed columnar cells located in the midline. This region has been called the node on the assumption that it is equivalent to Hensen's node in the chick. It has also been described as the archenteron, notochordal plate and head process mesoderm. I don't know the fate of the cells in the node, but they are likely to give rise to prechordal plate and the notochord.

Beddington: The pattern of recruitment of definitive endoderm provides a possible mechanism for preserving the primitive streak. The same population of primitive endoderm could underlie it until at least the late primitive streak stage; so a mesoderm-inducing signal initiated in that population would be present for some time.

Tam: That's right. I want to emphasize that there is a resident population of ancestral primitive endoderm underlying a major portion of the streak. New endodermal cells are not added to it and this population persists at least to the late primitive streak stage. If this population is the source of the mesoderm-inducing signal, then its presence may be required for a fairly long time during axis development.

Herrmann: I wonder about the time scale. Are the cells induced to form mesoderm until they begin to migrate?

Tam: I don't know. The induction has not been looked at in the mouse as extensively as Jim Smith has studied it in *Xenopus*. Everybody would like to repeat his experiments in the mouse, but I don't think there is a good equivalent of animal cap cells in the mouse embryo. The closest may be embryonic stem (ES) cells before they differentiate.

Hogan: I would argue that ES cells are not the closest equivalent of the animal cap because they are unpolarized—they have no organization into an epithelial layer or sheet. A closer equivalent might be the very anterior part of the egg cylinder that, from the fate maps, is going to form the ectoderm and neuroectoderm.

Tam: Cells in the anterior ectoderm do have an ectodermal fate but they may not be fully committed to it.

Hogan: No one has done the experiment. No one has added endoderm or mesoderm-inducing factors to isolated groups of anterior egg cylinder embryonic ectoderm cells.

Tam: The most relevant experiment has been done by Rosa Beddington. Heterotopic grafting of anterior ectoderm of late primitive streak stage embryos (7.5 days) showed that these cells are biased towards a neuroectodermal fate. No one has looked at earlier stages.

Beddington: The difficulty is that anterior ectoderm can still readily form mesoderm. If you explant, ectopically graft or heterotopically graft anterior ectoderm in the mouse embryo, it can make mesoderm under all those situations. There is no clear-cut tissue that will make ectoderm in isolation and make mesoderm only in the presence of an inducer.

Hogan: Rosa, did you actually separate the anterior epiblast or ectoderm from the underlying endoderm?

Beddington: Yes, and they both form mesoderm equally well under the testis capsule.

Hogan: Under the testis capsule they could be getting mesoderm-inducing factors.

Smith: Only at very low levels (Meunier et al 1988).

Beddington: Let's consider the reverse experiment. If you take epiblast from a region you would suppose to be induced in the egg cylinder, either at the posterior end of the streak or at the front of the streak, and put it into the anterior region, it happily forms ectoderm despite the fact that it should have already received a mesoderm-inducing signal. That is at the tissue level, not the single cell level. There seems to be considerably more lability in the mouse embryo during gastrulation than in the amphibian embryo, where the time points at which decisions are made are a bit clearer.

Herrmann: The midline of the primitive streak where cells are induced to form mesoderm is equivalent to the marginal zone in the *Xenopus* embryo. Lateral to the midline and the anterior half of the embryo would correspond to the animal cap. If you split the primitive streak along its midline and form a circle, you have a structure similar to a *Xenopus* embryo.

Smith: But in *Xenopus*, isolated animal caps only form epidermis. Analogous regions of the mouse form mesoderm.

Herrmann: But if you add inducing factors to the *Xenopus* animal cap cells, the cells will form mesoderm. In the mouse, if the cells that are induced to make mesoderm have ingressed and the others follow, the latter can also be induced and form mesoderm.

McLaren: You suggested that induction of the germ layers is likely to have taken place before the onset of gastrulation. What do you mean by that?

Tam: The process of gastrulation or germ layer formation is perhaps the earliest morphological manifestation of any spatial pattern of tissue determination. The pattern has to be specified slightly earlier than the event of morphogenesis leading to germ layer formation. This may involve germ layer induction, whereby specific populations of cells in the primitive ectoderm are biased towards different pathways. If you look at differentiation of embryonal carcinoma (EC) cells and ES cells, there can be germ layer-type cell differentiation *in vitro* without any of the major morphogenetic events of cell migration and primitive streak formation. The only prerequisite for ectodermal or mesodermal differentiation seems to be the formation of primitive endoderm. If induction does occur between the EC or ES cells and the newly generated endoderm, this may be sufficient to initiate phenotypic differentiation and the expression of collagen genes (E. Lau, K. S. E. Cheah, P. P. L. Tam, unpublished). These observations suggest that the specification of cell fate is independent of the act of gastrulation and that certain determinative events related to germ layer induction may have happened before that.

Evans: You suggested that the ability of ES and EC cells to differentiate without going through gastrulation proves that these cells do not need this type of influence. I think that's a misinterpretation, particularly of the EC cell type of differentiation. It is possible to get EC cells to differentiate *in vitro* in disorganized situations but one has to treat them with specific inducing agents or start with very specifically selected EC cell lines that are not in themselves normal.

The information about cell commitment that EC and ES cells provide comes from results that Gail Martin and I obtained a long time ago that the best differentiation occurs when the cells are allowed to go through a pseudo embryoid differentiation (Martin & Evans 1975, Evans & Martin 1975). This suggests to me that it is the topological orientation that's important in cell commitment and not the other way around. The cellular differentiation which it is possible to obtain *in vitro* is artifactual and the embryoid organization—with in good cases an ectoderm, mesoderm, endoderm type of organization—is the model of the normal process.

Hogan: In those experiments the mesoderm forms everywhere: it is nothing like a primitive streak, there is no polarity. There is no positional information in the mesoderm that is coming out. There is no anteroposterior or dorsoventral axis in those embryoid bodies. Maybe you need to implant the anterior end of a primitive streak into those bodies.

Evans: I think that's why you get disorganized differentiation and never get an embryo out of it.

Joyner: But the differentiation goes through a defined set of events: you don't get the mesoderm until a nice layer of endoderm has formed around the ectoderm.

McLaren: So it's organized but not positionally organized.

Beddington: It's almost uninterpretable at present. We have no early tissue-specific markers to say what type of cells we are looking at, unless it's primitive endoderm which we can be fairly confident about. For definitive endoderm, there is no assay to say that is being formed as opposed to primitive endoderm. The same applies to a lot of the experiments and lineage data we have heard about this morning. We desperately need markers that say we are looking specifically at one type of tissue rather than another.

McMahon: Gail Martin has reported that a small percentage of embryoid bodies do have a structure that closely resembles a primitive streak (Martin et al 1977).

Smith: Patrick, do you think the process of gastrulation is a playing out of existing anteroposterior information that these cells have or that the anteroposterior information arises as a result of the cell movements?

Tam: Kirstie Lawson's fate map for the early streak and pre-streak stage embryo shows that the mesodermal patch in the epiblast has an anteroposterior axis, but this is reversed—the anterior cells are towards the back. If that is a pre-pattern, then gastrulation is a realization of this pre-pattern achieved by organizing the cells into an anteroposterior pattern and turning them around during ingression.

Smith: But that isn't telling you anything about the way the cells are organized or even when it happens.

Lawson: Karen Downs is doing grafts in 6.5 day embryos to study the primordial germ cell problem. This approach may also give us information about pre-patterning or the lack of pre-patterning. She is able to take very small groups of cells and position them quite precisely, even in the 6.5 day embryo. So she will be able to see whether there is any difference between posterior and anterior regions at 6.5 days, and between distal and proximal, by making the appropriate orthotopic and complementary heterotopic grafts.

Smith: Patrick, you said that cells form various sorts of mesodermal cell types without passing through the primitive streak. It is not known whether those cells express any regional marker i.e. whether they have any anteroposterior positional information.

Tam: I separate the two events: the initial inductive event just tells the cell to become mesoderm; regional specification comes at the time of gastrulation. The situation in *Xenopus* may be different.

Smith: It's not known how much information there is at the time of mesoderm induction in *Xenopus*.

Beddington: Jim and I tried to see if *Xenopus* MIF (mesoderm-inducing factor) would induce mesoderm in mouse egg cylinders. In order to get the mesoderm-inducing activity to the appropriate surface of the ectoderm, we removed small patches of endoderm before there was any mesoderm present. We saw cells delaminating that looked exactly like mesoderm in the anterior region, in the presence or absence of mesoderm inducer. The problem is that rat serum is probably full of mesoderm-inducing factors, and there are no neutral culture conditions for mice embryos.

Balling: If gastrulation is required for positional information, wouldn't you predict that the cells that are derived by direct delamination wouldn't have positional information?

Tam: They may not have acquired any anteroposterior information. We have a similar problem in the formation of definitive ectoderm. Cells in this layer never go through the primitive streak and it remains unclear as to how they acquire the information to differentiate correctly into, say, forebrain and midbrain. One can get around this by postulating that the information is provided by the mesodermal cells that have ingressed through the primitive streak and come to lie under the ectoderm. The same signal may also be given to the endoderm cells derived by direct delamination.

Solter: We can imagine that during gastrulation the cells stream like a waterspout from the primitive streak toward the primitive endoderm. Some cells will be incorporated into primitive endoderm and become definitive endoderm; others will go under the endoderm and become mesoderm. Mesoderm differentiation is the default pathway and endoderm differentiation is an instructory pathway. That means the cells don't have to pick up any information going through the primitive streak.

Tam: The problem there is that only the cells that go through the anterior end of the streak will have a chance to become endoderm. The cells not being recruited at that site are likely to end up as mesoderm. You are proposing that the germ layer's allocation during ingression through the primitive streak is totally opportunistic—cells happen to be there and become mesoderm. I don't think the endoderm behaves like that: you have to look at the anterior end of the streak to see the definitive endoderm coming through. I have no evidence that cells at the posterior end of the streak will become endoderm. In the experiments I have done with Rosa Beddington, we never saw any colonization of the gut endoderm by labelled cells grafted into any regions of the streak, apart from the anterior region (Tam & Beddington 1987).

Joyner: Couldn't that be due to a defined set of cell movements that specifies which cells will go through the anterior streak and later become induced to form endoderm?

Tam: It could be, but then you have to ask why the hiatus is present only at the site where recruited cells are incorporated into endoderm. Why could there not be several sites of incorporation?

Saxén: Is any information available as to what kind of extracellular molecules might be involved in this guided migration of the cells here?

Tam: Takeichi (1988) has studied the expression of cadherins during embryogenesis. In the primitive streak embryonic cells express both mesoderm-specific (N-) and endoderm-specific (E-) cadherins. Mesodermal cells leaving the primitive streak then stop expressing E-cadherin. Once the cells have been sorted into mesoderm and endoderm, they express only the relevant types of cadherin.

McLaren: I would like to go back to a conceptually simpler problem, the growth of the egg cylinder. Do we know whether the mitotic rate is equal throughout the egg cylinder as it extends? Are those mitoses oriented or are they equally distributed in all dimensions? Is cell mixing equal in all dimensions? Is it equal within or between domains or is it restricted?

Beddington: It seems quite likely that cell mingling occurs as the pro-amniotic cavity is formed. That's a major horror show for the embryo, to have a hole made in the middle of it. If you inject single inner cell mass cells that express *lacZ* into the blastocyst and look at their distribution in the egg cylinder, even at 7.5 days they appear intimately intermixed. It doesn't answer the question, but mixing obviously continues within these fate domains—whether it's just within them or also across borders, I don't think we can say.

References

Evans MJ, Martin GR 1975 The differentiation of clonal teratocarcinoma cell culture in vitro. In: Solter D, Sherman M (eds) Roche symposium on teratomas and differentiation. Academic Press, New York

Gardner RL, Rossant J 1979 Investigation of the fate of 4.5 day post coitum mouse inner cell mass cells by blastocyst injection. J Embryol Exp Morphol 52:141–152

Martin GR, Evans MJ 1975 Differentiation of clonal lines of teratocarcinoma cells: formation of embryoid bodies in vitro. Proc Natl Acad Sci USA 72:1441–1445

Martin GR, Wiley LM, Damjanov I 1977 The development of cystic embryoid bodies *in vitro* from clonal teratocarcinoma stem cells. Dev Biol 61:230–244

Meunier H, Rivier C, Evans RM, Vale W 1988 Gonadal and extragonadal expression of inhibin α, βA and βB subunits in various tissues predicts diverse functions. Proc Natl Acad Sci USA 85:247–251

Takeichi M 1988 The cadherins: cell–cell adhesion molecules controlling animal morphogenesis. Development 102:639–655

Tam PPL, Beddington RSP 1987 The formation of mesodermal tissues in the mouse embryo during gastrulation and early organogenesis. Development 99:109–126

General discussion I

Problems with cell markers and embryo cultures

Buckingham: One problem with the fate mapping studies and cell lineage experiments in the mouse is the lack of early markers to distinguish the phenotypes of some of these cells. Are people willing to list the markers of this kind that they use at the moment to distinguish different cell types?

McMahon: This is difficult because a number of markers have been reported in the literature as being expressed in the primitive streak which may not be expressed in the primitive streak itself, but somewhat laterally. What this entails now is doing a close analysis with the various genes that have been described. These include genes that we will hear about later like *Brachyury*, growth factors like *int-2* and some of the other FGF family members, plus some of the *Hox* genes. To date there has been no comparative analysis of all of these, and I think that's what is called for. *Brachyury* would be a very useful point of reference.

Goodfellow: I haven't clearly understood what you need to do to resolve the questions that have been asked.

Beddington: We would like to know when any lineage becomes committed and self-contained as a population; first as a tissue type, secondly as a spatial domain. Those are the questions one wants to ask in pattern formation. There is no experimental evidence that says when notochord is committed as a self-contained population, which might give some clue to which gene is involved in the pathway associated with that decision. We haven't defined clearly enough in experimental terms what questions we can ask sensibly with the molecules that we are finding.

Goodfellow: If you could define a cell, do you have the rest of the technology to find out which are the descendants of that cell? If you were given enough molecular markers of cells, could you do the transplantation experiments that would answer the question?

Beddington: Only when we have very early tissue markers. The problem in the mouse is that all these things are happening after implantation. One is restricted to short-term experiments of no more than 48 hours. The decisive answer comes if you can look at all the descendants of a single cell and see what tissues they form when there is no ambiguity about the cell types produced. At present, there is ambiguity; one defines cell type by cell position rather than by tissue-specific markers.

Hastie: I take it part of the problem also is getting enough material from a particular embryonic cell type to allow you to carry out biochemical and biological studies. Are people going to be able to use systems such as the immortomouse, for example? (This is a transgenic mouse that carries a temperature-sensitive SV40 virus T antigen driven by a ubiquitously—in theory—expressed promoter. Stem cell lines from various tissues have been established from this mouse at the permissive temperature; Jat et al 1991). Will you be able to get transformed cell lines which represent different stem cells that you could incorporate into a system like this?

Hogan: We already have ES cells. Why do you need more?

Hastie: I am asking about something that has gone beyond that stage. Cell lines that are not totipotent but have some limited potency. For example, could we get kidney blastemal stem cells that give huge amounts of material, which we could manipulate genetically and put back into the *in vitro* nephrogenesis system of Lauri Saxén (this volume).

Hogan: The important thing we are talking about now is how positional information is established.

Goodfellow: But Rosa said you can't define the cell you start with, and you can't define the cell you end up with, which rather limits the interpretation of your experiments!

McLaren: That's absolutely right. At the stage when you want to start transplantation experiments—and you can't answer any of these questions if you leave cells where they are—you don't know which bit of the embryo is which. If you do transplant part of the embryo, you then want to know what cell types develop in it during the period that you are allowed by *in vitro* culture, and we lack the cell markers to identify those cell types.

Lawson: Apart from primordial germ cells, which are problematical, one lineage may be amenable and that's the heart lineage. When one finds clones in heart they are quite compact; they tend to come from a well defined area at the early streak stage . One could focus in on that area and we know more or less where it is (Fig. 8, Lawson, this volume). Specific heart markers are becoming available. It is no problem to find heart cells in the 8.5 day embryo.

Buckingham: The structural genes of the heart, cardiac actin and cardiac myosin, are expressed extremely early. One could go back and see if one could follow cells on that basis.

Hogan: Xenopus has no more markers than mouse: cardiac actin, N-CAM, maybe a few others, but people working on *Xenopus* have done some very beautiful experiments suggesting which molecules could be responsible for patterning in early embryos. The problem is that their experiments were achieved with an easy assay that could be quantified quickly.

People have been trying experiments to get mesodermal induction in ES cells by culturing ES cells with different factors, for example activin, and it doesn't seem to work. The question is what in the mouse embryo is substituting for the mesoderm-inducing factors in *Xenopus*. We could do the experiments in mouse with the tools we have, if we knew how to ask the question.

Joyner: Maybe ES cells aren't ready to respond to mesoderm-inducing agents.

Hogan: That's what I mean, it comes back to 'What is the equivalent of the animal cap in the mouse?'

McMahon: You can't do these experiments unless you can grow cells in simple defined medium. ES cells may be capable of responding but they may receive too many conflicting signals.

Goodfellow: This begs the question of what it's like inside a uterus.

Smith: What people who work on the mouse need is something larger, something which is accessible and something which will develop in a simple salt solution—an amphibian.

Beddington: But something that is not tetraploid!

Smith: In the amphibian there is evidence for the existence of inhibitors of induction as well. Perhaps when you remove cells from the mouse embryo, you are taking them away from the inhibitors and they form mesoderm as a result of that. We need to identify the inhibitors in the frog, so people can see if similar inhibitors are present in the mouse embryo.

Hogan: There is still controversy as to whether these things work *in vivo* for *Xenopus*.

McMahon: Clearly, what has been valuable in the frog experiments is the ability to add components. In the mouse, the strength is the ability to take away components using genetic techniques like homologous recombination. This is going to be one of the most important aspects of the mouse studies.

Hogan: An important difference between the egg cylinder and ES cells is that the egg cylinder is organized into an epithelial sheet. Some polarity may be needed in order to get a response. Maybe the cells don't switch on the receptors for the patterning molecules until they are organized into an epithelial layer with basal and apical surfaces.

Solter: At the time the primitive ectoderm is formed— and all of us would agree that primitive ectoderm is unorganized, radially symmetrical and consists of cells that are basically totipotent—the ectoderm is for a relatively short time in contact with another type of cell, the extraembryonic ectoderm. This type of contact never happens in *Xenopus* or *Drosophila*. What is the extraembryonic ectoderm doing during this brief period of contact? What kind of organization does it impart to the embryonic ectoderm? Extraembryonic ectoderm looks basically like a single sheet of cells. How does it get delaminated so precisely and then pushed away? Is there any evidence that extraembryonic ectoderm induces or affects primitive ectoderm?

Beddington: It inhibits α-fetoprotein synthesis in visceral endoderm.

Solter: Can one culture the embryonic ectoderm without it contacting extraembryonic ectoderm and see whether it will continue normally?

McLaren: The embryonic part of the egg cylinder is quite unhappy in culture without its extraembryonic ectoderm.

Beddington: It doesn't need Reichert's membrane, which will impair it's development because the membrane doesn't expand in culture.

Solter: I am talking about the place where yolk sac will subsequently form: at one stage it is basically a line of contact. The two completely different lineages meet briefly at that line and then split apart.

Beddington: They are not in very intimate contact. If you take an embryo out, once you have removed the endoderm it will fall apart naturally at the embryonic/extraembryonic ectoderm join. They are separate epithelia, not continuous.

Solter: But they touch each other very closely.

Hogan: Some time ago, we cultured immunosurgically isolated inner cell masses in suspension (Hogan & Tilly 1978). In most cases, they re-formed endoderm and then would differentiate like embryoid bodies with random mesoderm formation. If some trophoblast cells were left behind or, more likely, if some inner cell mass cells were still able to differentiate into trophoblast, then the inner cell mass formed something quite organized like a primitive streak (see Hogan & Tilly 1978, Figs 5 and 6), as though it was the interaction between the epiblast and the trophoblast that signalled 'Localize the production of mesoderm here'.

McLaren: How like a primitive streak?

Hogan: The mesoderm was formed in only one place rather than being all the way round the structure.

Joyner: Rosa, what are the prospects for extending the length of embryo culture, because that would clearly help determine the end-point, even if we don't know the starting point?

Beddington: I am not very optimistic. The growth medium is a complete soup. I think 2.5 days is about the maximum.

Joyner: Is it longer for rat embryos?

Beddington: It is not dramatically longer if you start with the early stages. A low percentage of rat embryos develop a bit further; you probably have another day.

Joyner: A day would get the embryos to a point where there would be more recognizable cell types.

Beddington: In the end we really want biochemical markers, so that we do not have just a subjective assessment.

McLaren: The culture period available at present should be enough to have the positional information all laid down, if only one had the cell markers to examine the tissues that are formed.

Beddington: It's probably going to become circular in that you define the marker by what you look at morphologically. Hopefully, you can then look at those markers a bit later and reassure yourself that they really are specific.

Goodfellow: The markers allow you to subdivide morphologically identical structures.

Hastie: Are people looking for new markers by the new high resolution methods? What are people doing about this?

Beddington: We are planning to do subtractive hybridization using germ layer-specific cDNA libraries with Davor Solter and Barbara Knowles.

Saxén: What is the restricting factor in the culture of early embryos? What happens after 2.5 days?

Beddington: After about a day and a half everything starts to go off: the division rate slows, growth decreases. Metabolism is affected: the embryos are not getting enough oxygen, not getting enough nutrients. After 2.5 days you are seeing the accumulation of all these effects and the result is often a pretty abnormal embryo.

Lawson: I agree, retardation of growth is very important, particularly for 6.5 day embryos. I suspect there is slight retardation anyway, which will build up and give abnormalities, especially in the head region and in the neural ectoderm. My impression for later stages is that functionally one of the things that slows development is abnormalities in circulation; the heart may keep on pumping but unless the circulation stays well open the embryo begins to suffer. There is clogging up of blood vessels in the yolk sac around where the placenta would normally form and there tends to be stagnation in the cranial region. I don't know what causes this. Perhaps we should look more closely at how the circulatory system is laid down. From other work on vasculogenesis and angiogenesis we know quite a lot about the factors involved.

Kaufman: There may be some clues from the culture of slightly later, say late egg cylinder or even primitive streak stage, embryos. If you leave them in culture for relatively short periods and then look at the ultrastructure of the yolk sac, it is very different from a freshly isolated yolk sac. Deterioration of the yolk sac is probably the restricting factor from the point of view of nutrients and oxygen going into the embryo. It is virtually impossible to maintain in culture an embryo in which the yolk sac looks reasonable for even a few hours. You have to put higher and higher oxygen levels into the culture system, which must become more and more toxic. The embryo is probably protected to a limited extent because it is more distant from the culture environment, but the yolk sac is at the dividing line with the outside environment and is inevitably exposed to a higher and higher, potentially toxic level of oxygen.

Lawson: What are these ultrastructural changes?

Kaufman: The microvilli on the surface of the endoderm cells look very abnormal; in some cases their distal region dilates and they have the appearance of a drumstick. The whole of the thickness of the yolk sac and the basement membrane is very different from normal.

Copp: One should bear in mind that the embryo is not normally rolling around in a bottle; some of the abnormalities may represent adaptation to an unusual environment.

In relation to extending the culture period, we are able to culture embryos to this stage because they depend on the yolk sac circulation and therefore have a large surface area. We are never going to be able to culture beyond the time at which the yolk sac needs to be superseded by the placenta, unless some new technology is developed such as Dennis New (1990) is trying, to replace the placenta.

Lovell-Badge: Can one culture whole embryos for a day or two then cut out little bits and culture them further?

Beddington: We tried that once with somite grafts, looking at colonization of the forelimb bud. Limb bud culture in mice is suboptimal compared with the chick anyway. It didn't improve things. If you strictly restricted yourself to one day in whole embryo culture and then tried organ culture, it might be alright. If you try to catch the embryo when it's already in decline, you may introduce a lot of artifacts.

Three-dimensional representation of gastrulation in the mouse

Beddington: The object of this discussion is to give some idea of the three-dimensional structure of the mouse embryo during gastrulation. This is intended to provide some guidance for interpreting the expression patterns of the ever-increasing number of genes being studied at these stages. The stages I am going to talk about are shown in Fig. 1 and illustrate pre-, early, mid- and late primitive streak stages, ending with the headfold stage. The fate maps of epiblast at these stages are shown at the bottom of Fig. 1, illustrating approximately which regions will give rise to which tissues during the course of gastrulation. Before I start I should emphasize that the age attributed to each developmental stage is very approximate and will vary not only between mouse strains but also within a single litter from a particular strain. I would also like to take this opportunity to re-christen the structure known as the 'archenteron' in the mouse. This is a misnomer since it is not equivalent to the archenteron of amphibians but, as far as we can tell, corresponds to the dorsal blastopore lip of *Xenopus* or Hensen's node of the chick. Therefore, I would suggest that we call it the 'node'.

The pre-streak egg cylinder is shown in Fig. 1A. It is probably simplest to think of the embryo as radially symmetrical about its proximodistal axis at this stage. Essentially, it is a closed bilaminar cylinder with the proamniotic cavity at its centre. It can be subdivided into two halves: proximally the extraembryonic region (where extraembryonic ectoderm constitutes the inner layer) and distally the embryonic region (where epiblast lines the proamniotic cavity). In all the diagrams a dashed line marks the junction between the embryonic and extraembryonic regions. Some asymmetries may exist at the pre-streak stage but at present there is no unequivocal evidence which correlates any described asymmetry with the future position of the primitive streak, and therefore the anteroposterior axis of the embryo can be defined only once the primitive streak forms at the posterior-most aspect of the embryo (Fig. 1B).

The initial formation of the primitive streak is signified by a thickening of the junction region owing to the emergence of mesoderm at what can now be called the posterior end of the embryo. This nascent mesoderm moves in two directions: proximally to push its way into the extraembryonic region (between the extraembryonic ectoderm and primitive endoderm) and laterally around the circumference of the cylinder in both the extraembryonic and embryonic regions. In effect, this means that mesoderm can reach the anterior extreme of the embryo without having to move along the length of the anteroposterior axis.

At the mid-streak stage (Fig. 1C) the primitive streak extends about two-thirds of the way down the posterior side of the embryo. If the primitive endoderm is enzymically removed at this stage (it has been removed from the posterior half of the cylinder in these diagrams), one can see two wings of mesoderm on either side of the embryonic region which remain physically attached to the epiblast only along the length of the primitive streak. Mesoderm emerging from the posterior part of the streak continues to move into the extraembryonic region. In the posterior extraembryonic region, and to a lesser extent anteriorly and laterally, the mesoderm accumulates and increases in volume because of the formation of coalescing intercellular lacunae—large spaces appear within the extraembryonic mesoderm contributing to its expansion. As a result of this, the most proximal rim of epiblast and the most distal part of the extraembryonic ectoderm are pushed towards the centre of the proamniotic cavity. This junctional bulging into the proamniotic cavity, which is most pronounced in the posterior region, is what makes up the amniotic folds. Eventually, these folds meet and fuse and a continuous layer of extraembryonic mesoderm, now with its own internal cavity (the exocoelom), separates extraembryonic ectoderm from epiblast. Thus, the chorion (extraembryonic ectoderm and extraembryonic mesoderm) and the amnion (epiblast and extraembryonic mesoderm) are formed.

By the late streak stage (Fig. 1D), expansion of the exocoelom results in a three-chambered egg cylinder. The embryonic part of the cylinder lies distally, surrounding the amniotic cavity and separated from the exocoelom, the cavity of the future visceral yolk sac, by the amnion. Proximally, the exocoelom is bounded by the chorion. In the embryonic region, the primitive streak now extends to the distal tip of the cylinder and the anterior extreme of the streak is morphologically recognizable as the node. The node is about 10–15 cells in diameter and is marked by a clear indentation at the tip of the cylinder, due partly to the epithelial-like morphology of node cells and partly to a hiatus in the outer endoderm layer in this region. Effectively, the node represents a region where delaminating epiblast has direct access to the outer surface of the cylinder. Emanating anteriorly from the node in the midline is the head process; lateral to this, definitive endoderm is emerging, also from the anterior end of the streak,

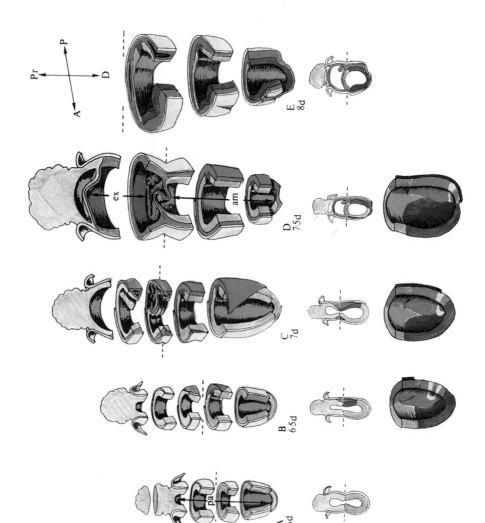

A
6d

B
6.5d

C
7d

D
7.5d

E
8d

Pr

P

A

D

pa

ex

am

and intercalating directly into the outermost epithelial layer. The head process will go on to form the prechordal plate and is continuous posteriorly with the notochord, which is also derived from recruitment of cells in the node region. Initially, the head process probably moves about 200–300 µm anteriorwards relative to the overlying ectoderm. This will bring its anterior limit in line with its future cranial boundary in the neurectoderm. Thereafter, the head process and notochord elongate dramatically in concert with the extensive expansion of the neuroectoderm but probably do not migrate relative to it.

By the late primitive streak stage mesoderm has formed a complete intermediate layer in the embryonic region and may well already be patterned into paraxial and lateral components. Extraembryonic mesoderm is still being produced at the posterior end of the streak and some, perhaps most, of this is forming a distinct structure, the allantois, situated in the midline emerging from the posterior end of the embryo. In due course, the allantois will traverse the exocoelom and fuse with the chorion to contribute a major component of the chorioallantoic placenta and to provide a direct link for nutrient exchange between the embryo and the placenta.

At the head fold stage (Fig. 1E) gastrulation begins to merge with organogenesis. Epiblast continues to invaginate through the streak to maintain a supply of posterior mesoderm while anteriorly differentiation of the future organ systems begins. By this stage neurectoderm formation is apparent and largely responsible for the formation of the head folds. Organization of the cardiac mesoderm is underway and the first pair of somites will shortly appear near the distal tip of the cylinder. Within 36 hours the cylindrical epithelial sheet of epiblast has given rise to at least four distinct categories of tissue: definitive ectoderm, definitive endoderm, mesoderm and the head process. These, by virtue of their own differentiation, morphogenetic movements and interactions,

---

FIG. 1. (*Beddington*) A diagram depicting the three-dimensional arrangement of tissues during gastrulation in the mouse. Midline sagittal sections of the five stages are drawn beneath each three-dimensional representation. The embryonic/extraembryonic junction is signified by a dashed line. The extent of the primitive streak is marked by a solid black line. The fate map of the epiblast is shown for the early (B), mid (C) and late (D) primitive streak stages at the bottom. The fate map at 6.5 days is based on clonal analysis (Lawson & Pedersen, this volume); that at 7.5 days is based on orthotopic grafting experiments (Beddington 1983, Tam & Beddington 1987, Tam 1989) and comparison of clonal analysis of embryos labelled at 6.5 days and cultured for 24 and 36 hours (Lawson & Pedersen, this volume). The map at 7.0 days has been inferred from those of the 6.5 and 7.5 day embryos. N.B. The boundaries shown are approximate and in reality there is considerable overlap between the presumptive regions. The proximal (Pr)/distal (D) and anterior (A)/posterior(P) coordinates are shown at the top right. pa, proamniotic cavity; ex, exocoelom; am, amnion. Grey, extraembryonic ectoderm/trophoblast; magnolia, primitive endoderm; turquoise, epiblast/uninvaginated ectoderm, surface ectoderm in fate maps; green, extraembryonic mesoderm; red, embryonic mesoderm; magenta, primitive streak; brown, node/head process/notochord; yellow, definitive endoderm; dark blue, neurectoderm.

establish both the basic building blocks and the appropriate floor plan for subsequent organogenesis. In addition, the mesoderm that moves into the extraembryonic region propels the critical partitioning of the conceptus and the development of the embryo's two major nutritive organs: the visceral yolk sac and the placenta.

Jaenisch: Could you put the notochord in?

Beddington: The origin of the notochord is precisely in the midline. The hiatus is about 10–15 cell diameters; it sits as a little patch in the bottom of the cylinder. At the mid-streak stage cells are coming out of the midline but there is no obvious external hiatus.

Smith: In *Xenopus*, the dorsal lip of the blastopore is where gastrulation begins: in the mouse, the equivalent region, that we have agreed to call the node, is where gastrulation finishes.

McLaren: Perhaps in the mouse there are early events of gastrulation, that we don't know about, that occur at that location.

Smith: Eddie de Robertis has isolated a gene from a *Xenopus* dorsal lip cDNA library (Blumberg et al 1991). The gene is called *goosecoid*, because it shows homologies with both *gooseberry* and *bicoid* in *Drosophila*. He has an *in situ* at 6.5 days in the mouse, which seems to show that *goosecoid* is expressed in what we are going to call the node (personal communication).

Lawson: I think the node is present in the mouse earlier than when we first see it—things are happening at the anterior end of the streak. The streak seems to be moving forward; it probably isn't in reality doing that.

Beddington: How can you say that the streak is not moving forwards?

Lawson: Because of the way the streak is forming behind its anterior end— all the material that is coming in from the proximal, the lateral and the anterior lateral areas is going in behind the anterior end of the streak. This is the simplest explanation that fits the results from the clonal distributions.

Beddington: You can't identify the anterior end of the streak as an entity when it starts.

Lawson: The very beginning is the time that we know so little about. Initially, the streak is difficult to identify; all you can see is that the mesoderm starts to form locally at the junction between the embryonic and extraembryonic regions. In any one litter, you will find embryos in which there is no indication of the streak, and also embryos in which the streak seems to be a quarter to half of the way down to the distal tip of the egg cylinder with a thickening of the epiblast and probably some mesoderm formation. These embryos overlap in size, suggesting that initially the streak develops by rapidly mobilizing cells over a limited distance, say about 80 μm or about 10 cell diameters.

If you label cells at the anterior end of the steak at the early streak stage, there are still labelled cells in the anterior end of the streak at the neural plate stage, whereas there are no labelled cells from the rest of the early streak still in the streak at the end of gastrulation. During that first day the streak is acting as a conduit; it lengthens by transitorily incorporating the expanding population from lateral and then anterolateral parts of the epiblast.

Beddington: So there has to be more growth proximally, in the streak, than distally?

Lawson: Yes, as far as the streak is concerned, but there is axial growth distally, anterior to the streak. There seem to be two main ways in which the ectodermal component of the axis lengthens. Using the zones defined earlier (Fig. 3, Lawson, this volume), the axis at the early streak stage can be divided into five equal lengths (Fig. 2). By the early neural plate stage in culture the axis has doubled its length, but the contribution of the original zones is not equal. Zone I spreads a little within the axis, but mainly expands laterally around the epiblast cup and so also contributes to the extreme posterior portion of the streak. Zone VI also extends laterally and makes little additional contribution to axial growth. Zone XI, however, makes a substantial contribution to the axis, mainly anterior to the streak but also to the ectodermal portion of the anterior part of the streak. Zone X has also expanded within the axis, and the relative position of the anterior end of the streak has altered little within this zone. The posterior one-third of the axis, which is the posterior two-thirds of the streak, is formed from originally non-axial material that has grown in from the more lateral and anterior zones VIII, VII, II and I.

Herrmann: If you think the streak is not elongating anteriorly, can you explain what I see in early streak embryos? The *T* gene is expressed in nascent mesoderm in the streak, at the posterior end of the embryo, and in some cells along the midline of the posterior half to the distal tip of the embryo. At 7.3 days the streak is fully elongated and *T* is expressed in many more cells along the midline of the streak. I interpreted this as an anterior extension of the streak. But, if Kirstie is right that the streak does extend only posteriorly, I do not know how to interpret my observation.

Lawson: In the pre-primitive streak stage fate map, the presumptive notochord and endoderm are in the position where the anterior end of the streak will be at

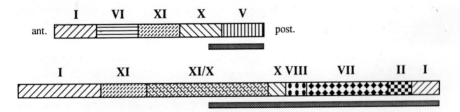

FIG. 2. (*Lawson*) Axial growth between the early streak stage (upper figure) and the early neural plate stage (lower figure). The linear contribution of the zones defined at the early streak stage (Fig. 3, Lawson, this volume) are shown. Zone VI at the neural plate stage was overlapped by zones I and XI and has been omitted for clarity. The length of the primitive streak at each stage is indicated by a shaded bar.

the early streak stage (Fig. 5, Lawson & Pedersen, this volume), i.e. about half way down the length of the epiblast cup, so it's perfectly feasible that in the very early development of the streak, T is expressed this far down. After the early streak stage, clonal analysis indicates streak growth occurs mainly by incorporation of new material posteriorly. The embryo is changing so rapidly at this time: we may only have a problem of staging, or a more fundamental difference in interpretation.

References

Beddington RSP 1983 The origin of the foetal tissues during gastrulation in the rodent. In: Johnson MH (ed) Development in mammals. Elsevier, Amsterdam p 1–32

Blumberg B, Wright CVE, de Robertis EM, Cho KWY 1991 Organizer-specific homeobox genes in *Xenopus laevis* embryos. Science (Wash DC) 253:194–196

Hogan B, Tilly R 1978 In vitro development of inner cell masses isolated immuno-surgically from mouse blastocysts. J Embryol Exp Morphol 45:93–105

Jat PS, Noble MD, Ataliotis P et al 1991 Direct derivation of conditionally immortal cell lines from an H-2Kb-tsA58 transgenic mouse. Proc Natl Acad Sci USA 88:5096–5101

Lawson KA, Pedersen RA 1992 Clonal analysis of cell fate during gastrulation and early neurulation in the mouse. In: Postimplantation development in the mouse. Wiley, Chichester (Ciba Found Symp 165) p 3–26

New DAT 1990 Introduction. In: Copp AJ, Cockcroft DL (eds) Postimplantation mammalian embryos: a practical approach. IRL Press, Oxford p 1–14

Saxén L, Thesleff I 1992 Epithelial–mesenchymal interactions in murine organogenesis. In: Postimplantation development in the mouse. Wiley, Chichester (Ciba Found Symp 165) p 183–198

Tam PPL 1989 Regionalisation of the mouse embryonic ectoderm: allocation of prospective ectodermal tissues during gastrulation. Development 107:55–67

Tam PPL, Beddington RSP 1987 The formation of mesodermal tissues in the mouse embryo during gastrulation and early organogenesis. Development 99:109–126

Use of chimeras to study gene function in mesodermal tissues during gastrulation and early organogenesis

Rosa S. P. Beddington*, Andreas W. Püschel† and P. Rashbass*

*AFRC Centre for Genome Research, King's Buildings, West Mains Road, Edinburgh EH9 3JQ, UK and †Institute of Neuroscience, 222 Huestis Hall, University of Oregon, Eugene, OR 97403, USA

Abstract. The origin of different mesodermal tissues during gastrulation and the developmental lability of mesodermal precursors can be mapped by transplanting marked epiblast cells to the same or a different position in a host egg cylinder, and assessing the subsequent fate of transplanted tissue. This information provides the context for assessing the role of particular patterns of gene expression during mesoderm formation and differentiation. For example, the stability of *Hox* gene expression can be examined by transplanting transgenically marked somites that express a particular *Hox* gene to a position in the somite file where it is not normally expressed. Such experiments can reveal not only the cues required for *Hox* gene expression but also the relevance of a circumscribed pattern of *Hox* gene expression to a specific developmental fate. A different approach to resolving gene function is to mix mutant cells known to affect mesoderm formation with normal cells and to determine the cell autonomy of mutant cells in a normal environment. Homozygous *Brachyury* (T/T) embryonic stem cell lines have been isolated and injected into normal blastocysts. The presence of T/T cells in chimeras results in mesodermal defects similar to those seen in the intact mutant.

1992 Postimplantation development in the mouse. Wiley, Chichester (Ciba Foundation Symposium 165) p 61–77

The mammalian fetus is derived from a single epithelial sheet, the epiblast (see Beddington 1986). Proliferation, differentiation and movement of the epithelial cells during gastrulation transform a simple monolayer into a complex trilaminar cylinder with a defined anteroposterior axis. On the outer surface the gut endoderm and notochord precursors are laid down. On the inner surface, where epithelial continuity is maintained except in the primitive streak region, are the neurectoderm and surface ectoderm rudiments. Between the two is an array of different mesodermal tissues, at the posterior end of which the primordial germ cells are first identified. In addition, mesoderm moves out of the embryonic area into the extraembryonic region to supply components of the visceral yolk sac, the amnion and the allantois. Consequently, within 48 hours of the onset of gastrulation, a single apparently homogeneous population of cells has generated at least twenty distinct tissues, laid down in a precise pattern, readily recognizable as the basic conformation of the final organism.

Clearly, gastrulation encompasses every fundamental developmental process and a complete molecular understanding of this symphony of morphogenesis, cytodifferentiation and pattern formation remains remote. However, the experimental tools are now available to provide, at least, a more detailed score of the cell movements, differentiation sequences and genetic players involved. This paper will examine some of the ways in which chimeras may help to identify the developmental role of genes expressed during gastrulation and early organogenesis. After discussing the contribution that postimplantation chimeras have made to charting the cellular origin of mesodermal tissues during gastrulation it will concentrate on methods for testing the function of two genes, *Hox-1.1* and *Brachyury*, that are thought to be involved in particular aspects of mesoderm formation and specification.

Fate mapping

A prerequisite for any mechanistic explanation of gastrulation is a reliable description of which epiblast cells, or epiblast regions, give rise to which of the various fetal primordia. Although no regionalized gene expression has yet been seen in the epiblast before gastrulation starts, one might assume, from the precedent set by studies on the *Drosophila* blastoderm, that some molecular heterogeneity precedes the onset of epiblast differentiation. If and when such heterogeneity within the epiblast is uncovered, it will be meaningless unless one knows the developmental fate of the subset of cells expressing such molecules. Without an accurate fate map there is no embryological context for molecular findings.

The first fate maps of mammalian epiblast were described only ten years ago (Beddington 1981, 1982, Tam & Beddington 1987); they still lag far behind those of lower vertebrates, in detail and in coverage of consecutive stages during gastrulation. The principal reason for this poverty of information is that both *in situ* labelling of cells and transplantation of tissue in the egg cylinder require direct access to the embryo, and mammalian gastrulation does not start until after implantation. Consequently, lineage studies had to await the development of an adequate culture system that could support normal postimplantation development long enough for the epiblast to differentiate into recognizable ectodermal, mesodermal and endodermal derivatives. Even now, few biochemical markers are available for distinguishing the earliest products of gastrulation; therefore, the greater the extent of development, the more reliable is the morphological classification of each derivative. While rat embryos had been successfully cultured from before gastrulation to quite advanced stages of organogenesis (New 1978), there were reasons why it was preferable to pursue lineage studies in the mouse. Firstly, the majority of available preimplantation lineage information was restricted to the mouse. More importantly, there is a comparative wealth of genetic resources in the mouse, as opposed to other non-human mammals. If the ultimate purpose of fate maps is to provide the basic description with which to test molecular function, then the availability of relevant mutants is of paramount importance.

The initial fate maps were created by transplanting [^{3}H]thymidine-labelled epiblast on the 8th day of development, when the primitive streak is fully extended to the tip of the egg cylinder. The differentiation seen after grafting into a position in the host embryo equivalent to that from which the donor tissue had been recovered (orthotopic grafts), should approximate the normal fate of the donor tissue. The fate map obtained by this method for the epiblast origin of mesodermal tissues during the later stages of gastrulation (Beddington 1981, 1982, Tam & Beddington 1987, Copp et al 1986) is summarized in Fig. 1. More recently, similar grafts have been done using a transgenic strain of mice producing bacterial β-galactosidase as a source of donor cells (Beddington et al 1989). The advantage of such a marker, which can be detected in intact embryos, is that the three-dimensional distribution of donor cells can be readily observed. Preliminary analysis of chimeras made using this transgenic marker reveals extensive cell mixing within nascent mesoderm and that epiblast originating on one side of the primitive streak can contribute progeny to both the right and left sides of the fetus (R. Beddington, unpublished observations 1990). More precise clonal fate mapping has been carried out, without resorting to the perturbation incurred by transplantation, by using iontophoretic marking of individual cells (Lawson et al 1986, 1987, Lawson & Pedersen, this volume). Despite the intrinsic superiority of the *in situ* labelling method for charting cell fate, both approaches produce similar results and, therefore, reinforce the validity of the current fate maps.

Developmental lability

Transplantation is essential for testing the stability of a particular developmental fate within the embryo. Only by moving a cell or cells to a different environment in the embryo can one test whether it is capable of forming more tissues than those described by its normal fate. Epiblast in the primitive streak region gives rise almost exclusively to either mesodermal or endodermal derivatives. Conversely, epiblast at the anterior extreme of the embryonic region generates predominantly surface ectoderm and neurectoderm. However, if epiblast from either the rostral or caudal end of the primitive streak is transferred to the anterior region, it changes its fate and differentiates into the ectodermal derivatives characteristic of its new position (Beddington 1982). Single cells have not been tested for their ability to change fate but, at least at the tissue level, it appears that commitment to form mesoderm or endoderm does not occur

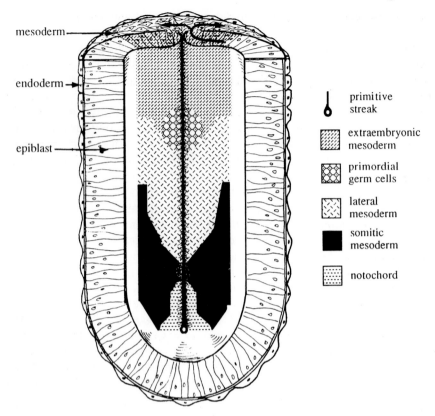

FIG. 1. Fate map of the epiblast adjacent to the primitive streak in 8th day late primitive streak stage embryo. Only the mesodermal derivatives are shown. The arrows show the general ingression movement of epiblast cells passing through the streak.

before ingression through the streak. Once it has left the streak, mesoderm can no longer be incorporated into the anterior epiblast when grafted, and forms only unassimilated clumps within the cranial region (R. Beddington, unpublished observations 1988). This makes it likely that mesodermal fate is sealed once the cells have left the epiblast epithelium. No experiments have been undertaken to see how interchangeable the different mesodermal derivatives may be during early organogenesis. Therefore, it is not known whether different mesodermal fates are fixed as the cells leave the streak or whether equivalent cells acquire different fates by virtue of where they move to. If the latter were true, the initial diversification of mesoderm may rely more on mechanical constraints affecting morphogenesis than on clear-cut genetic directives.

The relationship between gene expression and developmental fate

Transplantation can also be used, in principle, to examine the relationship between the expression of a particular gene and a particular developmental fate. Mouse homeobox-containing (*Hox*) genes show an intriguing correspondence between their 3' to 5' arrangement along the chromosome and their anterior expression boundaries in neurectoderm and somitic mesoderm along the anteroposterior axis of the embryo (Graham et al 1989, Duboule & Dolle 1989, Dressler & Gruss 1989). This is reminiscent of the chromosomal order and segmental expression pattern of their *Drosophila* counterparts. Consequently, *Hox* genes have been invoked, albeit tentatively, as the mouse equivalent of homeotic or selector genes. The combination of different *Hox* genes expressed in a particular axial region is proposed to provide a unique address specifying the characteristics, or identity, of that particular axial level. Although the parallel with *Drosophila* is intellectually compelling, direct evidence that *Hox* genes can dictate axial fate is lacking.

Transplantation of tissue with one axial fate expressing a particular repertoire of *Hox* genes to a region with a different fate and a different *Hox* profile might show whether *Hox* gene expression is autonomous and itself dictates axial identity, or whether it is regulated according to where cells lie along the axis and is therefore a secondary consequence of axial specification. Fig. 2 illustrates such an experiment. Taking into account that the stage at which grafts are undertaken may be critical, the only result compatible with *Hox* genes being selector genes is the demonstration that a particular fate is *invariably* associated with a defined combination of *Hox* gene expression.

We have investigated the stability of *Hox-1.1* expression in such relocation experiments. Initial attempts to examine endogenous *Hox-1.1* expression by *in situ* hybridization after relocation of somites transgenically marked with a ubiquitous *lacZ* reporter gene (Beddington et al 1989) were frustrated by the incompatibility of conditions for nucleic acid hybridization and β-galactosidase histochemistry. Detection of the endogenous gene remains our ultimate goal but, in the meantime, we have used *lacZ* as a reporter for *Hox-1.1* expression itself. Somites from a transgenic strain, Tg.m6lacZ1 (kindly supplied by P. Gruss, Göttingen; Püschel et al 1990, 1991), containing a *lacZ* gene regulated by the 5' non-coding region of the *Hox-1.1* gene were transplanted into wild-type hosts and the expression of *Hox-1.1* was assessed by X-Gal staining for β-galactosidase activity.

Hox-1.1 is normally expressed in somites 14 to 24, which correspond to thoracic vertebrae 3 to 13 (Mahon et al 1988). In the transgenic strain used for donor tissue the *lacZ* expression pattern shows that sufficient regulatory elements are present to replicate the anterior boundary of this somitic domain but the posterior boundary is missing; *lacZ* expression occurs more caudally than do transcripts of the endogenous *Hox-1.1* gene (Püschel et al 1991). Fig. 3 illustrates

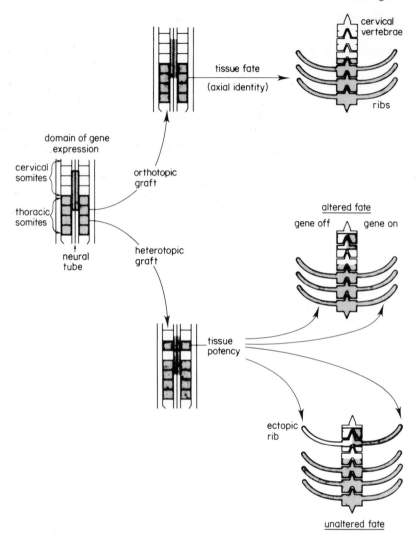

FIG. 2.   General scheme for assessing the relationship between a specific pattern of gene expression and a specific axial developmental fate. Orthotopic grafts establish the normal fate of thoracic somites (to produce vertebrae with rib appendages). Heterotopic grafts establish the stability of this developmental fate and its dependence on the initial pattern of gene expression. The stippled regions indicate maintenance of the gene expression pattern.

the three series of grafts undertaken. The donor somite and the position to which it was grafted were matched with respect to the stage of somite maturation. For example, somites from anterior non-expressing regions were recovered from the four most caudal somites of the somitic file in younger embryos (8.5 d) and

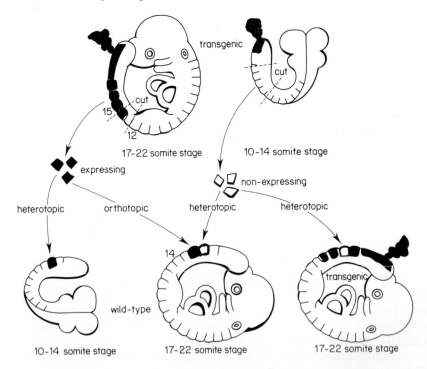

FIG. 3. Diagram summarizing the orthotopic and heterotopic grafts using transgenic somites recovered from regions either expressing or not expressing endogenous *Hox-1.1*. The shaded areas represent the expression of the reporter *Hox-1.1–lacZ* transgene as judged by β-galactosidase activity.

grafted in place of the penultimate somite in older embryos (9.5 d). After grafting, embryos were grown for 48 hours in culture and stained for *lacZ* expression.

In all types of graft there was no evidence that relocation to a new axial level at these developmental stages resulted in either induction or repression of *Hox-1.1* expression (Table 1, Fig. 3). In other words, once somites have formed *Hox-1.1* expression appears to be stable. It is possible that the stability of the reporter β-galactosidase protein obscures any decrease in expression; however, the equivalent intensity of staining following orthotopic and heterotopic grafts of somites expressing *Hox-1.1* argues against this (Fig. 4). Furthermore, if Tg.m6lacZ1 somites expressing *Hox-1.1* are explanted *in vitro*, at least 30% of the cells are negative for β-galactosidase activity after 48 hours (R. Beddington, unpublished observations 1990); thus down-regulation of *lacZ* expression can be detected in this time.

We have yet to apply tissue-specific markers to distinguish the fate of somites transplanted to different axial levels. In the most advanced embryos developed

TABLE 1   Expression of *lacZ* in orthotopically grafted Tg.m6lacZ1 somites

	Orthotopic grafts (expressing to expressing region)	Heterotopic grafts (expressing to non-expressing) region)	Heterotopic grafts (non-expressing to expressing region)
Number of grafts	13	9	6
Number incorporated	9	7	4
	69.2%	77.8%	66.7%
Number expressing *lacZ*	9	7	0
	100%	100%	

in vitro, it should be possible to discriminate between vertebrae forming lateral ribs, which would be the normal fate of somites expressing *Hox-1.1*, and vertebrae lacking rib appendages, a morphology characteristic of the more anterior cervical region. In the chick, relocation of thoracic somites, grafted to a more anterior position at a slightly earlier stage than the experiments described above, results in the formation of ectopic ribs (Kieny et al 1972), indicating that the axial fate of somitic sclerotomal cells is determined during somite formation. If the same proves to be true for the mouse, and earlier grafts of *Hox-1.1*-expressing tissue exhibit the same position-independent autonomy of gene expression, then the argument that *Hox* genes serve as axial determinants will become more substantial.

Chimeras and analysis of mutations

It is apparent from recent work in *Drosophila* that the availability of mutant embryos exhibiting a specific phenotype is a great asset, if not a necessity, in identifying genes which direct pattern formation and tissue diversification. Most mutations affecting early development are likely to be recessive and lethal during embryogenesis. Consequently, it was not until Nüsslein-Volhard & Wieschaus (1980) undertook a specific and comprehensive screen for early embryonic lethals, that *Drosophila* genetics became a truly systematic analytical tool for identifying developmental genes. In theory, it is now possible, although considerably more labour intensive and time consuming, to screen for early embryonic lethal mutations in mice created by random insertional mutagenesis in embryos or embryonic stem (ES) cells. However, it is clear from developmental mutants already known in the mouse that the phenotype of an abnormal, dying embryo is not always informative regarding primary gene function. Turning again to *Drosophila*, the mixing of mutant and wild-type cells in mosaics has proved uniquely instructive not only in determining the cell autonomy of gene function but also in defining the effective, as opposed to inessential, regions of gene expression (e.g. ommatidia development; Rubin 1989).

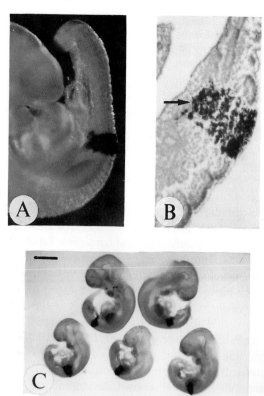

FIG. 4.   (A) Intact embryo orthotopically grafted with a Tg.*Hox-1.1–lacZ* somite
expressing *Hox-1.1* at 9.5 d and cultured for a further 24 hours before staining for bacterial
β-galactosidase activity. (B) Sagittal section of embryo depicted in A showing correct
incorporation of the donor somite (arrow). (C) Intact embryos stained for β-galactosidase
activity 48 hours after receiving either orthotopic (top row) or heterotopic (bottom row)
Tg.*Hox-1.1–lacZ* somite grafts at 9.5 d. Bar = 300 µm in A, 600 µm in B and 2 mm in C.

In the mouse such mixing of wild-type and mutant cells in embryonic or
haemopoietic chimeras or in tissue explants *in vitro* has helped to determine
the cell autonomy or otherwise of gene function in mutations such as *Dominant
Spotting* (*W*) and *Steel* (*Sl*) (Gordon 1977, Stephenson et al 1985, Harrison &
Astle 1976, Fleischman & Mintz 1979, Bannerman et al 1973). There is a problem
in creating chimeras from embryos homozygous for early embryonic lethal
mutations. Chimeras can be made only during preimplantation development,
either by aggregation of cleavage stage embryos or by injection of inner cell
mass cells into the blastocyst. Consequently, it is often impossible to identify
the homozygous embryos at the time the chimeras are made; therefore, the
genotype of the chimera has to be inferred retrospectively. Although polymerase

chain reaction techniques now make it possible to characterize preimplantation embryos genetically (Handyside et al 1990), it would be preferable to have a permanent source of mutant embryonic cells for such studies.

The derivation of ES cells from blastocysts obtained from matings between animals heterozygous for a particular mutation provides just such a resource (Martin & Lock 1983). After genetic characterization of such ES cell lines one can directly compare the development, *in vitro* or in intact chimeric embryos, of homozygous, heterozygous and wild-type cells of the same genetic background. ES cells also provide a continuous source of large numbers of mutant cells *in vitro*, which permits further directed genetic manipulation and selection for rare recombination events (Robertson & Bradley 1986). For example, a reporter gene, such as *lacZ*, can be introduced either to serve as a ubiquitous single cell marker for all mutant cells in chimeras or, by using an appropriate promoter, to monitor selectively the fate of those mutant cells which normally express the wild-type gene. Similar transgenic reporter gene constructs can be used to follow subpopulations of mutant cells expressing other genes involved in the same developmental pathway. Judicious choice of heterologous promoters allows the normal gene product, or genes acting downstream of it, to be expressed only in a specific subset of embryonic tissues or in precise spatial or temporal domains. It is also now possible to make specific double mutants by homologous recombination (Bradley et al, this volume) into other genes in the mutant ES cell line.

As a first step towards such a sophisticated chimeric analysis, we have isolated ES cells which are null for a well-known developmental deletion mutation, *Brachyury* (*T*). *Brachyury* is a semi-dominant mutation that was first identified because heterozygotes have shortened tails (Dobrovolskaia-Zavadskaia 1927). Homozygous embryos die at mid-gestation and characteristically have a degenerate or absent notochord, greatly reduced allantois, accumulation of mesoderm cells beneath the primitive streak and, probably as secondary effects, disrupted neural tube and somites (Chesley 1935, Glücksohn-Schönheimer 1944, Grüneberg 1958, Yanagisawa et al 1981). The origin of all these defects can be traced to the early stages of gastrulation; this implies a requirement for *T* in the generation of mesoderm emerging after the delamination of the earliest rostral and extraembryonic mesoderm.

The *T* gene has been cloned and its expression in early postimplantation embryos mapped by *in situ* hybridization (Herrmann et al 1990, Wilkinson et al 1990). Neither the sequence of the gene nor its expression pattern unequivocally defines its function. It is undoubtedly associated with mesoderm formation; the range in the severity of abnormality caused by different alleles indicates an increasing demand for the *T* gene product in mesoderm forming at later stages and contributing to more caudal structures. A more detailed discussion of the molecular and genetic evidence for the role of *T* in anteroposterior specification can be found in Herrmann (this volume).

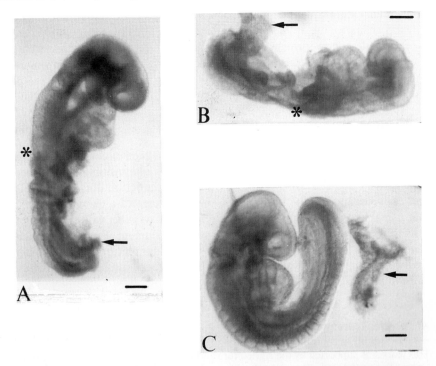

FIG. 5. Phenotype of T/T, $T/T \leftrightarrow {}' +/+$ and control ES cell chimeras on the 10th day of gestation. (A) Lateral view of a T/T embryo showing abnormal trunk region posterior to somite 7 (asterisk) and a minimal allantois (arrow). Bar = 200 µm. (B) Lateral view of $T/T \leftrightarrow {}' +/+$ chimera showing abnormal trunk region posterior to somite 7 (asterisk) and a greatly reduced allantois (arrow) which had not fused with the chorion. Bar = 175 µm. (C) $T/+ \leftrightarrow {}' +/+$ or $+/+ \leftrightarrow {}' +/+$ ES cell chimera that is entirely normal and has a well differentiated allantois (arrow) which has fused with the chorion. Bar = 300 µm.

Chimeras made from injecting T/T ES cells into blastocysts indicate that the gene acts cell autonomously in some tissues (Rashbass et al 1991). Using glucose phosphate isomerase allozymes to distinguish between host and donor ES cell contributions, we have shown that embryos containing T/T ES cells replicate the mid-gestation phenotype of the intact homozygous T/T mutant (Fig. 5). Conversely, wild-type or heterozygous ES cells derived from blastocysts obtained from the same strain of $T/+ \times T/+$ matings form morphologically normal chimeras at mid-gestation (Fig. 5C). Thus, the demonstration that the effects of the *Brachyury* deletion cannot be rescued in $T/T \leftrightarrow +/+$ chimeras provides the foundations for an excellent experimental system in which to analyse the function of the T gene, and possibly other genes acting downstream of it. Introduction of a transgenic single cell marker will enable us to monitor the behaviour of intermixed normal and mutant cells in affected and unaffected tissues. In addition, the *T/T* ES cells provide an ideal population for genetically manipulating certain aspects of mesoderm formation and will allow the effects of genetic perturbation, such as ectopic *T* expression, to be studied in parallel *in vitro* and in the intact embryo *in vivo*.

Conclusions

Three somewhat different chimeric strategies have been discussed in this paper, each of which can be used to shed light on the relationship between specific gene activity and cellular determination or differentiation during gastrulation and early organogenesis. In all cases, the advantage of a chimeric analysis is that it enables the development of a subpopulation of either wild-type or mutant cells to be assessed in a normal embryonic environment. Purely molecular or cell biological studies *in vitro* may identify particular features of developmental pathways, but their relevance to normal development can be tested only in the embryo itself.

Acknowledgements

This work was supported by the Agriculture and Food Research Council and the Imperial Cancer Research Fund. We would also like to thank Linda Manson and Louise Anderson for valuable technical assistance.

References

Bannerman RM, Edwards JA, Pinkerton PH 1973 Hereditary disorders of the red cell in animals. Prog Hematol 8:131–179

Beddington RSP 1981 An autoradiographic analysis of the potency of embryonic ectoderm in the 8th day postimplantation mouse embryo. J Embryol Exp Morphol 64:87–104

Beddington RSP 1982 An autoradiographic analysis of tissue potency in different regions of the embryonic ectoderm during gastrulation in the mouse. J Embryol Exp Morphol 69:265–285

Beddington RSP 1986 Analysis of tissue fate and prospective potency in the egg cylinder. In: Rossant J, Pedersen RA (eds) Experimental approaches to mammalian embryonic development. Cambridge University Press, New York p 121–147

Beddington RSP, Morgenstern J, Land H, Hogan A 1989 An in situ transgenic enzyme marker for the midgestation mouse embryo and the visualization of inner cell mass clones during early organogenesis. Development 106:37–46

Bradley A, Ramirez-Solis R, Zheng H, Hasty P, Davis A 1992 Genetic manipulation of the mouse via gene targeting in embryonic stem cells. In: Postimplantation development in the mouse. Wiley, Chichester (Ciba Found Symp 165) p 256–276

Chesley P 1935 Development of the short-tailed mutant in the house mouse. J Exp Zool 70:429–459

Copp AJ, Roberts HM, Polani PE 1986 Chimerism of primordial germ cells in the early postimplantation mouse embryo following microsurgical grafting of posterior primitive streak cells in vitro. J Embryol Exp Morphol 95:95–115

Dobrovolskaia-Zavadskaia N 1927 Sur la mortification spontanée de la queue chez la souris nouveau-née et sur l'existence d'un caractère hereditaire "non-viable". C R Soc Biol 97:114–116

Dressler GR, Gruss P 1989 Anterior boundaries of Hox gene expression in mesoderm-derived structures correlate with the linear gene order along the chromosome. Differentiaton 41:193–201

Duboule D, Dolle P 1989 The structural and functional organisation of the murine *HOX* gene family resembles that of *Drosophila* homeotic genes. EMBO (Eur Mol Biol Organ) J 8:1497–1505

Fleischman RA, Mintz B 1979 Prevention of genetic anaemia in mice by microinjection of normal haematopoietic stem cells into the fetal placenta. Proc Natl Acad Sci USA 76:5736–5740

Glücksohn-Schönheimer S 1944 The development of normal and homozygous *Brachy* (*T/T*) mouse embryos in the extraembryonic coelom of the chick. Proc Natl Acad Sci USA 30:134–140

Gordon J 1977 Modification of pigmentation patterns in allophenic mice by the *W* gene. Differentiation 9:19–27

Graham A, Papalopulu N, Krumlauf R 1989 The murine and *Drosophila* homeobox gene complexes have common features of organization and expression. Cell 57:367–378

Grüneberg H 1958 Genetical studies on the skeleton of the mouse. XXIII. The development of *Brachury* and *Anury*. J Embryol Exp Morphol 6:424–443

Handyside AH, Kontogianni EH, Hardy K, Winston RML 1990 Pregnancies from biopsied human preimplantation embryos sexed by Y-specific DNA amplification. Nature (Lond) 344:768–770

Harrison D, Astle CM 1976 Population of lymphoid tissue in cured *W*-anemic mice by donor cells. Transplantation 22:42–46

Herrmann BG, Labeit S, Poustka A, King TR, Lehrach H 1990 Cloning of the *T* gene required in mesoderm formation in the mouse. Nature (Lond) 343:617–622

Herrmann BG 1992 Action of the *Brachyury* gene in mouse embryogenesis. In: Postimplantation development in the mouse. Wiley, Chichester (Ciba Found Symp 165) p 78–91

Kieny M, Mauger A, Sengel P 1972 Early regionalisation of the somitic mesoderm as studied by the development of the axial skeleton of the chick embryo. Dev Biol 28:142–161

Lawson KA, Pedersen RA 1992 Clonal analysis of cell fate during gastrulation and early neurulation in the mouse. In: Postimplantation development in the mouse. Wiley, Chichester (Ciba Found Symp 165) p 3–26

Lawson KA, Meneses JJ, Pedersen RA 1986 Cell fate and cell lineage in the endoderm of the presomite mouse embryo, studied with an intracellular tracer. Dev Biol 115:325–329

Lawson KA, Pedersen RA, van de Geer S 1987 Cell fate, morphogenetic movement and population kinetics of embryonic endoderm at the time of germ layer formation in the mouse. Development 101:627–652

Mahon KA, Westphal H, Gruss P 1988 Expression of homeobox gene *Hox-1.1* during mouse embryogenesis. Development (Suppl) 104:167–174

Martin GR, Lock LF 1983 Pluripotent cell lines derived from early mouse embryos cultured in medium conditioned by teratocarcinoma stem cells. In: Silver LM, Martin GR, Strickland S (eds) Cold Spring Harbor Conferences on Cell Proliferation. Cold Spring Harbor Laboratory, Cold Spring Harbor, p 635–646

New DAT 1978 Whole embryo culture and the study of mammalian embryos during organogenesis. Biol Rev Camb Philos Soc 53:81–122

Nüsslein-Volhard C, Wieschaus E 1980 Mutations affecting segment number and polarity in *Drosophila*. Nature (Lond) 287:795–801

Püschel AW, Balling R, Gruss P 1990 Position-specific activity of the Hox-1.1 promoter in transgenic mice. Development 108:435–442

Püschel AW, Balling R, Gruss P 1991 Separate elements cause lineage restriction and specify boundaries of Hox 1.1 expression. Development 112:279–289

Rashbass P, Cooke LA, Herrmann BG, Beddington RSP 1991 A cell autonomous function of *Brachyury* in *T/T* embryonic stem cell chimaeras. Nature (Lond) 353:348–350

Robertson EJ, Bradley A 1986 Production of permanent cell lines from early embryos and their use in studying developmental problems. In: Rossant J, Pedersen RA (eds) Experimental approaches to mammalian embryonic development. Cambridge University Press, New York p 475–508

Rubin G 1989 Development of the *Drosophila* retina: inductive events studied at single cell resolution. Cell 57:519–520

Stephenson RA, Glenister PH, Hornby JE 1985 Site of beige (*bg*) and leaden (*ln*) pigment gene expression determined by recombinant embryonic skin grafts and aggregation mouse chimaeras employing sash (*W^{sh}*) homozygotes. Genet Res 46:193–205

Tam PPL, Beddington RSP 1987 The formation of mesodermal tissues in the mouse embryo during gastrulation and early organogenesis. Development 99:109–126

Wilkinson DG, Bhatt S, Herrmann BG 1990 Expression pattern of the mouse *T* gene and its role in mesoderm formation. Nature (Lond) 343:657–659

Yanagisawa KO, Fujimoto H, Urushihara H 1981 Effects of the *Brachyury* (*T*) mutation on morphogenetic movement in the mouse embryo. Dev Biol 87:242–248

DISCUSSION

Balling: In some of your *Brachyury* chimeras there were dead cells in the notochord area. Are those notochord-derived cells or paraxial mesoderm-derived cells?

Beddington: My suspicion is that they are notochord because paraxial mesoderm is clearly present: one sees disorganized somites, they look quite healthy but they are not organized in the correct way. Cell death always occurs on the ventral aspect of the neural tube. It may be quite an enlarged area but it's always where you would expect to find a notochord. We don't see dead cells more dorsally where paraxial mesoderm is located.

Brown: What were the percentages of chimerism in the *T/T* ↔ *+/+* embryos?

Beddington: It is complicated. The 8.5 day embryos were divided into head, trunk, tail and allantois. There was essentially little difference between head, trunk and tail; there is a marginal tendency, which may not stand up statistically, for a higher contribution of *T/T* cells in the allantois.

For the 9.5 day embryos, we analysed the amnion or the allantois for glucose phosphate isomerase activity and did histology on the rest of the embryo.

Tam: In the *Brachyury* chimeras, is the severity of the abnormalities related to the overall somatic chimerism or more specifically to the notochordal chimerism?

Beddington: I predict that it may be related specifically to the notochordal chimerism. But until we can examine the distribution with a single cell marker, I can't answer that.

Balling: How do you account for the fact that when you explant blue somites expressing *Hox-1.1–lacZ* into an organ culture about 30% of those cells lose expression? Does that mean the expression isn't cell autonomous or did the cells lose the chromosome carrying the transgene, for example?

Beddington: I would consider elimination of the chromosome as highly unlikely and that it is actually loss of expression.

Balling: But wouldn't this indicate that the expression is not really cell autonomous, if a certain cellular environment is still needed?

Beddington: Hox-1.1 expression eventually becomes sclerotome specific and we see that to some extent in the grafts. In culture, we may be seeing switching off of *Hox-1.1* expression in somite derivatives that aren't going to make sclerotome. I can't answer your question properly. The important point is that it demonstrates that, despite the stability of bacterial β-galactosidase, we are not being totally misled by using *lacZ* as a reporter.

Joyner: Did you do the reverse grafts of putting somites that would never express *Hox-1.1* into regions that would normally express *Hox-1.1*?

Beddington: I have done. The experiments are more difficult to interpret categorically because a non-stained somite is put into a very densely stained region. As far as I can see, *Hox-1.1* is not switched on. At the boundary regions, where there is some diffusion of the staining product when there is the very high level of *lacZ* expression, the staining of the cells is somewhat ambiguous. In the middle of the somite expression is still negative.

Joyner: Can you take somites from the region that will express *Hox-1.1* before they start expressing and do the heterotopic transplant?

Beddington: It is difficult to know what that region is in the embryo. *Hox-1.1* expression comes on at the posterior end of the streak, so it's basically on in the presomitic mesoderm and subsequent somites from the moment it's turned on.

Goodfellow: If you take out a somite and look for endogenous *Hox-1.1* expression by PCR (polymerase chain reaction), can you detect it?

Beddington: We have not done that. One problem is that the enzyme treatment used to isolate somites also degrades RNA.

Goodfellow: One interpretation is that the construct can switch on *Hox-1.1* expression but it can't switch it off because the construct is being expressed inappropriately at the end of the tail. The PCR experiment could check at least whether the endogenous gene was behaving normally.

Beddington: I agree. The only way to do the experiment properly is to look at endogenous gene expression, either by *in situ* hybridization or by PCR.

It would probably be best to look for the endogenous *Hox-1.1* protein using an antibody.

Goodfellow: I don't understand what you meant by specification with respect to the defective allantois in $T/T \leftrightarrow +/+$ chimeras.

Beddington: Presumably, at some point cells, either before or after they have passed through the streak, get some information that tells them to go to the extraembryonic region and make allantois, to go to lateral regions and make lateral mesochyme or whatever. My use of specification here implies that some signal is missing as the allantois forms or some signal is missing to direct cells properly to become allantois. Cells are present in the allantoic region but they appear to be confused in that many start to spread over the amnion.

Goodfellow: Can you distinguish between there being a general growth factor which is needed for allantois to survive and something which changes or controls the fate of cells that are going to give rise to the allantois?

Beddington: I would not necessarily expect to see abnormalities in the allantois, just fewer cells, if a factor affecting only growth or cell maintenance was lacking. That doesn't appear to be the case. Where the allantois is stunted, there is actually a higher density of cells. I haven't done cell counts but it doesn't look as though it's simply a lack of proliferation and there is no obvious increase in cell death.

Hogan: Do you find the right number of primordial germ cells in the *T* mutant?

Beddington: Primordial germ cells have never previously been looked at in the *T* mutant. The mutation seems to affect midline structures; the allantois is in the midline, the notochord is in the midline. The other cells that come out in the midline are primordial germ cells. We have looked at 9.5 days using whole-mount alkaline phosphatase staining. That's too late to get a clear answer because there is a rather retarded posterior region which is strongly alkaline phosphatase positive. So although you can see one or two individual positive cells that may be primordial germ cells, the general picture is obscured by very strong staining. Val Wilson is repeating these experiments on 8.5 day embryos, which should give a clear answer to whether there are germ cells present.

Lee: There is a lectin from *Lotus tetragonolobus* which is specific for germ cells.

Beddington: And it does not label epiblast or neuroectoderm?

Lee: It doesn't label anything else in the embryo.

Hogan: Do you think the extraembryonic mesoderm goes through a regular sort of streak with a Hensen's node-type structure? Is it formed prior to notochord formation?

Beddington: Much extraembryonic mesoderm is certainly formed prior to notochord formation, but it is formed at the opposite end of the streak to Hensen's node, at the very posterior extreme. If you graft epiblast cells into the posterior region, they will make extraembryonic mesoderm and those cells have to go through the streak.

Lawson: It is difficult to define the posterior end of the streak, so perhaps it is rather academic just how the cells move into the extraembryonic mesoderm.

Hogan: Could there be a different mechanism for induction of extraembryonic mesoderm than for embryonic mesoderm? What is the equivalent in chick or in *Xenopus* of extraembryonic mesoderm? Isolated *Xenopus* animal caps exposed to certain mesodermal inducing factors, e.g. FGF, BMP-4, give rise to mesothelium and blood islands. Are these tissues equivalent to extraembryonic mesoderm?

Smith: Xenopus doesn't have extraembryonic mesoderm.

I think that for the mouse we should avoid saying mesoderm induction: the phrase has been used a lot, but the induction has not been shown at all.

Krumlauf: The mouse might not be the best system in which to do these experiments. The cell autonomy of the somites has been examined in the chick by Drew Noden's neural crest grafts and Andrew Lumsden's study of *Hox* gene expression. How safe would you feel extrapolating those results back to the mouse? If the mouse isn't the best system, can we just do the experiments in another organism and apply the results to the mouse?

Beddington: It's a bit of an impasse. I would hope that some of the mutational analysis in the mouse will complement the grafting experiments in the chick. There is no reason to do third-rate experiments in the mouse for the sake of saying they have been done in the mouse.

Hogan: I am still not satisfied about what happens during the formation of the extraembryonic mesoderm in the chick.

Tam: In the mouse, the progenitor cells of the extraembryonic mesoderm are localized in the posterior region of the primitive streak. Presumably, they have to go through the primitive streak to reach their destination in the yolk sac and amnion.

Beddington: You can map quite lateral epiblast posteriorly in the chick which will end up in extraembryonic mesoderm.

Lawson: That is early on. In the area opacum there is a lot of presumptive yolk sac mesoderm—it's very anterior and must all come down and go through the streak.

Action of the *Brachyury* gene in mouse embryogenesis

Bernhard G. Herrmann

Max-Planck-Institut für Entwicklungsbiologie, Abteilung Biochemie, Spemannstraße 35/II, 7400 Tübingen, Germany

Abstract. The murine developmental mutation *T* identifies a gene required in mesoderm formation. *T/T* mutant embryos develop normally to the primitive streak stage; during early organogenesis they show insufficient mesoderm and absence of the notochord. The mutants die at around 10 days of gestation because of the lack of the allantois. We have localized the *T* mutation relative to DNA markers and used a combination of genetic and molecular techniques to clone the *T* gene. Expression of the *T* gene is restricted to nascent mesoderm and to the notochord, the tissues most strongly affected by the mutation. Recent results suggest that the *T* gene encodes a nuclear factor involved in establishing notochord cell identity and differentiation, and is directly or indirectly involved in the organization of axial development.

1992 Postimplantation development in the mouse. Wiley, Chichester (Ciba Foundation Symposium 165) p 78–91

Mesoderm formation is a fundamental process in vertebrate development, establishing many features of the body (Slack 1983, Gilbert 1988). The *T* mutation identifies one of several murine developmental loci involved in mesoderm formation that have been investigated by genetic and embryological techniques (Green 1989). But so far, genes identified only by mutations are not generally accessible to cloning.

The *T* mutation was initially identified by its effect on tail length. *T/+* animals have short tails (Dobrovolskaia-Zavadskaia 1927). This is due to haplo-insufficiency—the phenotype can be complemented by a chromosomal duplication including the *T* gene (Styrna & Klein 1981). *T/t* mice are tailless, probably because of reduced activity of the *t* chromosome allele of the *T* gene. A tailless phenotype and skeletal malformations are observed in mice heterozygous for the antimorphic *T* allele *T^c* (Searle 1966, MacMurray & Shin 1988).

T/T embryos develop an apparently normal head but axial elongation ceases during early organogenesis, at the same time as defects in mesoderm formation become apparent (Chesley 1935, Grüneberg 1958, Yanagisawa et al 1981). The notochordal plate forms, but fails to differentiate into a notochord. As a result of the absence of a notochord the somites and neural tube are abnormal. The embryos die at 10 days of gestation owing to lack of the allantois (Glücksohn-Schönheimer 1944). Homozygous T^c/T^c embryos are slightly more affected than T/T embryos (Searle 1966).

In all cases, the most severely affected structure is the notochord (Chesley 1935, Yanagisawa 1990). The graded severity of the phenotype along the anteroposterior axis correlates with gene dosage (MacMurray & Shin 1988), suggesting an increasing demand for the T gene product as axial elongation proceeds (Yanagisawa 1990).

Positioning and cloning of the T gene

The T gene has been cloned via a multistep process (Herrmann et al 1990). Random DNA markers isolated from the t complex region of chromosome 17 by microdissection (Röhme et al 1984) were mapped to subregions of the chromosome (Herrmann et al 1986, 1987) (Fig. 1A). Two duplicated markers, 119I and 119II, 66E and 66EII, proved to be closely linked to the T mutation, since neither marker could be separated from the mutation in 380 meioses scored. Both 119I and 119II are deleted in the T alleles T^{tOrl} and T^{Or4}, whereas 119II but not 119I is absent in the T deletion T^{Hp} (Fig. 1B). This suggested that T was located near 119II. In addition, the centromeric limit of the chromosomal duplication t^{Tu3} was mapped to within the cloned region 119II. Since t^{Tu3} duplicates the T gene and complements the T mutant phenotype (Styrna & Klein 1981), this demonstrated that the T gene must lie on the telomeric side of 119II. An accurate physical and genetic map of the T region was thus established.

Chromosome walking and jumping were used to bridge a region of approximately 400 kb on the telomeric side of 119II (Fig. 1B). The end fragment of the second jumping clone identified deletions in the original T allele and in T^{2J}. Their sizes were estimated as 160–200 kb and 80–110 kb, respectively, using pulsed-field gradient gel electrophoresis. This localized the T gene to a genomic region of no more than 100 kb. Isolation of cosmids from this region facilitated the screening of a cDNA library prepared from 8.5 day embryos (Fahrner et al 1987) and several cDNA clones were isolated. The analysis of a number of T alleles using the cDNA pme75 as a probe revealed the insertion of a transposable element in the T allele T^{Wis} (Shedlovsky et al 1988, Herrmann et al 1990) which results in the alteration of the T gene product and pinpoints the locus of the T gene. T^{Wis} produces a phenotype which completely lacks a tail and is thus more severe than the deletion mutant.

Expression pattern of T

Northern blotting revealed a transcript of 2.1 kb in early embryos, but no T transcripts in any of the adult organs examined (not shown). The temporal and

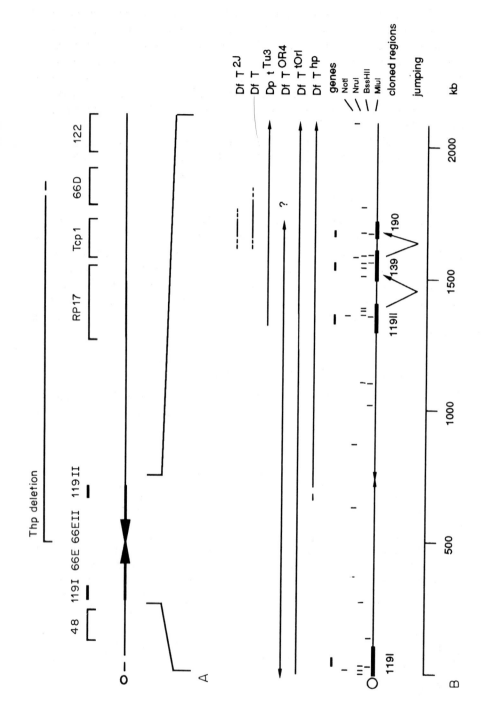

spatial distribution of *T* transcripts was determined by *in situ* hybridization of embryo sections (Wilkinson et al 1990). *T* is expressed in nascent mesoderm of the primitive streak of 7, 8.5 and 9.5 day embryos, and in the notochordal plate and notochord of 8.5 day and older embryos (Fig. 2). The expression is down-regulated or turned off in migrating presomitic and lateral plate mesoderm; in the notochord it persists at least until 17.5 days post coitum (p.c.). Notochordal cells are integrated into the intervertebral discs where expression can still be detected at 17.5 days p.c..

The expression pattern correlates with the mutant phenotype, suggesting a direct role for the *T* gene in the formation of mesoderm and of the notochord.

Molecular properties of *T*

The hybridization of the *T* cDNA to genomic DNA of different vertebrate species demonstrates a high degree of sequence conservation during evolution (Herrmann et al 1990). Its nucleotide and predicted amino acid sequence do not reveal significant homologies to any known gene products. The absence of a signal peptide is consistent with an intracellular localization of the *T* gene product. These limited data and the pleiotropic phenotype of the *T* mutation suggest a regulatory role of *T* in the formation of mesoderm and the differentiation of the notochord.

What is the role of the *T* gene in embryogenesis?

The *Xenopus* and zebrafish homologues of the *T* gene were cloned using the mouse gene as a probe and their expression patterns were examined. Analysis of

---

FIG. 1. Genetic and molecular maps of the proximal part of chromosome 17 and the region near the *T* gene. (A) An overview over the entire region analysed. The centromere is represented by a circle. The locations of several DNA markers are indicated above the chromosome. A 650 kb inverted duplication (Herrmann et al 1987) is shown in bold by two facing arrows; the thick bars on top represent cloned regions. The deletion in the deficiency T^{hp} starts near the centre of the inverted duplication and covers a large part of chromosome 17 near the centromere. (B) The inverted duplication (119I–119II) and adjacent regions are shown at the resolution of a (rare cutter enzyme) restriction map providing details on the extent of deficiency (Df) and duplication (Dp) alleles of the *T* mutation. Restriction sites are represented by vertical lines. The start and finish of the deletions of *T* and T^{2J} were not precisely localized, symbolized by broken lines. Arrow heads indicate the continuation of the deficiencies or duplication; in T^{Or4} the ? means that the telomeric extension of the deficiency beyond region 190 is uncertain. Jumping clones (Poustka & Lehrach 1986) are represented by a V with an arrowhead facing the end point fragment. Jumps start and end in *Bss*HII/*Eco*RI fragments isolated in phage clones. Reprinted by permission from Nature 343:618. Copyright© 1991 MacMillan Magazines Ltd.

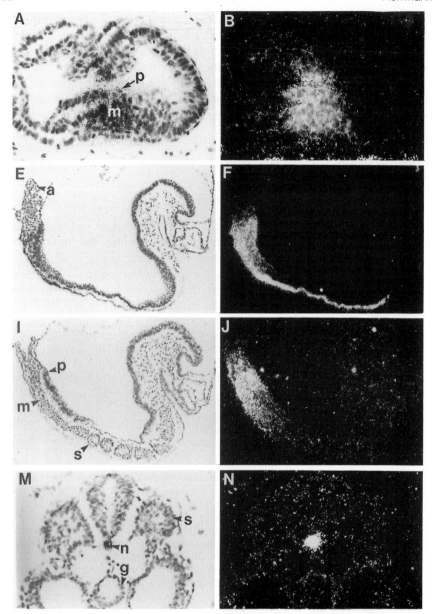

FIG. 2.   Spatial distribution of *T* gene transcripts in mouse embryos. Sections of mouse embryos were used for *in situ* hybridization with the *T* gene probe. (A,B) Longitudinal section through 7 day embryo. (C,D) Transverse section through 7 day embryo. (E,F) (I,J) Saggital and parasaggital longitudinal sections, respectively, of the same 8.5 day embryo. (G,H) Higher magnification view of E and F. (K,L) Transverse section through an 8.5 day embryo. (M,N) Transverse section through the spinal region of a 9.5 day embryo. (O,P) Longi-

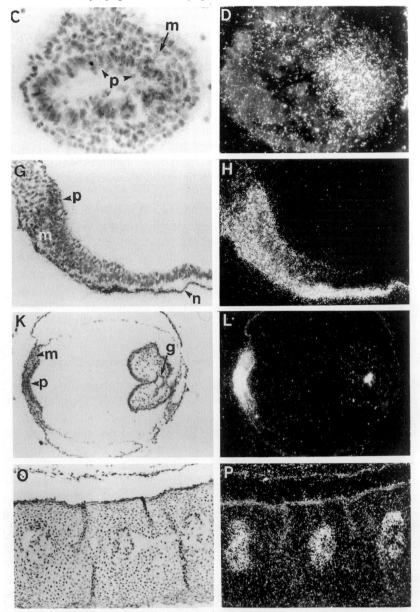

tudinal section through the vertebral column of a 17.5 day fetus. The pictures A, C, E, G, I, K, M, and O are bright field images corresponding to dark field photographs of B, D, F, H, J, L, N and P. Abbreviations: p, primitive ectoderm; m, early mesoderm; a, allantois; s, somite; n, notochord; g, gut. The anterior notochordal plate is intercalated into gut at 8.5 days of development (K,L), but has separated from the gut by 9.5 days (M,N). Reprinted by permission from Nature 343:658. Copyright©1991 Macmillan Magazines Ltd.

T/T embryonic stem cell mouse chimeras and of the *T* gene in *T^{Wis}* mutant mice, combined with the results from other species, has given a fairly accurate picture of the action of the *T* gene.

The *T* gene of *Xenopus* is expressed in response to mesoderm-inducing signals, even when protein synthesis is blocked, and is restricted specifically to mesoderm and the notochord (Smith et al 1991). Studies in the zebrafish revealed a nuclear localization of the *T* gene product (S. Schulte-Merker, B. G. Herrmann, C. Nüsslein-Volhard, unpublished). Chimeric embryos derived from blastocysts injected with *T/T* mutant embryonic stem cells demonstrate a cell-autonomous action of the *T* gene; mutant cells integrated in the notochord cannot be rescued by normal cells present in the same notochord and become pycnotic (Rashbass et al 1991).

T^{Wis}/T^{Wis} embryos develop normally during the primitive streak stages. *T* is expressed in the nascent mesoderm of the primitive streak, in the head process and in its posterior extension, the notochord precursor. However, during early organogenesis *T* transcripts disappear prematurely in these mutant embryos (Herrmann 1991). The abolition of *T* expression in *T^{Wis}/T^{Wis}* embryos parallels an arrest of mesoderm formation that correlated with an abnormally high cell number in the primitive ectoderm described in *T/T* embryos (Yanagisawa et al 1981).

A hypothesis for the action of the *T* gene product during embryogenesis

Mesoderm may be induced during the primitive streak stage in the mouse by specific signals, as described for *Xenopus* (Slack et al 1987, Smith et al 1988). *T* is expressed as an immediate response to mesoderm induction. The expression persists in mesoderm which becomes incorporated into the head process/notochord precursor, but is switched off in all other mesoderm subtypes. *T* is involved in, but is not essential for, early mesoderm formation, since early embryogenesis proceeds apparently normally without the *T* gene product.

During early organogenesis, however, the *T* gene product is required for the formation of notochord precursor cells. *T* may act as a regulator of genes involved in notochord establishment and differentiation. The notochord precursor, at its posterior end (which is similar to the Hensen's node region in the chick) may provide a signal required to induce primitive ectoderm cells to form mesoderm, and a signal for mesoderm specification and axial organization. The amount of *T* gene product per cell and of signal secreted by the notochord precursor may increase as axial elongation proceeds.

The *T* gene product is directly or indirectly involved in providing the signal. Thus, a feedback loop of increasing signal and increasing *T* gene expression could be established.

In embryos with a mutant *T* gene, normal development proceeds along the anteroposterior axis to an axial position where the amount of the speculative signal compound no longer suffices for notochord formation and organization of further axial development. Consequently, there must be at least one other (*T*-like) gene that supports early axial development. There is genetic evidence for a gene similar to *T* (Lyon & Meredith 1964, Nadeau et al 1989, Herrmann et al 1990).

In summary, the *T* gene as a tool allows molecular studies of the processes organizing axial development of the mouse embryo.

Acknowledgements

I thank U. Schwarz and A. Kispert for critical comments on the manuscript, and D. G. Wilkinson for Fig. 2.

References

Chesley P 1935 Development of the short-tailed mutant in the house mouse. J Exp Zool 70:429–435

Dobrovolskaia-Zavadskaia N 1927 Sur la mortification spontanée de la queue chez la souris nouveau-née et sur l'existence d'un caractère hereditaire "non viable". C R Soc Biol 97:114–116

Fahrner K, Hogan BLM, Flavell RA 1987 Transcription of H-2 and Qa genes in embryonic and adult mice. EMBO (Eur Mol Biol Organ) J 6:1265–1271

Gilbert SF 1988 Developmental Biology, 2nd edn. Sinauer Associates Inc, Sunderland, MA

Glücksohn-Schönheimer S 1944 The development of normal and homozygous *Brachy* (*T/T*) mouse embryos in the extraembryonic coelom of the chick. Proc Natl Acad Sci USA 30:134–140

Green MC 1989 Catalog of mutant genes and polymorphic loci. In: Lyon MF, Searle AG (eds) Genetic variants and strains of the laboratory mouse, 2nd edn. Oxford University Press, Oxford

Grüneberg H 1958 Genetical studies on the skeleton of the mouse. J Embryol Exp Morphol 6:424–443

Herrmann BG 1991 Expression pattern of the *Brachyury* gene in whole-mount T^{wis}/T^{wis} mutant embryos. Development 113:913–917

Herrmann B, Bucan M, Mains PE, Frischauf A-M, Silver L-M, Lehrach H 1986 Genetical analysis of the proximal portion of the mouse t complex: evidence for a second inversion within t haplotypes. Cell 44:469–476

Herrmann BG, Barlow DP, Lehrach H 1987 A large inverted duplication allows homologous recombination between chromosomes heterozygous for the proximal t complex inversion. Cell 48:813–825

Herrmann BG, Labeit S, Poustka A, King TR, Lehrach H 1990 Cloning of the *T* gene required in mesoderm formation in the mouse. Nature (Lond) 343:617–622

Lyon MF, Meredith R 1964 Investigations of the nature of t-alleles in the mouse. Heredity 19:313–325

MacMurray A, Shin H-S 1988 The antimorphic nature of the T^c allele at the mouse T locus. Genetics 120:545–550

Nadeau JH, Varnum D, Burkart D 1989 Genetic evidence for two t complex tail interaction (tct) loci in t haplotypes. Genetics 122:895–903

Poustka A, Lehrach H 1986 Jumping libraries and linking libraries: the next generation of molecular tools in mammalian genetics. Trends Genet 2:174–179

Rashbass PR, Cooke LA, Herrmann BG, Beddington RSP 1991 A cell autonomous function of *Brachyury* in *T/T* embryonic stem cell chimaeras. Nature (Lond) 353:348–351

Röhme D, Fox H, Herrmann B et al 1984 Molecular clones of the mouse t complex derived from microdissected metaphase chromosomes. Cell 36:783–788

Schulte-Merker S, Herrmann BG, Nüsslein-Volhard C 1992 The zebrafish homologue of the mouse *Brachyury* gene encodes a nuclear protein. Submitted

Searle AG 1966 Curtailed, a new dominant T-allele in the house mouse. Genet Res 7:86–95

Shedlovsky A, King TR, Dove WF 1988 Saturation germ line mutagenesis of the murine t region including a lethal allele at the quaking locus. Proc Natl Acad Sci USA 85:180–184

Slack JMW 1983 From egg to embryo. Cambridge University Press, Cambridge

Slack JMW, Darlington BG, Heath JK, Godsave SF 1987 Mesoderm induction in early *Xenopus* embryos by heparin-binding growth factors. Nature (Lond) 326:197–200

Smith JC, Yaqoob M, Symes K 1988 Purification, partial characterization and biological properties of the XTC mesoderm-inducing factor. Development 103:591–600

Smith JC, Price BMJ, Green JBA, Weigel D, Herrmann BG 1991 Expression of a Xenopus homologue of *Brachyury* (*T*) is an immediate-early response to mesoderm induction. Cell 67:79–87

Styrna J, Klein J 1981 Evidence for two regions in the mouse t complex controlling transmission ratios. Genet Res 38:315–325

Wilkinson DG, Bhatt S, Herrmann BG 1990 Expression pattern of the mouse T gene and its role in mesoderm formation. Nature (Lond) 343:657–659

Yanagisawa KO 1990 Does the T gene determine the anteroposterior axis of a mouse embryo? Jpn J Genet 65:287–297

Yanagisawa KO, Fujimoto H, Urushihara H 1981 Effects of the brachyury (T) mutation on morphogenetic movement in the mouse embryo. Dev Biol 87:242–248

DISCUSSION

Smith: I would like to describe the expression of the *Xenopus* homologue of *Brachyury*. It is first expressed when the frog gastrulates. *In situ* hybridization shows that it is expressed in the frog equivalent of the primitive streak and in the mesoderm, exactly the same as in the mouse. At the early neurula stage, there is expression in the notochord but not in the somites or in the lateral plate.

Expression of *Brachyury* in *Xenopus* is inducible. It is not expressed in cells of the animal cap when these are cultured alone or in the vegetal cells; it is expressed in the mesoderm. If the animal cap cells are put into physical contact with cells of the vegetal hemisphere, they begin to express *Brachyury*. This expression is also switched on if animal cap cells are exposed to activin or to FGF. It is a very rapid response and will occur in dispersed cells, which implies there is no need for cell–cell contact. It is expressed in the presence of inhibitors of protein synthesis.

In *Xenopus* we are now looking at, for example, whether the expression in notochord is switched on in response to factors like activin while expression in the primitive streak is induced by FGF. We can inject the message into the egg and ask what happens if it is expressed in the wrong place.

Buckingham: You said it is a rapid response; how does the timing compare with the induction of *MyoD* or actin?

Smith: It depends on how you detect the message. Using RNase protection, expression of actin occurs between five and seven hours after induction (Gurdon et al 1985); *MyoD* is probably switched on about two hours earlier than that (Hopwood et al 1989). *Brachyury* can be expressed after as little as 90 minutes, but it varies between experiments.

Buckingham: Is it still being expressed when *MyoD* is switched on?

Smith: Yes, but by then it may be being down-regulated in the somites.

McLaren: Bernhard, in the homozygous mutant, *T* is expressed early, at 6.5–7 days, but not later; mesoderm is made early and is normal but is defective later; yet you say that *T* is not required for early formation of mesoderm. How do you know? Why shouldn't the early *T* expression be required for the early mesoderm?

Herrmann: It is not essential. I think that at some point during organogenesis there is a limiting amount of some product which depends on the presence of the *T* gene product. At earlier stages some of this product is present despite the absence of *T*. There may be another gene involved in that process, for example the *T*-like gene that has been described which is probably linked to the *T* allele (Lyon & Meredith 1964, Nadeau et al 1989), and the additive action of at least two genes may be necessary for axial development at the end of the tail. When only one of these genes is functional, the other is sufficient for some early embryogenesis and axis formation but not all the way to the posterior end of the axis.

Balling: In the *T* mutants that make at least some somites, do those somites have any anteroposterior polarity?

Herrmann: I don't think anybody has looked at that in detail. Dorothea Bennett (1958) showed that the somites cannot respond to cartilage-inducing signals.

Evans: Is there any prediction about the function of the *T* gene product that you can make from the amino acid sequence? Secondly, is there any homology with genes in *Drosophila* or other invertebrates?

Herrmann: There is no homology to *Drosophila* that we can detect by hybridization. It may be possible to design oligos to pick up conserved domains.

There is no clear prediction about the protein structure. The N-terminal is highly conserved; the C-terminal half is less conserved. The protein is found in the nucleus (S. Schultz-Merker, B. G. Herrmann and C. Nüsslein-Volhard, unpublished work), so I like to speculate that it is a transcription factor. In the homeobox proteins, for example, the most conserved region is the DNA-binding domain and we are investigating whether the *T* gene product can bind DNA.

Hastie: What are the phenotypes of the heterozygotes in the gain-of-function mutants?

Herrmann: Heterozygote gain-of-function mutants are tailless; loss-of-function mutants have short tails of varying lengths. The effects of T^C on malformation of vertebrae have been described by Tony Searle (1966).

I like to think they are antimorphic. MacMurray & Shin (1988) postulated that T^C is an antimorphic allele and I think they are right. By providing more activity on a duplication chromosome, you can complement the phenotype of T^C and restore the tail length. We have some evidence in transgenic mice that tail lengths can be restored, at least to some degree, when there is a normal transgene in a $T^C/+$ mouse. In the homozygous state of T^C or T^{Wis} the phenotype is only slightly more severe than in the loss-of-function mutants.

Basically, the changed product antagonizes the function of the normal allele: it removes functional product from the pool within the cell. This gives a more severe phenotype in the heterozygotes, but in the homozygotes the phenotype is almost identical because there is no functional product present.

Joyner: The three dominant-negative mutants or antimorphs are lacking the C-terminus, which is the non-conserved region. In transcription factors, often the domains of the protein that interact with other proteins are not highly conserved, although they tend to have certain characteristics, such as being rich in alanine or proline. Does the T protein have any such regions?

Herrmann: It is proline rich and serine rich.

Joyner: Is the proline-rich region missing in the mutants?

Herrmann: I haven't looked specifically at the different domains. I would like to speculate that the N- terminal region which is strongly conserved is the DNA-binding domain and the C-terminal region is involved in protein–protein interactions, but there is no evidence for this.

Goodfellow: What's the prediction for phenotype of ectopic expression or just overexpression?

Herrmann: One would hope that high ectopic expression of T would cause formation of a notochord in a second axis. I am afraid things will probably not turn out so simple.

Jaenisch: Concerning the first signal for mesoderm induction in ES cells, do you know whether and what type of embryoid bodies are formed by homozygous T/T ES cells?

Beddington: We have started to look at differentiation *in vitro* but we don't yet have any conclusive results. When we make experimental teratomas, the homozygous line forms less well differentiated teratomas than the heterozygous line. But we have started to look at a second homozygous line and that seems to be differentiating quite well into tissues like cartilage and skeletal muscle.

McLaren: Ephrussi (1935) explanted bits of early homozygous T/T embryos and they made muscle and cartilage.

Beddington: Fujimoto & Yanagisawa (1979) have also done that.

Balling: You mentioned an increasing requirement for T towards the caudal end. Could this be an increasing requirement for downstream elements of the

T signalling pathway rather than for *T* itself? The requirement for *T* could be constant along the axis and the receptor for a notochord signal, for example, could be needed at increased levels at the caudal end.

Herrmann: One would have to postulate an in-built clock in the cells so that cells formed very late in gestation have an increased requirement for a signal to make any mesoderm at all. In my model, the increase of signal is transferred along the axis by an increasing amount of *T* product from one cell to the next.

Balling: What's the experimental evidence for that?

Herrmann: The evidence is that at the posterior end there is a higher requirement for *T*. This is clear from the genetic results; if you take out one gene then you don't get a complete axis formed. If you take out more of the gene product, more anterior regions are affected.

Balling: That does not allow you to discriminate whether it's the signal or the receiver.

Goodfellow: Bernhard measures the signal. Your statement would be true only if you couldn't see expression of *T*, but he has shown expression of *T*.

Balling: Not at increasing levels towards the posterior end. Can you induce autoinduction of *T* by transfecting the *T* gene transiently into mesodermal cells, for example?

Herrmann: We haven't done those experiments.

Smith: Have you looked at the very early expression of *T* in the primitive ectoderm to ask whether it matches with the fate map? Is it expressed where you would expect it to be early on or is there general expression followed by cell sorting?

Herrmann: The early expression is in the posterior end of the axis; in the region where the primitive streak is formed initially, at 6.5 days approximately. I don't see any expression outside that, but the sensitivity of whole-mount *in situ* hybridization might not be sufficiently high. The signal is fairly weak; I don't think many transcripts are present.

Copp: Dorothea Bennett reported that in *T* homozygotes the extracellular matrix is abnormal; this was the interstitial matrix between the paraxial mesenchyme cells (Jacobs-Cohen et al 1983). Would you see this as just the secondary effect of cell-autonomous expression? Taken alone, matrix abnormalities are more reminiscent of a non-cell-autonomous type of defect.

Herrmann: The cell autonomous action of *T* is fairly well established (Rashbass et al 1991, Beddington et al, this volume). But there are also defects in *T* mutants suggesting a signal-mediated action of *T*. The abnormality of the extracellular matrix in *T* homozygotes might be signal mediated or a secondary effect of the lack of *T* product in the paraxial mesenchyme.

Buckingham: You are implying that the *T* gene is involved in a mesodermal signalling mechanism, perhaps inducing the signal as a transcription factor acting on another gene. Do you think that at some stage the *T* gene product is expressed in all mesodermal cells?

Herrmann: Yes.

Buckingham: *T* continues to be expressed in the notochord. You were suggesting that there again the *T* gene product acts as a transcription factor inducing a signalling mechanism coming from the notochord. Do you regard this as a second distinct function of *T*?

Herrmann: I don't know if I would call it a second distinct function. The two events are apparently closely connected. If *T* acts on the signal molecules that induce mesoderm which is recruited into the extending notochord precursor, then there is a direct downstream effect and a direct upstream effect of *T* expression on cell fate.

Buckingham: The notochord is going to emit a particular kind of signal that will affect, for example, somite development. It seems to be distinct from what is going on earlier.

Wilkinson: *T* continues to be expressed at 17.5 days, well after these events you are talking about. What do you think it might be doing then?

Herrmann: I am not sure it is doing anything then. In zebrafish and in *Xenopus* it is turned off at that stage.

Brown: Several *T/T* mutant embryos have sinistral hearts, suggesting inverted heart looping. Is that an established effect of the *T* mutation?

Beddington: I haven't examined the hearts carefully but they clearly are slightly abnormal.

References

Beddington RSP, Püschel AW, Rashbass P 1992 Use of chimeras to study gene function in mesodermal tissues during gastrulation and early organogenesis. In: Postimplantation development in the mouse. Wiley, Chichester (Ciba Found Symp 165) p 61–77

Bennett D 1958 *In vitro* studies on cartilage induction in *T/T* mice. Nature (Lond) 181:1286

Ephrussi B 1935 The behaviour in vitro of tissues from lethal embryos. J Exp Zool 70:197–204

Fujimoto H, Yanagisawa KO 1979 Effects of the T mutation on histogenesis of the mouse embryo under the testis capsule. J Embryol Exp Morphol 50:21–30

Gurdon JB, Fairman S, Mohun TJ, Brennan S 1985 Activation of muscle-specific actin genes in Xenopus development by an induction between animal and vegetal cells of a blastula. Cell 41:913–922

Hopwood ND, Pluck A, Gurdon JB 1989 MyoD expression in the forming somites is an early response to mesoderm induction in *Xenopus* embryos. EMBO (Eur Mol Biol Organ) J 8:3409–3417

Jacobs-Cohen RJ, Spiegelman M, Bennett D 1983 Abnormalities of cells and extracellular matrix of T/T embryos. Differentiation 25:48–55

Lyon MF, Meredith R 1964 The nature of t alleles in the mouse. II. Genetic analysis of an unusual mutant allele and its derivatives. Heredity 19:313–325

MacMurray A, Shin H-S 1988 The antimorphic nature of the T^c allele at the mouse T locus. Genetics 120:545–550

Nadeau JH, Varnum D, Burkart D 1989 Genetic evidence for two t complex tail interaction (tct) loci in t haplotypes. Genetics 122:895–903

Rashbass PR, Cooke LA, Herrmann BG, Beddington RSP 1991 A cell autonomous function of *Brachyury* in T/T embryonic stem cell chimaeras. Nature (Lond) 353:348–351

Searle AG 1966 Curtailed, a new dominant T-allele in the house mouse. Genet Res 7:86–95

Molecular mechanisms of pattern formation in the vertebrate hindbrain

M. A. Nieto, L. C. Bradley, P. Hunt, R. Das Gupta, R. Krumlauf and D. G. Wilkinson

Laboratory of Eukaryotic Molecular Genetics, National Institute for Medical Research, The Ridgeway, Mill Hill, London NW7 1AA, UK

Abstract. During early stages of neural development a series of repeated bulges, termed rhombomeres, form in the vertebrate hindbrain. Studies in the chick have shown that rhombomeres are segments that underlie the patterning of nerves in the hindbrain, and this raises the question of the molecular basis of segment development. Several genes have been found with expression patterns consistent with roles in the formation or differentiation of rhombomeres. The zinc finger gene *Krox-20* is expressed in two alternating rhombomeres, r3 and r5, in the mouse hindbrain; these stripes of gene expression are established prior to the morphological appearance of segments. *Krox-20* is also expressed in this pattern in the chick and *Xenopus*, suggesting that it has a conserved role, possibly in the formation of rhombomeres. Four members of the *Hox-2* homeobox gene cluster have limits of expression at rhombomere boundaries. Three genes, *Hox-2.6*, *-2.7* and *-2.8* have progressively more anterior limits of expression at two-segment intervals, whereas expression of *Hox-2.9* is restricted to one rhombomere, r4. The *Hox-2* genes are expressed in spatially restricted patterns in early neural crest cells. These findings suggest that the *Hox* genes have roles in specifying the identity of rhombomeres and of neural crest.

1992 Postimplantation development in the mouse. Wiley, Chichester (Ciba Foundation Symposium 165) p 92–107

One of the more challenging problems in developmental biology is to understand the cellular and molecular mechanisms underlying pattern formation in the most complex tissue, the vertebrate nervous system. As for other tissues discussed in this volume, little is known of how the diverse cell types are generated and how they become organized in region-specific patterns at appropriate locations in the developing embryo. In addition, the nervous system poses a special problem: how are the extraordinarily specific projections of axons to neural and peripheral targets established? Although elucidation of how the structure of the mature nervous system arises is a daunting prospect, the final complexities are built upon simple beginnings, and it seems likely that the early developmental decisions will prove to be more amenable to molecular and genetic analysis than later events.

The central nervous system of vertebrates arises from the neural plate, a thickened epithelium that is induced in dorsal ectoderm by signals from the underlying mesoderm. The lateral edges of the neural plate fold up, meet and then fuse along the dorsal midline to form an enclosed cylinder, the neural tube. As the neural tube is forming, certain cells detach and migrate away from its dorsal edge. These cells constitute the neural crest and contribute to several tissues, including the peripheral nervous system and, in the head, cartilage and bone. The neural plate and early neural tube consist of a one cell thick layer of dividing cells; postmitotic differentiating neurons are not generated until some time after closure. Postmitotic neurons migrate radially away from the proliferating epithelium, which persists during subsequent development as a stem cell layer adjacent to the lumen of the neural tube. The spectrum of neural cell types and the pattern in which they are organized varies along the anteroposterior axis, with particularly dramatic variations in different parts of the brain. The regionalization of the nervous system is manifested before and during neural tube closure, well before neurogenesis starts. Constrictions appear in the early neural epithelium, subdividing it along the anteroposterior axis into the primary brain vesicles (the forebrain, midbrain and hindbrain) and the spinal cord. Subsequently, the neural epithelium is further subdivided by the formation of a series of bulges which are particularly prominent in the brain (reviewed by Vaage 1969). In the hindbrain, these bulges are known as rhombomeres, and their conservation throughout vertebrates suggests that they are of developmental significance.

Recent studies in the chick have shown that rhombomeres are segments that underlie the patterning of the hindbrain. Clonal analysis indicates that the boundaries between rhombomeres restrict cell movement, and thus subdivide the hindbrain into a series of compartments (Fraser et al 1990). These compartments generate specific populations of neurons, including the branchial motor nerves, whose cell bodies are located in adjacent pairs of rhombomeres (Fig. 1; Lumsden & Keynes 1989). Intriguingly, the formation of these and certain other nerves is first initiated in alternating rhombomeres, implying the existence of cellular mechanisms with a two-segment periodicity (Lumsden & Keynes 1989). A potential broader significance for hindbrain segmentation is suggested by the anatomical relationships between rhombomeres, cranial ganglia and the branchial arches (Fig. 1). As will be discussed later, neural crest arising from specific rhombomeres contributes to these latter tissues and may be imprinted according to its segmental origin.

These findings raise the question of the genetic basis of the segmental patterning of the hindbrain. Certain of the genes involved in the formation and differentiation of rhombomeres would be expected to have segment-restricted expression. Several genes encoding putative transcription factors have been found to be expressed in segmental domains in the mouse hindbrain. We will discuss the implications of these observations for the function of these genes.

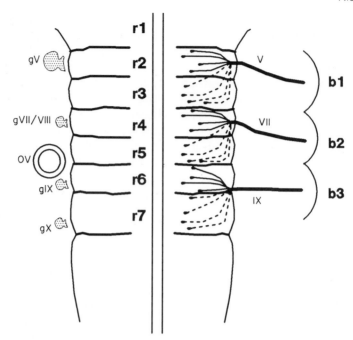

FIG. 1.   Cellular aspects of hindbrain segmentation. The left hand side depicts the anatomical arrangement of rhombomeres (r1–r7), cranial ganglia (gV–gX) and the otic vesicle (OV) found in all vertebrates. On the right is depicted the segmental arrangement of the branchial motor nerves (V–IX) which innervate the branchial arches (b1–b3). Early-differentiating branchial motor neurons are drawn as solid lines and later-differentiating neurons as dashed lines. For clarity, the Xth nerve which is located in r7 and r8 is not shown.

Expression of *Krox-20* in alternating rhombomeres

Krox-20 was first identified as a gene that is rapidly up-regulated when quiescent mouse fibroblasts are treated with serum or purified growth factors (Chavrier et al 1988). The *Krox-20* gene encodes a protein with three zinc fingers, structural motifs that are characteristic of a large family of DNA-binding proteins, and several lines of evidence support the idea that it acts as a transcription factor (Chavrier et al 1990).

In situ hybridization analysis of the expression of *Krox-20* in mouse embryos revealed that transcripts are restricted to specific regions of the neural epithelium and neural crest (Wilkinson et al 1989a,b). Expression is first detected in the early neural plate, prior to the formation of rhombomeres. A single domain of *Krox-20* expression is found in the prospective hindbrain of the 8 day mouse embryo (Fig. 2a,b); by 8.5 days a second, more posterior, domain has formed. At 9.5 days, these stripes correspond to two alternating rhombomeres, r3 and r5,

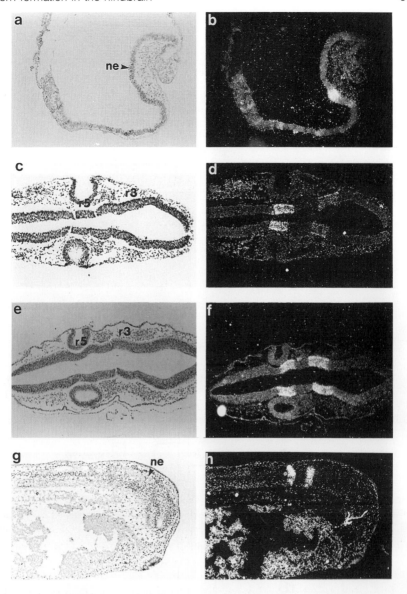

FIG. 2. Expression of *Krox-20* in hindbrain segments. The pattern of *Krox-20* expression is shown in the (a,b) 8 day mouse embryo, (c,d) 9.5 day mouse embryo, (e,f) stage 15 chick embryo, (g,h) stage 28 *Xenopus* embryo. a,c,e,g are bright field photographs corresponding to the dark field images shown in b,d,f,h, respectively. The apparent signal in the ventral part of the *Xenopus* embryo is not silver grains, but due to the refraction of light by yolky endoderm cells. ne, neural epithelium; r, rhombomere. Data from Wilkinson et al (1989b) and Nieto et al (1991).

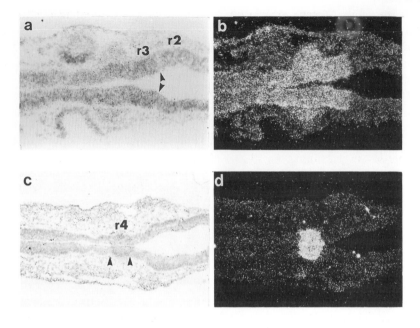

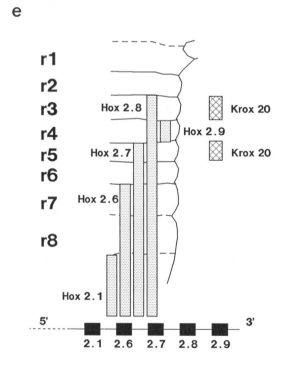

with the limits of gene expression coincident with rhombomere boundaries (Fig. 2c,d). Expression of *Krox-20* is sharply down-regulated in the hindbrain during the next day of development, first in r3 and then in r5. These data indicate that rhombomeres are domains of gene expression and suggest that *Krox-20* might play a role in early aspects of the segmental patterning of the hindbrain in the mouse. This role is likely to be conserved in birds and amphibians, as *Krox-20* is also expressed in this alternating pattern in the chick and *Xenopus* hindbrain (Fig. 2e–h).

The alternating expression pattern of *Krox-20* correlates with several cellular properties of rhombomeres. r3 and r5 have a delayed onset of neurogenesis relative to r2, r4 and r6 (Lumsden & Keynes 1989), and differ in that they do not generate migratory neural crest (Lumsden et al 1991). Grafting experiments suggest that an alternating property of rhombomeres underlies the formation of boundaries; morphological boundaries do not form when r3 and r5 are juxtaposed, but do form when either of these are grafted adjacent to any of the even-numbered rhombomeres (Guthrie & Lumsden 1991). A simple explanation for this behaviour would be that cells in r3 and r5 express adhesion molecules that render them miscible with each other, but immiscible with cells in r2, r4 and r6. It is therefore intriguing that a cell surface glycoprotein, recognized by the HNK-1 antibody, is specifically expressed in r3 and r5 in the chick hindbrain (Kuratani 1990), although its expression occurs too late to have a role in establishing lineage restriction. In contrast, *Krox-20* is expressed before overt segmentation and prior to lineage restriction in the chick hindbrain (Nieto et al 1991). Thus, regardless of the nature of the alternating cellular property underlying the formation of boundaries, it is possible that *Krox-20* is involved in regulating this process.

Segmental expression of *Hox-2* homeobox genes

The genetic analysis of *Drosophila* development identified clustered genes, the homeotic genes, involved in specifying variations in pattern along the anteroposterior axis (Gehring 1987). The early domains of expression of these genes are coupled to parasegments and correlate with their function in determining segmental identity (Akam 1987, Ingham 1988). The presence of

---

FIG. 3. Segmental expression of *Hox-2* genes. Examples of the expression patterns of *Hox-2* genes in the 9.5 day mouse hindbrain are shown: (a,b) *Hox-2.8*, (c,d) *Hox-2.9*. a,c are bright field photographs corresponding to the dark field images shown in b,d, respectively. The arrows indicate limits of gene expression at rhombomere boundaries. r, rhombomere. (e) Summary of the expression of *Hox-2* genes and *Krox-20* in hindbrain segments. The 3′ end of the *Hox-2* cluster on the chromosome is depicted at the bottom. Modified from Wilkinson et al (1989b).

a highly conserved sequence encoding a DNA-binding domain, the *Antp*-like homeodomain, in the homeotic genes defines them as a family of closely related genes. Four clusters of *Antp*-like genes, termed the *Hox* genes, exist in the mouse, and a number of striking parallels have been found between them and their *Drosophila* counterparts (Graham et al 1989, Duboule & Dolle 1989, Akam 1989, Kessel & Gruss 1990). Sequence comparisons indicate that the gene complexes of these diverse organisms are homologous with each other, indicating that they arose from a common ancestor. Moreover, there is a similar correlation between the physical order of the *Antp*-like genes and their expression along the anteroposterior axis. In both *Drosophila* and the mouse the genes within each complex are in the same transcriptional orientation, and successively more 3' genes have increasingly anterior domains of expression. These results provide circumstantial evidence that the *Hox* genes may have analogous roles to the *Drosophila* homeotic genes in specifying differences in pattern along the anteroposterior axis.

Analysis of the expression pattern of 3' members of the *Hox-2* cluster in 9.5 day mouse embryos showed that four of them have limits of expression coincident with rhombomere boundaries (Wilkinson et al 1989b, Murphy et al 1989). The successively linked *Hox-2.6*, *-2.7*, and *-2.8* genes are expressed from the caudal end of the spinal cord to an anterior limit at the r6/7, r4/5 and r2/3 boundaries, respectively (Fig. 3). Expression of the 3'-most gene, *Hox-2.9*, differs in two respects from these other *Hox-2* genes. Although expression initially occurs in a continuous domain extending from the caudal end of the spinal cord to the r3/4 boundary, it then becomes restricted in the hindbrain to a single rhombomere, r4 (Sundin & Eichele 1990, Murphy & Hill 1991). The anterior limit of this expression domain is caudal to that of its 5' neighbour, thus the *Hox-2.9* gene defies the correlation between position within the cluster and expression along the anteroposterior axis.

The segmental expression patterns of *Hox-2* genes in the hindbrain are reminiscent of the early metameric expression of the *Drosophila* homeotic genes, which suggests that they have roles in specifying the identity of rhombomeres. The expression of *Hox-2.6*, *-2.7* and *-2.8* with boundaries at two-segment intervals could specify pairs of rhombomeres. In principle, members of other *Hox* clusters could be expressed in distinct segmental patterns from the *Hox-2* genes, so that each rhombomere has a unique *Hox* code, as proposed by Kessel & Gruss (1990) for the pre-vertebrae. However, all of the genes in other clusters analysed thus far have the same anterior limits as their *Hox-2* counterparts (P. Hunt, unpublished work). These results seem to preclude a simple model in which the combinatorial expression of *Hox* genes directly specifies branchial motor nerves, as the latter arise from pairs of rhombomeres out of phase with the *Hox* gene expression patterns (compare Fig. 1 and Fig. 3).

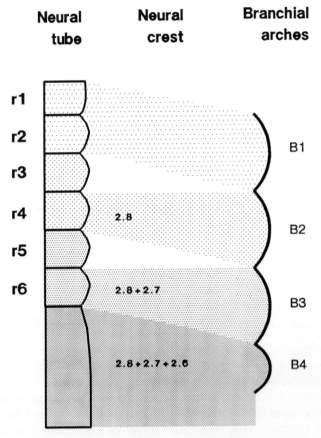

FIG. 4.   Summary of *Hox-2* gene expression in rhombomeres and neural crest. The shading indicates the combinatorial expression of *Hox-2* genes and the migratory pathways of neural crest from the hindbrain into the branchial arches. The segment-restricted expression of *Hox-2* genes is also imparted to early neural crest which emigrates from specific rhombomeres into the branchial arches. These results suggest that *Hox-2* genes may be involved in patterning neural crest according to its site of origin in the neural epithelium. Data from Lumsden et al (1991) and Hunt et al (1991).

Expression of *Hox-2* genes in neural crest

There is a conserved anatomical relationship between rhombomeres and other structures of the head (Fig. 1). Cranial ganglia are associated with specific rhombomeres (Vaage 1969) and the branchial arches are aligned with, and are innervated by the branchial motor nerves arising from, pairs of rhombomeres (Lumsden & Keynes 1989). A developmental basis for these relationships could lie in the neural crest, which contributes to the cranial ganglia and forms the skeletal components of the branchial arches. Grafting experiments suggest that neural crest destined to enter the branchial arches is imprinted prior to its emigration from the neural epithelium and imposes pattern on mesodermal tissue; when neural crest destined to enter the first arch was transplanted so that it entered the second arch, it formed not only the bones typical of the first arch, but also the appropriate set of mesodermally derived muscles (Noden 1983).

In view of the expression of *Hox* genes in segment-restricted patterns, it was pertinent to examine whether they are expressed in a corresponding pattern in neural crest derived from the hindbrain. Spatially restricted expression of *Hox-2* genes does occur in early neural crest, and is maintained as these cells migrate into the branchial arches (Fig. 4; Hunt et al 1991). Fate mapping has shown that neural crest from each of r2, r4 and r6 enters only a specific branchial arch, the first, second and third, respectively; r3 and r5 seem not to form neural crest (Lumsden et al 1991). Thus, the combinatorial expression pattern of *Hox-2.6*, *-2.7* and *-2.8* in rhombomeres correlates precisely with their expression in the neural crest of the branchial arches (Fig. 4). On the basis of this, we suggest the segment-restricted expression of *Hox-2* genes in the neural epithelium is also imposed upon early neural crest and might contribute to the imprinting of these cells prior to their migration.

Conclusions and future prospects

Analysis at the cellular and molecular level has given us some clues regarding mechanisms that pattern the early hindbrain. The neural epithelium of the hindbrain becomes subdivided into a series of compartments, possibly through the generation of an alternating pattern of cellular properties. These compartments underlie the patterning of the branchial motor nerves, and possibly also of neural crest that enters the branchial arches. The expression pattern of *Krox-20* correlates with several alternating properties of rhombomeres, including cellular differences that lead to boundary formation, the relative timing of neurogenesis and the formation of neural crest. In view of the very early establishment of the *Krox-20* expression pattern, this gene could be involved in any of these events. The segment-restricted expression patterns of *Hox-2* genes suggest that they may be involved in specifying segment identity. Moreover, the expression of these genes in neural crest derived from rhombomeres suggests that *Hox-2* genes may also be involved in the patterning of the branchial arches.

These findings present a number of challenges. Clearly, it is essential to test directly the functions of the *Hox-2* and *Krox-20* genes by manipulative experiments. It is important to examine whether they are linked in a regulatory cascade; this will require finding other genes that act upstream or are potential targets of these transcription factors. Good candidates are certain retinoic acid receptors and binding proteins that have segment-restricted expression (Maden et al 1991, Ruberte et al 1991). Perhaps the most challenging problem is to forge links between these genes and the cellular events of hindbrain segmentation and neurogenesis. The key to this may be the identification of genes with potential roles in cellular interactions that pattern the hindbrain, such as cell adhesion or signal transduction.

References

Akam M 1987 The molecular basis for metameric pattern in the *Drosophila* embryo. Development 101:1–22

Akam M 1989 Homologous gene clusters in insects and vertebrates. Cell 57:347–349

Chavrier P, Zerial M, Lemaire P, Almendral J, Bravo R, Charnay P 1988 A gene encoding a protein with zinc fingers is activated during G_0/G_1 transition in cultured cells. EMBO (Eur Mol Biol Organ) J 7:29–35

Chavrier P, Vesque C, Galliot B et al 1990 The segment-specific gene *Krox-20* encodes a transcription factor with binding sites in the promoter region of the *Hox-1.4* gene. EMBO (Eur Mol Biol Organ) J 9:1209–1218

Duboule D, Dolle P 1989 The murine *Hox* network: its structural and functional organisation resembles that of *Drosophila* homeotic genes. EMBO (Eur Mol Biol Organ) J 8:1497–1505

Fraser S, Keynes R, Lumsden A 1990 Segmentation in the chick embryo hindbrain is defined by cell lineage restrictions. Nature (Lond) 344:431–435

Gehring WJ 1987 Homeo boxes in the study of development. Science (Wash DC) 236:1245–1252

Graham A, Papalopulu N, Krumlauf R 1989 The murine and *Drosophila* homeobox gene complexes have common features of organisation and expression. Cell 57:367–378

Guthrie S, Lumsden A 1991 Formation and regeneration of rhombomere boundaries in the developing chick hindbrain. Development 112:221–229

Hunt P, Wilkinson D, Krumlauf R 1991 Patterning the vertebrate head: murine *Hox-2* genes mark distinct subpopulations of premigratory and migrating cranial neural crest. Development 112:43–50

Ingham PW 1988 The molecular genetics of embryonic pattern formation in *Drosophila*. Nature (Lond) 335:25–34

Kessel M, Gruss P 1990 Murine developmental control genes. Science (Wash DC) 249:374–379

Kuratani SC 1991 Alternate expression of the HNK-1 epitope in rhombomeres of the chick embryo. Dev Biol 144:215–219

Lumsden A, Keynes R 1989 Segmental patterns of neuronal development in the chick hindbrain. Nature (Lond) 337:424–428

Lumsden A, Sprawson N, Graham A 1991 Segmental origin and migration of neural crest cells in the hindbrain region of the chick embryo. Development 113:1281–1291

Maden M, Hunt P, Eriksson U, Kuroiwa A, Krumlauf R, Summerbell D 1991 Retinoic acid-binding protein, rhombomeres and the neural crest. Development 111:35–44

Murphy P, Hill RE 1991 Expression of the mouse *labial*-like homeobox-containing genes, *Hox*-2.9 and *Hox*-1.6, during segmentation of the hindbrain. Development 111:61–74

Murphy P, Davidson DR, Hill RE 1989 Segment-specific expression of a homeobox-containing gene in the mouse hindbrain. Nature (Lond) 341:156–159

Nieto MA, Bradley LC, Wilkinson DG 1991 Conserved patterns of segmental expression of the zinc-finger gene *Krox*-20 in the vertebrate hindbrain. Development 113, in press

Noden DM 1983 The role of the neural crest in patterning of avian cranial, skeletal, connective and muscle tissues. Dev Biol 96:144–165

Ruberte E, Dolle P, Chambon P, Morriss-Kay G 1991 Retinoic acid receptors and cellular retinoid binding proteins. Development 111:45–60.

Sundin OH, Eichele G 1990 A homeo domain protein reveals the metameric nature of the developing chick hindbrain. Genes & Dev 4:1267–1276

Vaage S 1969 The segmentation of the primitive neural tube in chick embryos (*Gallus domesticus*). Adv Anat Embryol Cell Biol 41:1–88

Wilkinson DG, Bhatt S, Chavrier P, Bravo R, Charnay P 1989a Segment-specific expression of a zinc finger gene in the developing nervous system of the mouse. Nature (Lond) 337:461–465

Wilkinson DG, Bhatt S, Cook M, Boncinelli E, Krumlauf R 1989b Segmental expression of *Hox*-2 homeobox-containing genes in the developing mouse hindbrain. Nature (Lond) 341:405–409

DISCUSSION

Goodfellow: Could your compartments be parasegments?

Wilkinson: I think parasegments are a red herring because they were defined in *Drosophila* and my belief is that the segments in the hindbrain are not homologous with those in flies. The extrapolation is therefore not valid.

Goodfellow: The word was just used as an example. Perhaps the boundary you see isn't going to form at the precise rhombomere boundary but is related to where that boundary will form. In which case, Andrew Lumsden's marking experiment may have worked before the rhombomeres had formed but the cells were marking a different compartment.

Wilkinson: Fraser et al (1990) found clones that span from one boundary to the other of a single rhombomere. As far as they can tell there is no subcompartmentalization.

Goodfellow: Can you tell us anything about the expression of the proteins?

Wilkinson: I can't, because we don't have any antibodies and that's a weakness of all our arguments.

Buckingham: In experiments where manipulations in the chick juxtapose even-numbered rhombomeres with even-numbered rhombomeres or odd with odd, what happens to the expression of genes like *Krox-20* or *Hox-2*? And what happens to the development of the branchial arches?

Wilkinson: We don't know what happens to the gene expression; we are collaborating with Andrew Lumsden to look at that.

The embryos haven't been left to develop long enough to find out what happens to the branchial arches. There is an interesting question as to whether the neural crest coming from those transplanted rhombomeres does something to the pattern of the branchial arches.

Krumlauf: Andrew is unable to look at the connective tissue but if he looks at the neurogenic component, the position of the nerves seems to be associated with the origin of the rhombomere after it is grafted to a new place. He is letting those embryos develop longer to see whether he can repeat Drew Noden's experiments with an isolated rhombomere before neural crest migration.

Wilkinson: It will then be interesting to look at the *Hox* gene expression patterns in the neural crest to see whether the correlations with phenotype hold.

McMahon: There seems to be a lack of positional information in both the mandibular arch and branchial arch in terms of *Hox* gene expression, so how is the pattern of those structures determined?

Wilkinson: They could be the default pathway or there could be other genes involved in patterning. There is evidence that *engrailed*-like genes are expressed in neural crest in these arches in a number of species.

Joyner: The *engrailed* genes may be expressed in neural crest cells in the mouse (Davis et al 1991).

Hastie: How does the knock out of *Hox-1.5* fit in with all this?

Krumlauf: There is an interesting dilemma. In the mice with a disrupted *Hox-1.5* gene that Mario Capecchi has reported, there are phenotypes more anterior than the patterns of *Hox-1.5* expression would predict (Chisaka & Capecchi 1991). One explanation is that if you insert a construct with gene regulatory regions into the genome, you alter the expression of surrounding genes.

A consequence of rhombomeres 3 and 5 not generating neural crest is that in transgenic mice we have made marked with *lacZ*, we would predict that *Hox-2.8* is expressed only in the second branchial arch not in the first. But we do see a few labelled cells in the first arch. It is possible that some neural crest cells coming out of r4 migrate into the first arch and pattern a very small percentage of the arch. This would not have been seen by *in situ* hybridization. It is dangerous to make absolute correlations based on *in situ* analysis, and we need to look at the protein to be sure.

Pierre Chambon has a *Hox-1.6* knock out in which the phenotypes are restricted to arches where *Hox-1.6* is expressed in the corresponding rhombomeric 4-6 regions (Lufkin et al 1991). It is not expressed in the mesenchymal crest, only in the neurogenic crest, and those components are missing or altered in the mutant mice. The correlations that David Wilkinson is making hold well with the knock out data.

Hogan: In both those knock outs, *Hox-1.6* and *Hox-1.5*, the rhombomeric pattern was unaltered. This again would suggest that *Hox* genes regulate the character of the rhombomeres rather than their initial formation.

Joyner: Does that correlate with the timing of expression?

Wilkinson: It correlates with what we believe to be the timing of expression relative to other markers of segments.

Kaufman: Is there an incompatibility on some of your diagrams, particularly relating to rhombomeres 2 and 3? You indicated that the Vth ganglion comes from r2 (Fig. 1), yet another slide (not reproduced here) showed neural components additionally coming from r3. Is there a possibility that the Vth nerve is a complex nerve, like the XIIth nerve? Could it be two fused together? It is surprising that the Vth nerve plays so many different roles, and it does have a mandibular and a maxillary component.

Wilkinson: The Vth branchial motor nerve does arise from r2 and r3. The trigeminal (Vth) ganglion is associated with only r2. We are talking about two different things, the sensory ganglion and the motor nerve.

There is an idea from comparative anatomy that the trigeminal is a composite of two nerves, if you look at the structures that it innervates. It is possible that in some primitive organism the segmentation of the hindbrain extended more anteriorly and some modification of that occurred very early in vertebrate evolution. That would explain the anatomy of the trigeminal nerve and might explain why *Hox-2.9* has this apparently aberrant expression pattern. You could hypothesize that in some primitive organism it followed the same pattern of successively more anterior expression with successively more 3′ position along the chromosome as do the other *Hox* genes. The current expression pattern of *Hox-2.9* in a more caudal domain than *Hox-2.8* may reflect some modification at the anterior end of the segmented region of the hindbrain.

Thesleff: Do neural crest cells migrate only from the even-numbered rhombomeres? Have neural crest cells been seen migrating from neural tubes in culture?

Wilkinson: Migration from chick neural tube in culture just produces a halo of cells around the explant; they don't migrate in a straight line. *In vivo* in the chick the neural crest streams out to the branchial arches in a more or less straight line. It is these crest-free zones which may keep the streams separate from each other.

Tam: Do you see a similar pattern in the rodent embryo? Is there a neural crest-free region in the mouse hindbrain?

Wilkinson: There is evidence for a crest-free region around r3 and r5 (Tan & Morriss-Kay 1985). My expectation is that the patterns will be conserved. Those observations were from scanning electron micrographs.

Tam: They have also studied the migration of cranial neural crest in rat embryos by orthotopic grafting (Tan & Morriss-Kay 1986).

Wilkinson: Andrew Lumsden believes his technique gives much finer resolution than do grafting experiments. He found the crest-free zones by marking very small populations of cells using DiI. He marked cells at the tip of the neural epithelium: when he marked cells in r5, he never found crest cells coming from r5 and entering the branchial arches. The resolution of grafting experiments may not be sufficient to pinpoint the crest-free zones.

Krumlauf: One problem is that because of steric hindrance from the otic vesicle it is very difficult to say if crest migrating around it has come from r5 or r6. Andrew Lumsden has done some analysis with stains that detect cell death. With r3 and r5 there seems to be a recurrent pattern that could be associated with cell death, so perhaps crest tries to form and just doesn't migrate very far.

Drew Noden, with his chick–quail chimera studies felt that crest did come from r5 but it didn't migrate very far. It was difficult to tell because of the position of the otic vesicle. Andrew Lumsden says his DiI injections wouldn't have been able to distinguish whether the crest doesn't migrate very far or doesn't survive because he was looking for lateral migration from r5.

Saxén: Wouldn't Noden's transplantation experiments indicate that there is a difference between the cephalic and the trunk neural crest? Similar experiments with the trunk neural crest show that the cells are determined according to their final position.

Wilkinson: Many people in the field believe that there is a fundamental difference between the head and the trunk. There is a body of evidence that in the head the neural crest is the dominant patterning tissue, whereas in the trunk the mesoderm is.

Smith: How are you going to find out how the rhombomere-specific expression patterns are controlled?

Wilkinson: There are two general ways. One is through molecular biology: you can start to define the transcriptional regulatory elements of the *Krox-20* gene which lead to its segmentally restricted expression pattern and work backwards from there to find the transcription factors which bind to these elements.

The other generalized approach relates to what I see as a major missing link in this whole field. All the markers we have are transcription factors. Cell–cell interactions must be crucial in this whole process: we really need to find genes involved in cell–cell interactions, which might be involved in the formation of rhombomeres and might be talking to transcription factors.

Herrmann: Haven't you missed an important point, namely making use of the mutants to show what happens if an embryo doesn't have the gene?

Wilkinson: Of course, there's homologous recombination knock out. If we had an embryo in which the *Krox-20* gene was knocked out, that could provide very direct evidence for its function. There is also the approach of mis-regulating the genes in transgenic mice. It would be nice to express *Krox-20*, for example, off a *Hox-2.9* promoter.

Hogan: Gail Martin has worked on a mutant called *Kreisler* which shows an altered pattern of expression of *Hox* genes in the hindbrain.

You say *Krox-20* expression extends laterally. Do you think that reflects that information is passing from the mesoderm into the neuroectoderm? Is there local induction from the paraxial mesoderm?

Wilkinson: Yes, our recent whole-mount *in situ* hybridization data show that in some embryos *Krox-20* expression is switched on in a medial-to-lateral direction. It's conceivable that the signal is from paraxial mesoderm but we have no experimental data to say where the signal is coming from.

Joyner: You see early expression of *Krox-20* in two little patches of cells in the neural plate on either side of the midline and the expression seems to fill in across the midline and laterally. That's exactly what we see for the *engrailed* genes being turned on in mouse and in frog (Davis et al 1991). This is intriguing, since it is at the same time. Is there any evidence in *Xenopus* where the signals that induce regionalization of the nervous system are coming from?

Smith: It's quite complicated. In Switzerland (May 1991) Ariel Ruiz i Altaba reported experiments on the expression of neural markers in *Xenopus* exogastrulae. Exogastrulae are made by allowing embryos to develop in hypertonic salt solutions. In these embryos the mesoderm does not move underneath the ectoderm but away from it, and the ectoderm forms an empty sac. The first experiments on exogastrulae were by Holtfreter in 1933, using urodeles, who found that the ectodermal sac developed as epidermis, indicating that during normal gastrulation a neural-inducing signal travels radially from the involuting dorsal mesoderm towards the overlying dorsal ectoderm. More recently, however, experiments with *Xenopus* have shown that exogastrulae can form neural structures (Kintner & Melton 1987, Ruiz i Altaba 1990), and that neural-inducing signals can travel tangentially, within the plane of the ectoderm (Dixon & Kintner 1989, Ruiz i Altaba 1990). Furthermore, these tangential signals seem to be able to pattern the neural ectoderm, as well as simply induce it to form neurectoderm; the gene *Xhox3*, for example, seems to be expressed in the right position (Ruiz i Altaba 1990). Using neural-specific antibodies, Ruiz i Altaba has shown that the neural pattern in exogastrulae is remarkably well-organized and quite complete, lacking only the forebrain and some ventral neuronal cell types such as secondary motoneurons. Not only is *Xhox3* expressed in the right place, *engrailed-2* is as well. Ruiz i Altaba suggests that the main inducing signal derives not from axial mesoderm but from the 'dorsal non-involuting marginal zone' or notoplate. Head mesoderm is required, however, for the induction of the forebrain, and notochord is required for induction of ventral neuronal cell types. For mouse and chick embryos this might mean that the anterior primitive streak plays a role in induction.

The question remains: why didn't Holtfreter see neural structures? One possibility is that he didn't have the right markers. Another is that urodele embryos, which he used, are different from *Xenopus*. They certainly gastrulate differently. Someone should repeat Ruiz i Altaba's experiments using newt embryos!

Saxén: I can confirm Holtfreter's finding, as we have never seen neural structures in *Triturus* exogastrulae. What were the neural markers which you referred to?

Smith: Ruiz i Altaba used four new monoclonal antibodies that recognize general neural antigens. Then he used an antibody that recognizes *engrailed* protein which showed expression at what seemed to be the midbrain/hindbrain junction.

Solter: But you could explain the induction and patterning of the neural structures in exogastrulae by the interaction between the posterior end of the ectoderm and the mesoderm coming out of the streak. This signal would form a gradient, thereby explaining why neural development is more and more deficient as one moves anteriorly.

Lawson: Is there not a controversy over homeogenetic induction in urodeles and *Xenopus*—that it occurs in one but not the other?

Smith: There is some controversy and some confusion. In this context homeogenetic induction of neural tissue can be viewed as 'tangential induction'; that is, as a neural-inducing signal spreading in the plane of the ectoderm. As I said, results from exogastrulae suggest that *Xenopus* shows tangential induction whereas newts do not. However, results implying the opposite have also been obtained. Pieter Nieuwkoop used urodele embryos to show that a neural-inducing signal can travel from the neural plate into flaps of ectoderm grafted on to it (Nieuwkoop et al 1952), while more recently Jones & Woodland (1989) were unable to demonstrate homeogenetic induction in *Xenopus*. In addition, Sharpe & Gurdon (1990) failed to detect tangential induction in *Xenopus*. The reasons for these differences may depend on the neural markers used, or on subtle differences in dissection. On the whole, I think a positive result beats a negative one and that homeogenetic induction probably occurs in both types of amphibian embryo.

Saxén: There is another difference between urodeles and *Xenopus*. One of the experimental nuisances in *Triturus* is their so-called autoneuralization. To my knowledge this has not been reported in *Xenopus*.

References

Chisaka O, Capecchi M 1991 Regionally restricted developmental defects resulting from targeted disruption of the mouse homeobox gene *hox*-1.5. Nature (Lond) 350:473–479

Davis CA, Holmyard DP, Millen KJ, Joyner AL 1991 Examining pattern formation in mouse, chicken and frog embryos with an *En*-specific antiserum. Development 111:287–298

Dixon JE, Kintner CR 1989 Cellular contacts required for neural induction in *Xenopus*: evidence for two signals. Development 106:749–757

Fraser S, Keynes R, Lumsden A 1990 Segmentation in the chick hindbrain is defined by cell lineage restrictions. Nature (Lond) 344:431–435

Jones EA, Woodland HR 1989 Spatial aspects of neural induction in *Xenopus laevis*. Development 107:785–791

Kintner CR, Melton DA 1987 Expression of *Xenopus* N-CAM RNA in ectoderm is an early response to neural induction. Development 3:311–325

Lufkin T, Dierich A, LeMeur M, Mark M, Chambon P 1991 Disruption of the *Hox*-1.6 homeobox gene results in defects in a region corresponding to its rostral domain of expression. Cell 66:1105–1119

Nieuwkoop PD, Boterenbrood EC, Kremer A et al 1952 Activation and organization of the central nervous system in amphibians. J Exp Zool 120:1–108

Ruiz i Altaba A 1990 Neural expression of the *Xenopus* homeobox gene Xhox3: evidence for a patterning neural signal that spreads through the ectoderm. Development 108:595–604

Sharpe CR, Gurdon JB 1990 The induction of anterior and posterior neural genes in *Xenopus laevis*. Development 109:765–774

Tan SS, Morriss-Kay G 1985 The development and distribution of the cranial neural crest in the rat embryo. Cell Tissue Res 240:403–416

Tan SS, Morriss-Kay GM 1986 Analysis of cranial neural crest cell migration and early fates in postimplantation rat chimaeras. J Embryol Exp Morphol 98:21–58

General discussion II

Developmental abnormalities in tetraploid mouse embryos

Kaufman: I have been working on a simple system that generates homozygous tetraploid mouse embryos (Kaufman & Webb 1990, Kaufman 1991). I isolate normal 2-cell fertilized eggs, electrofuse them to reconstitute a 1-cell tetraploid embryo, then transfer them to the oviducts of pseudopregnant recipients. These are then autopsied at different stages during the postimplantation period. The incidence of homozygous tetraploids that successfully develop to 10 days p.c. is about 70%, but the incidence of viable embryos decreases quite rapidly thereafter. The embryos are isolated for histological analysis, and the extraembryonic membranes are analysed to check the chromosome constitution of these embryos.

Because of the method employed, the embryos have either an XXYY or XXXX sex chromosome constitution; we also use male mice bearing a Robertsonian translocation to check the origin of these embryos. The females were mated to homozygous Rb(1.3)1Bnr male mice which carry a pair of 'marker' chromosomes. Cytogenetic analysis of the extraembryonic membranes of all the embryos isolated in this study revealed that they were all homozygous tetraploids. Each of the mitotic preparations contained two paternally derived Rb(1.3) 'marker' chromosomes. In the preliminary experiments, the most advanced embryo obtained was morphologically intermediate between a control 12.5 day and a control 13.5 day embryo. From the state of development of the organs, however, it was equivalent to a normal embryo of about 13.5 to 14 days of gestation. It had an abnormal craniofacial region, and similar features were seen in all of the other tetraploids isolated in this study. The postcranial axis of this embryo was close to normal, and the limbs also looked completely normal.

At about Day 12.5, approximately half the tetraploid embryos isolated appeared to have a normal postcranial axis, whereas the others have vertebral axial abnormalities and invariably body wall closure defects involving the abdomen and often the lower thoracic region. There is frequently lateral deviation of the vertebral axis at the level of the mid-trunk region. Duplications of the neural tube, such as forking, are also occasionally encountered. The forebrain region is invariably abnormal: there is incomplete separation of the telencephalic hemispheres, and there are often associated abnormalities of the diencephalon. For example, there may be absence of eyes on both sides, or a minute eye on one side and no eye on the other side. In the majority of cases, the pituitary gland is abnormal, and in some the pituitary is entirely absent. The midbrain generally looks pretty normal, as does the hindbrain.

The most advanced stage I have so far obtained is equivalent to about 14.5 days. Three embryos had a normal postcranial axis but the craniofacial features were quite characteristic and consistent with an incomplete separation of the forebrain-derived telencephalic hemispheres, while four additional embryos had an abnormal postcranial vertebral axis as described above, associated with a gross omphalocoele, but their craniofacial features were almost identical to those with a normal postcranial axis. Very often the eyelids are present but there is no eye subjacent to it; in most embryos there is just a stump of optic nerve. The extrinsic ocular muscle precursors are often present, although there is usually no optic nerve present beyond the region of the optic chiasma. If there is an eye present, there is often herniation or fragmentation of the lens. While the two telencephalic hemispheres are invariably abnormal, the hindbrain region usually has a normal morphology. All of the cranial nerve ganglia are recognizable and appear to be normal.

In the four embryos that had a normal postcranial axial morphology, the crown-rump lengths were only slightly smaller (81–91%) than those of developmentally matched control diploid embryos. In all seven of these embryos, gonadal differentiation was consistent with their developmental age. More interestingly, three of these embryos possessed abnormalities of the aortic arch arterial system: two had vascular rings around the oesophagus and trachea, while a third had a persistent right dorsal aorta. The other embryos each had one or more morphological abnormalities present, such as a left-sided congenital diaphragmatic hernia, and herniation of part of the liver into the physiological umbilical hernia. None of these abnormalities would be considered life-threatening, should these embryos have survived to birth. It is possible that the placenta and extraembryonic tissues were abnormal, and we hope to study this aspect in the near future.

In summary, roughly half of the tetraploids that do survive to Day 14–14.5 have vertebral abnormalities; the other half seem reasonably normal. In about half of those that survive, irrespective of whether they have a normal or abnormal vertebral axis, an abnormal arrangement of the aortic arch arteries is observed and all of these had an XXYY sex chromosome constitution.

McLaren: What strain of mouse was that? Was it all one strain?

Kaufman: We used as female a C57BLxCBA Fl hybrid. We tried a variety of males and observed that tetraploid development was very strain dependent. When F1 hybrid males were used, the tetraploids progressed beyond Day 10 only rarely. When homozygous Rb(1.3) males were used, considerably more advanced development was usually achieved.

References

Kaufman MH 1991 New insights into triploidy and tetraploidy, from an analysis of model systems for these conditions. Hum Reprod 6:8–16

Kaufman MH, Webb S 1990 Postimplantation development of tetraploid mouse embryos produced by electrofusion. Development 110:1121–1132

Myogenesis in the mouse

Margaret E. Buckingham, Gary E. Lyons*, Marie-Odile Ott and David A. Sassoon**

Department of Molecular Biology, CNRS URA 1148, Pasteur Institute, 25 rue du Dr. Roux, 75724 Paris Cedex 15, France

Abstract. The first striated muscle to form during mouse embryogenesis is the heart followed by skeletal muscle which is derived from the somites. The expression of genes encoding muscle structural proteins and myogenic regulatory sequences of the *MyoD1* family has been examined using ³⁵S-labelled riboprobes. In the cardiac tube, actin and myosin genes are expressed together from an early stage, whereas in the myotome, the earliest skeletal muscle, they are activated asynchronously over days. They are not expressed in the somite prior to myotome formation. One potential muscle marker, carbonic anhydrase III, is expressed in early mesoderm and subsequently in the notochord, similarly to the *Brachyury* gene. The myogenic sequences are not detectable in the heart. In the myotome they show distinct patterns of expression; this is discussed in the context of their role as muscle transcription factors. *myf-5* is the only myogenic factor sequence present in the somite prior to muscle formation and thus is potentially involved in an earlier step of muscle determination. It is also present in the early limb bud, but the status of myogenic precursor cells in the limb in this context is less clear.

1992 Postimplantation development in the mouse. Wiley, Chichester (Ciba Foundation Symposium 165) p 111–131

The formation of striated muscle during mammalian embryogenesis involves the activation of batteries of genes encoding muscle isoforms of the contractile proteins and of enzymes involved in muscle metabolism (Buckingham 1985). Much of our understanding of myogenesis is derived from experiments with cultured muscle cells. Mononucleated myoblasts, usually obtained from late fetal or new-born muscle, will undergo cell fusion to form muscle fibres and this morphological differentiation is accompanied by the onset of muscle protein synthesis. This process takes place spontaneously when the myoblasts reach a certain cell density. In addition to primary cultures, a number of mammalian muscle cell lines exist, again mostly derived from late fetal myoblasts or even adult muscle satellite cells. Such cell lines or primary cultures therefore provide model systems for examining the regulation of muscle differentiation.

*Current address: Department of Anatomy, University of Wisconsin Medical School, 1300 University Avenue, Madison, WI 53706, USA.
**Current address: Department of Biochemistry, Boston University School of Medicine, 80 East Concord Street, Boston MA 02118, USA.

Using muscle cell cultures, W. Wright cloned the myogenin regulatory sequence (Wright et al 1989) by subtractive hybridization of sequences from differentiating cells against those from myoblasts. The C3H 10T½ cell line represents a model system for examining earlier stages of myogenesis (see Konieczny & Emerson 1984). These cells resemble embryonic fibroblasts and have the notable property of becoming myogenic when treated with the drug 5-azacytidine. This cell system was used to clone the *MyoD* regulatory sequence by subtractive hybridization of sequences from myoblasts against those from untreated 10T½ cells (Davis et al 1987). Since then, two other sequences of this family have been identified by cross-hybridization to *MyoD*, namely *myf-5* (Braun et al 1989) and *MRF4* (Rhodes & Koznieczny 1989), also known as *myf-6* (Braun et al 1990) or *herculin* (Miner & Wold 1990). These sequences all contain a basic DNA-binding domain and a helix-loop-helix motif involved in their dimerization and for several muscle structural genes it has been shown that they act as muscle-specific transcription factors. In addition, they all demonstrate the striking property of myogenic conversion. When transfected into the 10T½ cell line or indeed into many other cell types, including primary cultures of diverse origin, the cells begin to express muscle-specific genes and in many cases form muscle fibres. It is on this basis that members of the *MyoD* family of genes have been called myogenic determination factors (see Weintraub et al 1991).

In this paper I shall address the expression of muscle structural and regulatory genes *in vivo* during mouse embryogenesis. We have developed a series of specific probes for different mouse muscle genes. These have been employed as ^{35}S-labelled riboprobes in *in situ* hybridization experiments on paraffin sections of mouse embryos, using a procedure originally described by David Wilkinson and modified in our laboratory (see Lyons et al 1990a). By this approach we have been able to describe the temporal and spatial patterns of expression of different muscle structural genes and of the myogenic regulatory sequences during striated muscle formation and maturation in the mouse.

The first striated muscle to form in the postimplantation mouse embryo is the heart; the cardiac tube is clearly visible at 7.5–8 days p.c. (post coitum). Subsequently, from about 8–8.5 days p.c. the first somites form in a rostrocaudal gradient over several days (see Theiler 1989). The skeletal muscles of the body are derived from precursor mesodermal cells present in the somite (Milaire 1976, Christ et al 1977, Chevallier et al 1977) that either migrate out to form muscle masses elsewhere or contribute to the formation of the central compartment of the somite, the myotome, which is the first skeletal muscle to form during embryogenesis.

Carbonic anhydrase III

The carbonic anhydrase gene family codes for enzymes involved in maintaining cellular pH, and the isoform carbonic anhydrase III was thought to be expressed

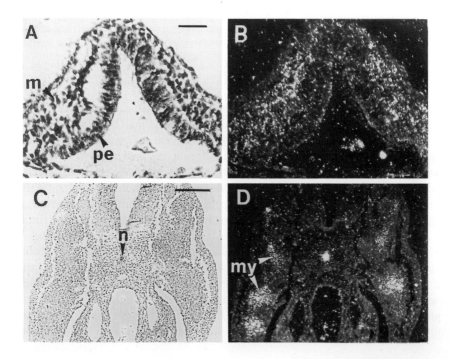

FIG. 1.   Carbonic anhydrase III transcripts in the mouse embryo. (A,B) Phase contrast and dark field micrographs through the caudal end of an 8 day p.c. mouse embryo hybridized with a probe specific for carbonic anhydrase III. Expression is seen in the caudal mesoderm (m) but not in the primitive ectoderm (pe). Bar = 50 μm. (C,D) Phase contrast and dark field micrographs of a transverse section of an 11.5 day p.c. mouse embryo hybridized with the carbonic anhydrase III probe, showing expression of the gene in the developing myotomes (my) and notochord (n). Bar = 200 μm.

only in skeletal muscle. It interested us as a potential early marker since it is already expressed in myoblasts of the muscle C2 line. In collaboration with Yvonne Edwards, we examined the expression of the gene in the mouse embryo (Lyons et al 1991a). As expected, the mRNA is present in skeletal muscle, although it is first detected in the myotome only from 10 days p.c. and is therefore not a particularly early marker of skeletal myogenesis *in vivo*; at later stages it is present in slow skeletal muscle fibres. However, it is also expressed in other tissues, for example brown fat and kidney tubules. Most interestingly for embryologists, carbonic anhydrase III transcripts are present at high levels in the notochord and in the prenotochordal mesoderm with a pattern of expression which is similar, although not identical, to that of the *Brachyury* gene (Herrmann, this volume) (Fig. 1). It therefore provides an early mesodermal marker in the embryo; it would be interesting to analyse more closely its expression compared with that of other genes expressed at this stage, in order to see whether subpopulations of early mesodermal cells can be defined on this basis.

Actin and myosin gene expression during myogenesis *in vivo*

The muscle structural genes that we have studied most extensively are those coding for the actins and myosins (myosin heavy and alkali light chains). These represent multigene families (see Table 1) in which cardiac and skeletal isoforms can be distinguished: a cardiac isoform is frequently expressed in developing skeletal muscle, for example cardiac actin, the myosin light chain MLC1A or the myosin heavy chain βMHC. In the case of the myosin heavy chains, further skeletal muscle-specific isoforms are expressed during muscle development, in addition to the isoforms which are characteristic of adult skeletal muscle fibres. It is thus a complicated scenario which tends to deter those not specifically interested in muscle. However, the regulation of these different genes during muscle formation and maturation poses an interesting problem, as does that of the function of the different isoforms which, in many cases, is far from clear (see Buckingham 1985). As a first step towards addressing these fundamental questions, we have described the distinct patterns of expression of these genes during embryogenesis.

In the heart

In the cardiac tube (Fig. 2), at the earliest stages that we have examined (7.5–8 days p.c.), major myosin sequences are already expressed (Lyons et al 1990b). Cardiac actin and, to a lesser extent, skeletal actin transcripts are also present (Sassoon et al 1988). These genes are among the first characteristic of an adult

TABLE 1 Actin and myosin genes expressed in striated muscle in the mouse

	Heart	Skeletal muscle	Fetal/embryonic skeletal muscle
Actins	α-Cardiac (α-Skeletal)	α-Skeletal	α-Cardiac α-Skeletal
Myosin heavy chains (MHC)	αMHC βMHC	MHCIIA MHCIIB MHCIIX βMHC	MHCemb βMHC MHCpn
Myosin alkali light chains (MLC)	MLC1A MLV1V	MLC1F/MLC3F MLC1V	MLC1A MLC1F/(MLC3F) (MLC1V)

The experiments on which this table is based are described in Buckingham (1985), Lyons et al (1990a,b). A, atrial; V, ventricular; pn, perinatal; emb, embryonic; F, fast.

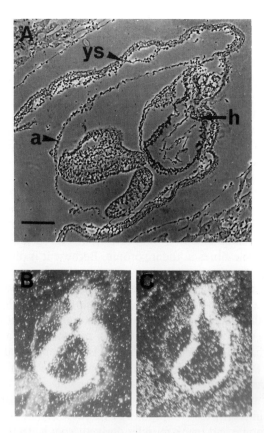

FIG. 2.   Expression of cardiac actin and myosin transcripts in the tubular embryonic heart. (A) Phase contrast, (B,C) dark field micrographs of a 7.5 day p.c. embryo *in decidua* hybridized with (A,B) a myosin heavy chain probe which recognizes all the striated muscle sequences of this family, notably cardiac α- and β-MHCs, and (C) with a cardiac actin-specific probe. h, heart; a, amnion; ys, yolk sac. Bar = 500 μm.

phenotype to be expressed during embryogenesis, and cardiac actin, for example, is a potentially interesting early marker for tracing cardiac cells as they emerge from the precardiac mesoderm. Subsequently, as the heart acquires its mature morphology, different myosin sequences become restricted to atrial or ventricular compartments. These restrictions occur asynchronously; the βMHC, for example, is present only in the ventricle by 10.5 days p.c., whereas the ventricular MLC1V form is still present in the atria at this time and is not confined to the ventricular compartment until 15.5 days p.c. (Lyons et al 1990b).

In skeletal muscle

In contrast to the situation in cardiac muscle, during formation of the first skeletal muscle, the myotome, transcripts encoding different sarcomeric proteins first accumulate very asynchronously over several days. Subsequently, as in the heart, different genes show different patterns of expression, and either cease to be transcribed or become restricted to a particular fibre type as skeletal muscles mature.

Cardiac actin transcripts appear very early (from about 8.5 days p.c.) in the myotome, when the few cells present still have a rounded morphology (Sassoon et al 1988). They are followed by transcripts of the myosin light chains, and again it is a cardiac isoform which predominates at this early stage. Myosin heavy chain transcripts are first detectable one day later, when the muscle cells begin to accumulate myofibrils and to acquire a fusiform shape. Initially, such cells remain mononucleated and it is only later that cell fusion takes place with formation of muscle fibres in the myotome. Because it is difficult to envisage a functional role for the sarcomeric α-actins in the absence of myosin heavy chains, we checked that the protein was in fact present, in a collaboration with Hans Mannhertz who has developed specific antibodies to these proteins. As shown in Fig. 3 the α-actin protein, as well as the transcript, accumulates (Lyons et al 1991b). The only exception to this, that we have noted so far, is the perinatal myosin heavy chain, for which the transcript is present several days before the protein (Lyons et al 1990a). However, this may be due to modification of an epitope recognized by the monoclonal antibody or there may be more than one isoform of this protein. Information on the onset of expression of different actin and myosin sequences is summarized in Table 2. The results shown here are for the myotome and trunk muscles derived from it. In the muscle masses of the limbs, the pattern of actin and myosin gene expression is essentially similar, with some differences in the relative timing for different genes. In general the process appears to be accelerated.

Some genes, such as that for MLC1V, are expressed only later in development, probably under the influence of innervation which begins at about 14 days p.c. in the mouse (Ontell & Kozeka 1984). Innervation is also probably responsible for the onset of spatial restrictions in the expression of myosin genes, which ultimately lead to the distinct protein composition of different adult fibre types. At earlier stages of myogenesis, the neural tube and indeed the notochord probably release growth factors which affect myotome development. Certainly, in birds they are both essential for somite maturation (Teillet & Le Douarin 1983). In the mouse, the neural tube appears to be necessary for the maturation of myotomal cells during a critical period around 10.5 days (Vivarelli & Cossu 1986). The nature of this process and other putative physiological factors affecting myogenesis in the embryo remain ill defined and these important aspects will not be discussed further here.

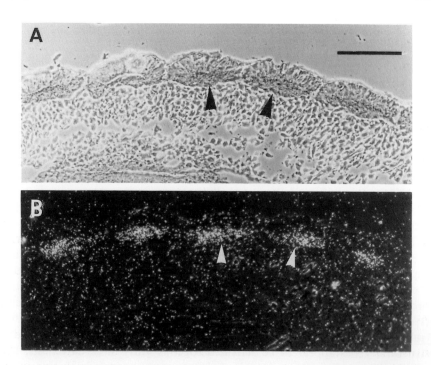

FIG. 3.   Co-localization of α-actin protein and mRNA in developing myotomes of the mouse embryo. Parallel transverse sections of a 10.5 day p.c. embryo (A) reacted with an α-striated actin-specific antibody, revealed by peroxidase staining, and (B) hybridized with an α-cardiac actin-specific probe. The arrowheads point to the myotomes of somites. Bar = 100 μm.

TABLE 2   Expression of actin and myosin genes in embryonic mouse skeletal muscle

Days post coitum	8.5	9.5	10.5	11.5	12.5	14.5	15.5
α-Cardiac actin	+	+ +	+ + +	+ + +	+ + +	+ + +	+ + +
α-Skeletal actin	±	+	+ +	+ +	+ + +	+ + +	+ + +
MLC1A	−	+	+ + +	+ + +	+ + +	+ + +	+ + +
MLC1F	−	±	+ +	+ +	+ + +	+ + +	+ + +
MLC3F	−	−	−	−	−	±	+
MLC1V	−	−	−	−	−	−	+
MHCemb	−	±	+ +	+ + +	+ + +	+ + +	+ +
βMHC	−	±	+	+	+	+	+
MHCpn	−	−	+	+ +	+ + +	+ + +	+ + +

These results are based on *in situ* hybridization experiments (Sassoon et al 1988, Lyons et al 1990a). A, atrial; V, ventricular; pn, perinatal; emb, embryonic; F, fast, MLC, myosin alkali light chain; MHC, myosin heavy chain.

Myogenic regulatory sequences

As transcription factors

Members of the MyoD family of proteins are clearly transcriptional regulators of muscle structural genes (see Olson 1990, Weintraub et al 1991). To understand more about how the myogenic programme is established, we have examined the expression of this gene family in the mouse embryo (in collaboration with E. Bober & H. Arnold, A. Lassar & H. Weintraub, W. Wright). In cultured muscle cells, different combinations of these sequences may be present and they appear to perform similar functional roles. It is not clear whether the four sequences do in fact have distinct functions during myogenesis. The analysis of their spatial and temporal patterns of expression during embryogenesis would suggest that they may. Each sequence has a distinct pattern of expression (see Table 3).

Transcripts of the MyoD family are detected only in skeletal muscle or its precursors (Fig. 4). At no stage examined (from 7.5 days p.c.) do we see any expression in the heart, despite the fact that many other genes expressed in skeletal muscle are also expressed here. Different regulatory sequences are probably involved in cardiac versus skeletal muscle transcription of these genes (see Ordahl 1992). It has not proved possible to detect any *MyoD*-related transcript in the adult heart by cross-hybridization. Given that many cardiac structural genes appear to be activated together early in cardiogenesis, it is possible that genes implicated in this phenomenon are expressed only transitorily

TABLE 3 Expression of myogenic factors in the embryo

Age of embryo (days post coitum)	Rostral myotomes				Forelimb buds and visceral arches			
	myf-5	*myogenin*	*myf-6*	*MyoD*	*myf-5*	*myogenin*	*myf-6*	*MyoD*
8.0	+	−	−	−				
8.5	+ +	+ +	−	−				
9.0	+ +	+ + +	+	−				
10.5	+ +	+ + +	+ +	+	+	−	−	−
11.0	+ +	+ + +	+	+ +	+ +	+ +	−	+ +
11.5	+	+ + +	−	+ + +	+	+ +	−	+ +
12.0	+	+ + +	−	+ + +	+	+ + +	−	+ + +
13.0	−	+ + +	−	+ + +	−	+ + +	−	+ + +
14.0	−	+ + +	−	+ + +	−	+ + +	−	+ + +

These results are based on *in situ* hybridization experiments (Sassoon et al 1989, Ott et al 1991, Bober et al 1991). At times ≤9.5 days no signal was observed for the myogenic factors in the forelimb buds or visceral arches. Before formation of the first somite (>8 days) no signal was detected for *myf-5* or the other myogenic sequences in the embryo.

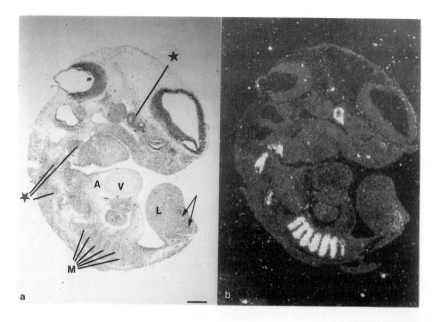

FIG. 4. *Myo-D* transcripts in the mouse embryo. (a) Phase contrast and (b) dark field micrographs of an 11.5 day p.c. mouse embryo hybridized with a *MyoD*-specific probe. Signal is seen in the myotomes (M), head and neck musculature (★) and limb (L, indicated by arrows), but not in the heart (A, atria; V, ventricle). Bar = 400 μm.

in the embryo. It is also possible, of course, that different classes of cardiac-specific regulators exist.

In skeletal muscle, as soon as a myotome forms, *myogenin* (Sassoon et al 1989) and *myf-5* (Ott et al 1991) transcripts are present, shortly followed by those of *myf-6* (Bober et al 1991). *MyoD* transcripts begin to be detectable at 10.5 days p.c. in the myotome of rostral somites, two days after the appearance of *myogenin* transcripts. This is in contrast to many tissue culture lines where expression of *MyoD* precedes that of *myogenin*. In cultured cells, as *in vivo*, *myogenin* transcripts are always present when muscle cells begin to differentiate, as defined by the expression of muscle structural genes, suggesting that they may play an essential role in this process, possibly in addition to that of the other myogenic factors (F. Catala & M. E. Buckingham, unpublished observations). Both *myf-5* and *myf-6* are expressed transitorily in myotomes. However, while *myf-5* transcripts are detectable only at early stages, *myf-6* reaccumulates in fetal muscle and is the major myogenic sequence in post-natal and adult muscle fibres (Rhodes & Konieczny 1989, Bober et al 1991). By this time both *myogenin* and *MyoD* have ceased to be major transcripts

(G. E. Lyons & M. E. Buckingham, unpublished observations). The continuing expression of *myf-6* at later stages suggests that this sequence, at least, is involved in the maintenance as well as the activation of muscle gene expression.

Comparison of the onset of expression of different muscle structural genes (Table 2) with that of the myogenic sequences (Table 3) suggests a correlation in a few cases (e.g. *MyoD* or the perinatal isoform *MHCpn*), but in general this is not obvious. Indeed, transcripts of the muscle isoform of creatine phosphokinase, which has been used extensively as a model gene to look at the mechanism by which a myogenic factor such as *MyoD* activates transcription, begin to be detectable only at 13 days p.c., two days after the onset of *MyoD* expression (Lyons et al 1991c). In attempting to draw such correlations, it is of course important to know whether transcripts, as detected by *in situ* hybridization, do reflect gene activation. For the myogenic sequences, analysis using the polymerase chain reaction (PCR) confirms the timing of activation given in Table 3 (Montarras et al 1991). It is not yet clear whether the corresponding proteins are present, although for *MyoD*, transcript and protein are detectable at about the same time (M.-O. Ott & M. E. Buckingham, unpublished). For the structural genes, transcriptional run-on or PCR analyses have not been performed at early stages. At later stages (from 14.5 days p.c.) there appears to be a good correlation between the *in situ* data and transcriptional activation as detected by nuclear run-on experiments (Cox & Buckingham 1992).

Assuming that the observations presented here can be taken to reflect transcriptional events, how can the asynchronous activation of muscle structural genes be explained in terms of transcriptional regulation by the myogenic factors? One explanation may be that each gene has a different threshold requirement for the concentration of a particular combination of myogenic factors. Myogenic sequences bind DNA most efficiently as heterodimers with the ubiquitous E12 helix-loop-helix factor and, probably, with other members of this superfamily. There is thus considerable scope for variation in the preferred heterodimer (see Blau & Baltimore 1991). In addition, Id-related sequences (Benezra et al 1990, see Weintraub et al 1991), which do not themselves bind DNA but are also helix-loop-helix proteins that can form heterodimers with the myogenic factors, may play a role in regulating availability of the latter. That quantitative effects may be very important is suggested by observations on the limb, where the pattern of expression of myogenic factors is different. *MyoD* and *myogenin* transcripts accumulate together and *myf-6* is not present in the pre-muscle masses (Sassoon et al 1989, Bober et al 1991). Nevertheless, most structural genes expressed in the myotome are also activated here. This would suggest that other combinations of myogenic factors at appropriate concentrations can also activate these genes.

As determination factors

If myogenic factors do play a role at earlier stages of myogenesis before activation of the muscle structural genes, as suggested by the demonstration of

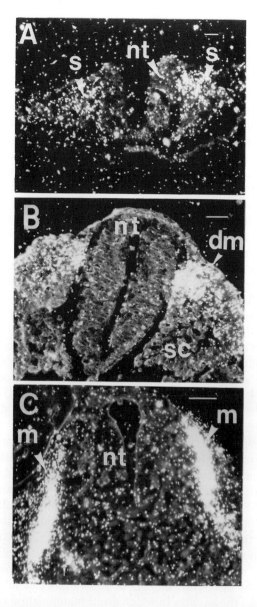

FIG. 5. *myf-5* transcripts in the developing somite. (A) Dark field micrograph of a section of a 4-somite embryo at about 8 days p.c., showing two early somites (s) which already express the *myf-5* sequence. (B) A more caudal section of a 9.5 day p.c. (21–24 somites) embryo, showing somites which now consist of dermomyotome (dm) and sclerotome (sc). (C) A transverse section of an 11 day p.c. (40–44 somites) embryo, showing somites at the level of the hind limbs. *myf-5* is expressed in the myotome (m) of these more mature somites. nt, neural tube. Bar A = 20 μm; B,C = 50 μm.

myogenic conversion *in vitro*, one would predict that they are expressed in the dermomyotome of the somite prior to myotome formation. The only myogenic transcript that is detectable in the mouse embryo before myotome formation is *myf-5* (Fig. 5, Ott et al 1991). It is first detected in the early somites when these are composed of a ball of epithelial-like cells. Transcripts are already partially concentrated in cells adjacent to the neural tube; this concentration becomes more pronounced as the dermomyotome forms and transcripts of *myf-5* accumulate in this region. According to the classic chick/quail experiments, cells will migrate from the dorsal-cranial edge of the dermomyotome to form the myotome (see Kaehn et al 1988). It thus seems probable that *myf-5* transcripts mark myotomal precursor cells and expression of this gene may be implicated in an earlier step in their commitment to myogenesis.

It is less clear whether *myf-5* is also expressed in myogenic precursors in the dermomyotome that will migrate out to form muscle masses elsewhere (Christ et al 1977, Chevallier et al 1977). Again, according to the chick/quail experiments, these come from the ventral-lateral edge of the dermomyotome and migrate before the myotome forms (see Bellairs et al 1986). No migrating *myf-5*-positive cells are detectable at any time, although the *in situ* technique with radioactive probes is not sensitive enough to detect low level expression in isolated cells. Some *myf-5*-positive cells are present throughout the dermomyotome. Techniques with higher resolution will be necessary to examine whether these cells do migrate. *myf-5* is the first myogenic transcript to be detectable in the pre-muscle masses of the limbs or the visceral arches (see Table 3), where it is transitorily expressed. However, *myf-5* transcripts are detectable in the forelimb bud, for example, only from 10 days p.c., when committed myogenic precursors should already be present.

The question then is whether other myogenic-specific regulatory genes, such as the putative *myd* gene (Pinney et al 1988), are expressed upstream of *myf-5*. It is almost certain that there is a cascade of mesodermal regulatory genes; the *Brachyury* gene (Herrmann et al 1990), for example, is probably several steps upstream of *myf-5*. The downstream genes are probably regulated directly by *myf-5* and each other, since autoregulation is a feature of this gene family. Again, different threshold levels of activation can be invoked to explain the temporal differences in the onset of their expression in the embryo. Whether *myf-5* is the only myogenic regulatory sequence which can play an upstream role or whether the different members of this family can substitute for each other are questions the answers to which await manipulation of these genes *in vivo*. Knock out experiments on each myogenic regulatory gene should also clarify the question of the transcriptional requirements of individual muscle structural genes. Such experiments are under way in several laboratories. We are attempting to knock out the *myf-5* gene (in collaboration with H. Arnold, E. Bober), to see if this eliminates all skeletal muscle formation. By inserting a β-galactosidase marker sequence, at the same time, we hope to be able to identify *myf-5*-expressing cells in the embryo more precisely and to follow their migration.

Acknowledgements

This work was supported by grants from the Pasteur Institute, the C.N.R.S., I.N.S.E.R.M., A.R.C. and the A.F.M. During the period when they carried out the work cited here in M.B.'s laboratory, D. S. was supported by a grant from the American Cancer Society and G. L. from the Fogarty Foundation and the A.F.M..

References

Bellairs R, Ede DA, Lash JW (eds) 1986 Somites in developing embryos. Plenum Press, New York p 320

Benezra R, Davis RL, Lockshon D, Turner DL, Weintraub H 1990 The protein *Id*: a negative regulator of helix-loop-helix DNA binding proteins. Cell 61:49–59

Blau HM, Baltimore D 1991 Differentiation requires continuous regulation. J Cell Biol 112:781–783

Bober E, Lyons GE, Braun T, Cossu G, Buckingham M, Arnold H 1991 The muscle regulatory gene, myf-6, has a biphasic pattern of expression during early mouse development. J Cell Biol 113:1255–1265

Braun T, Buschhausen-Denker G, Bober E, Tannich E, Arnold HH 1989 A novel human muscle factor related to but distinct from MyoD1 induces myogenic conversion in 10T½ fibroblasts. EMBO (Eur Mol Biol Organ) J 8:701–709

Braun T, Bober E, Winter B, Rosenthal N, Arnold HH 1990 Myf-6, a new member of the human gene family of myogenic determination factors: evidence for a gene cluster on chromosome 12. EMBO (Eur Mol Biol Organ) J 9:821–831

Buckingham ME 1985 Actin and myosin multigene families: their expression during the formation of skeletal muscle. In: Campbell PN, Marshall RD (eds) Essays in biochemistry. Academic Press, New York vol 20:77–109

Chevallier A, Kieny M, Manger A, Sengel P 1977 Developmental fate of the somitic mesoderm in the chick embryo. In: Ede DA, Hinchcliffe JR, Balls M (eds) Vertebrate limb and somite morphogenesis. Cambridge University Press, Cambridge p 421–432

Christ B, Jacob HJ, Jacob M 1977 Experimental analysis of the origin of the wing musculature in avian embryos. Anat Embryol 150:171–186

Cox R, Buckingham M 1992 Actin and myosin genes are transcriptionally regulated during mouse skeletal muscle development. Dev Biol 149:228–234

Davis RL, Weintraub H, Lassar AB 1987 Expression of a single transfected cDNA converts fibroblasts to myoblasts. Cell 51:987–1000

Herrmann BG 1992 Action of the *Brachyury* gene in mouse embryogenesis. In: Postimplantation development in the mouse. Wiley, Chichester (Ciba Found Symp 165) p 78–91

Herrmann BG, Labeit S, Poustka A-M, King TR, Lehrach H 1990 Cloning of the T gene required in mesoderm formation in the mouse. Nature (Lond) 343:617–622

Kaehn K, Jacob HJ, Christ B, Hinricksen K, Poelmann RE 1988 The onset of myotome formation in the chick. Anat Embryol 177:191–201

Konieczny SF, Emerson CP Jr 1984 5-Azacytidine induction of stable mesodermal stem cell lineages from 10T½ cells: evidence for regulatory genes controlling determination. Cell 38:791–800

Lyons GE, Ontell M, Cox R, Sassoon D, Buckingham M 1990a The expression of myosin genes in developing skeletal muscle in the mouse embryo. J Cell Biol 111:1465–1476

Lyons GE, Schiaffino S, Sassoon D, Barton P, Buckingham M 1990b Developmental regulation of myosin gene expression in mouse cardiac muscle. J Cell Biol 111:2427–2436

Lyons GE, Buckingham M, Tweedie S, Edwards Y 1991a Carbonic anhydrase III, an early mesodermal marker, is expressed in embryonic mouse skeletal muscle and notochord. Development 111:233–244

Lyons GE, Buckingham ME, Mannherz HG 1991b α-actin proteins and gene transcripts are colocalized in embryonic mouse muscle. Development 111:451–454

Lyons GE, Mühlebach S, Moser A et al 1991c Developmental regulation of creatine kinase gene expression by myogenic factors in embryonic mouse and chick skeletal muscle. Development 113:1017–1030

Milaire J 1976 Contribution cellulaire des somites à la genèse des bourgeons de membres postérieurs chez la souris. Arch Biol 87:315–343

Miner JH, Wold B 1990 Herculin, a fourth member of the MyoD family of myogenic regulatory genes. Proc Natl Acad Sci USA 87:1089–1093

Montarras D, Chelly J, Bober E et al 1991 Developmental patterns in the expression of myf-5, MyoD, myogenin and MRF4 during myogenesis. New Biologist 3:1–10

Olson E 1990 MyoD family: a paradigm for development? Genes & Dev 4:1454–1461

Ontell M, Kozeka K 1984 The organogenesis of murine striated muscle: a cytoarchitectural study. Am J Anat 171:133–148

Ordahl CP 1992 Developmental regulation of sarcoplasmic actin gene expression. Curr Top Dev Biol 26:145–168

Ott MO, Bober E, Lyons G, Arnold H, Buckingham M 1991 Early expression of the myogenic regulatory gene, myf-5, in precursor cells of skeletal muscle in the mouse embryo. Development 111:1097–1107

Pinney DF, Pearson-White SH, Konieczny SF, Latham KE, Emerson CPE Jr 1988 Myogenic lineage determination and differentiation: evidence for a regulatory gene pathway. Cell 53:781–793

Rhodes SJ, Konieczny SF 1989 Identification of MFR4, a new member of the muscle regulatory factor gene family. Genes & Dev 3:2050–2061

Sassoon D, Garner I, Buckingham M 1988 Transcripts of α-cardiac and α-skeletal actins are early markers for myogenesis in the mouse embryo. Development 104:155–164

Sassoon D, Lyons G, Wright W et al 1989 Expression of two myogenic regulatory factors: myogenin and MyoD1 during mouse embryogenesis. Nature (Lond) 341:303–307

Teillet MA, Le Douarin N 1983 Consequences of neural tube and notochord excision on the development of the peripheral nervous system in the chick embryo. Dev Biol 98:192–211

Theiler K 1989 The house mouse: atlas of embryonic development. Springer-Verlag, New York

Vivarelli E, Cossu G 1986 Neural control of early myogenic differentiation in cultures of mouse somites. Dev Biol 17:319–325

Weintraub H, Davis R, Tapscott S et al 1991 The *myoD* gene family: nodal point during specification of the muscle cell lineage. Science (Wash DC) 251:761–766

Wright WE, Sassoon DA, Lin VK 1989 Myogenin, a factor regulating myogenesis, has a domain homologous to MyoD1. Cell 56:607–617

DISCUSSION

Tam: Do you know which isoforms of the actin and myosin genes are being expressed in your early mouse embryo? Are they the fast form or the slow form?

Buckingham: All these probes are isoform specific so we know exactly which transcript is expressed when. The mouse has a whole battery of muscle contractile protein genes: embryonic, fetal, adult, fast and slow. The earliest genes to be expressed are mostly the embryonic fast isoforms, together with cardiac slow isoforms. In the case of the myosin heavy chains, for example, the first sequences to be expressed are the embryonic skeletal and the cardiac β slow; the latter remains a minor myosin heavy chain component during embryogenesis.

Tam: Is this related to the eventual formation of fast and slow muscle?

Buckingham: No, initially the genes are expressed uniformly throughout the muscle mass. It is only much later when innervation sets in that there is restriction of isoform expression to a particular muscle type.

Goodfellow: I am confused with the general hypothesis. I don't understand why the embryo switches on one gene which has the ability to turn the cell into a muscle cell, then switches that gene off and switches on another gene that has the same ability.

Buckingham: I think there are two things going on. One is the conversion of cells to become muscle, the other is the activation of the muscle structural genes. I was suggesting that *myf-5* may be concerned with converting cells to muscle initially, but that subsequently all these genes are involved at different times in activating muscle structural genes.

Goodfellow: What is the biology of a cell which has expressed *myf-5* but now has *myf-5* turned off?

Buckingham: *In vivo* such a cell would be expressing certain muscle structural genes and not others. *In vitro* there are cell lines that express various combinations of these factors. They all express *myogenin*, which seems to be essential for muscle differentiation. There are cell lines which express no *MyoD*, for example, but do express *myf-5*; they go through to myogenesis and express a subgroup of muscle structural genes.

Goodfellow: Would you predict that every cell in the muscle lineage expresses at least one of these four myogenic genes?

Buckingham: I would predict that in order to be a skeletal muscle cell a cell must express a combination of these factors.

Hogan: How do heart cells compare with skeletal muscle in their differentiation?

Buckingham: Heart seems to differ in that it initially expresses all the major myosin and actin genes together. We need to look at earlier stages to be sure that's really true, but at least by 7.5 days all these genes are expressed; that's different from the situation in skeletal muscle. If something is responsible for kicking off this expression, one might expect it to be expressed at only an early stage, e.g. as the cardiac tube forms. Most people have looked in heart at much later stages, when they have tried to isolate the *MyoD* homologue.

David Bader has used an antibody to look at the early chick heart (Bader et al 1991). In the chick heart, at the cardiac tube stage slightly earlier than the equivalent to the stage that I was showing, there is expression of a homologue of *MyoD*. It is present only transiently and is not expressed at later stages of development. So there may be a *MyoD*-like gene expressed in the heart that is responsible for initiating transcription of all these genes, but whose expression isn't maintained; however, that's preliminary.

Hastie: Could you discuss the ability of these myogenic genes to transform other cell types into muscle? Are all the genes of the family able to do that? What range of cell types can they transform and how far does the phenotype go?

Buckingham: The standard system initially used was the 10T½ embryonic fibroblast cell line; each of these genes individually will convert those cells to muscle. *MyoD* has probably been studied most extensively with other cell types (see Weintraub et al 1989). In most cases there is partial or complete conversion to muscle, with or without retention of expression of other differentiated genes. There are exceptions: for example, Helen Blau has shown that the hepatoma cell line HepG2 is not converted to myogenesis by introduction of a *MyoD*-expressing sequence (Schäfer et al 1990). One can speculate that this may be because it has a high level of negative regulators, such as the helix-loop-helix protein Id.

Experiments with *MyoD* have also been done on primary cultures and there is some conversion to myogenesis in all the cases that have been studied. For example, H. Holtzer looked at differentiated cell types derived from different embryonic germ layers in chick primary cultures (Choi et al 1990).

Jaenisch: What about EC cells like P19?

Buckingham: Moshe Shani has reported that *MyoD* expression in ES cells does not cause direct conversion to myogenesis (Shani et al 1991). I don't know of any report on EC cells.

Balling: When you treat 10T½ cells with 5-azacytidine, chondroblasts and adipoblasts are also formed. Is there any way to shift this ratio away from the muscle cell lineage into the chondroblast cell lineage or into the adipoblast cell lineage, for example by adding growth factors?

Buckingham: Not to my knowledge. When *MyoD* was cloned from that system, people immediately tried to do the same kind of subtraction hybridization to isolate genes from the chondroblastic lineage and didn't succeed in pulling anything out.

One should bear in mind that most of the experiments are transitory transfections and the cells are being blasted with enormous quantities of the sequence. Under these conditions we can push different cell types to myogenesis. As far as I know, people who have tried these experiments with other tissue-specific transcription factors, such as the liver-specific transcription factor HNF-1, cannot get the cells to form the relevant tissue type. That may reflect the fact that in the muscle system, one factor is sufficient, whereas in liver, for example, you need several factors. It may also be related to the autoactivation property of the *MyoD* family of genes.

McMahon: What's known about the differentiation of the somite? Is it autonomous to the somite or does it require information from the surrounding cells?

Buckingham: Nicole le Douarin's lab have done experiments on quail embryos which suggest that both the neural tube and the notochord exert an important influence on somite maturation. The most recent observations (C. Ziller, personal communication) are intriguing because in the absence of either neural tube or notochord or both, the somite doesn't mature correctly and the myotome doesn't form correctly, but at the stages they examined the limb muscles appear to form.

McMahon: When you say the myotome doesn't form correctly, does it express the myogenic genes?

Buckingham: They haven't looked at gene expression, but morphologically there is no maturation of myotome to form a muscle mass. It would be interesting to examine the early expression of myogenic factors.

Balling: What happens if you invert a somite so that the region that usually becomes dermomyotome is closer to the notochord?

Buckingham: Rosa Beddington and I have talked about doing that but haven't done so yet. It would be important to know whether the juxtaposition of the neural tube or notochord is responsible for the high level of expression of *myf-5*, which we find in the adjacent region of the dermomyotome.

Beddington: I think it's easier to move the notochord.

Smith: Could you say a little more about the migration of muscle cells into the mouse limb?

Buckingham: The precise information that we have about formation of muscle from the somite comes from chick/quail chimera experiments. That's because you can follow the quail cells and know exactly when and where they are going. In the mouse it is not possible to do those kinds of experiments so there isn't the same precise information about when these events take place.

Lee: Solursh & Meier (1986) used anti-desmin antibodies to trace the myogenic patterning from the dermomyotome into the limb bud in the chick. This hasn't been done in the mouse.

Buckingham: My understanding is that in the mouse, desmin is not expressed in the dermomyotome, it is expressed only in the myotome.

Smith: Is it correct that the sequence of expression of muscle regulatory genes in the mouse isn't the same as that in the quail?

Buckingham: The sequence of expression of the myogenic factors in the somites of the mouse embryo is clearly *myf-5*, *myogenin*, *myf-6* and *MyoD* considerably later. The work on the avian factors was done by Charles de la Brousse & Emerson (1990) on the quail. They found that *MyoD* expression in birds is early. We have also found that *MyoD* is expressed earlier than the myogenin gene in the chick, unlike the situation in the mouse (Lyons et al 1991).

Does that mean that these factors are really interchangeable and different strategies have been adopted in different systems but it doesn't really matter what the strategy is? Or does it mean that the gene family in the chick has evolved independently from an ancestral gene and one shouldn't make a precise equivalence between *MyoD* in the chick and *MyoD* in the mouse? We don't know yet.

Krumlauf: If you do cell transfection experiments, can you reveal any hierarchy of gene expression? If you transfect with *myf-5*, do you see activation of the downstream genes?

Buckingham: One very important feature of this system is that these genes autoactivate. When you put in *MyoD*, the endogenous *MyoD* gene is switched on and you also activate other genes in the pathway. There is a little evidence to suggest that, for example, when you put *MyoD* into a cell you don't activate the endogenous *myf-5* gene. That would support the idea that *myf-5* is upstream of *MyoD*. *In vivo* in the mouse, expression of one of these genes doesn't instantaneously activate the other myogenic factor genes; they are expressed at different times.

Jaenisch: Does this mean there are some other transcription activators in between or how do you interpret this?

Buckingham: I would try and argue in terms of thresholds for the myogenic sequences, as for the structural genes. Maybe there's not enough *MyoD* protein accumulated until 13 days p.c., which is when the gene for creatine protein kinase is switched on, whereas there is enough *MyoD* protein to initiate transcription of the myosin light chain gene MLC1A, for example, at an earlier stage. Or combinations of these factors may be needed. But there may be other factors required as well.

Jaenisch: The 10T½ cell line seems to need just a little kick to proceed into myogenic differentiation. Transfection of these cells with *MyoD* leads to autoactivation of the endogenous gene or of other myogenic genes. What about a cell which cannot differentiate into the muscle lineage, like the hepatoma cell line? If you transfected *MyoD* into hepatoma cells, would you activate the endogenous *MyoD* or any of the other myogenic genes?

Buckingham: In Helen Blau's experiments with the hepatoma cell line HepG2, it was not clear whether the endogenous myogenic factor genes were being activated; however, the high levels of transfected *MyoD* transcript and protein present were clearly not capable of producing the muscle phenotype (Schäfer et al 1990). It was proposed that this was due to the presence of negative regulators whose effect could be titrated out in cell fusion experiments with a *MyoD*-producing fibroblast, for example. In such heterokaryons, the liver cell transcribes muscle genes. Experiments with somatic cell hybrids containing single chromosomes have also led Harold Weintraub to postulate myogenic repressor genes (Thayer & Weintraub 1990).

Smith: Nick Hopwood and John Gurdon (1990) have injected *MyoD* mRNA into *Xenopus* embryos and then dissected animal pole regions from them. They find that such injections cause muscle-specific actin genes to be expressed in the caps, but only transiently. Recently, however, Hopwood has found that if he injects concentrations of RNA that are a few-fold higher, he gets much more prolonged expression of muscle-specific actin genes, as well as some other muscle markers. So this is quite a sharp threshold, although its physiological relevance is unclear.

Lovell-Badge: What's known about the distribution of Id in the mouse embryo?

Buckingham: Id was originally identified in the haemopoietic lineage. It turned out to be also present in myoblast cells (Benezra et al 1990). At least two other Id-like sequences have been identified, so it's probably a big family. One Id-like sequence is present in the heart. David Sassoon is looking with R. Benezra at the expression of Id in the embryo.

Hastie: In several systems that Weintraub has looked at where there is differentiation, such as F9 EC cells, Id or one of its relatives is expressed at high levels and then turned down on differentiation.

Buckingham: In the muscle cell systems examined, the idea is that Id does not bind most efficiently to the myogenic helix-loop-helix sequence but to the E12 helix-loop-helix sequence. Weintraub proposes that in order to get efficient myogenesis you need an E12–*MyoD* protein heterodimer and that Id messes things up by interfering with formation of this heterodimer (Weintraub et al 1991).

Wilkinson: What is known about expression of the E12 family members?

Buckingham: The E2A gene produces E12 and E47: those two proteins, as far as people have looked, seem to be universally present. But again, it's probably the tip of an iceberg and there are many more potential helix-loop-helix partners for the myogenic factors.

McLaren: We have these four regulatory genes and the various structural genes and we know that the regulatory genes encode DNA-binding proteins. Is it known what sequences those DNA-binding proteins are binding to? Do you envisage that they might be binding to the regulatory regions of some of the structural genes? Or do you envisage a long series of intermediate genes?

Buckingham: For some of the structural genes, for example creatine phosphokinase, all four of the regulatory factors in an *in vitro* system can bind to a consensus CANNTG motif. This is found in the muscle-specific enhancer of the creatine phosphokinase gene and is essential for its muscle-specific regulation. The more genes we look at, the more this seems to be a general principle.

McLaren: You described a temporal sequence of switching on and off of the structural genes. Do you know whether the later ones are in any sense dependent on the expression of the earlier ones? Can you test that by somehow inhibiting expression of the earlier ones and seeing whether the later ones are still switched on?

Buckingham: In vivo at the moment one can't really answer that. The way to do it would be to knock out the earlier gene and see if that affected the pattern of expression of the later gene.

Herrmann: How many genes do you think there are operating upstream of *myf-5*? From mesoderm induction to *myf-5* expression?

Buckingham: How far from *myf-5* to the *T* gene, for example? I have no idea. Unfortunately, no gene has been identified yet which falls between those two.

Herrmann: If one had a good idea of the timing from mesoderm induction to *myf-5* expression, for instance, one would have a better idea.

Buckingham: By PCR analysis, *myf-5* transcripts are accumulated in the somite as soon as it forms and perhaps even prior to that (Montarras et al 1991).

Balling: The analysis of the development of the skeleton is facilitated by the fact that the vertebrae have such different characteristics along the anteroposterior axis. Are there similar markers for the muscular system along this axis that would help decide whether homeotic transformations had occurred?

Buckingham: The first muscles to form are the myotomes and I don't know of any indication that an anterior myotome expresses different markers from those expressed by a posterior myotome. Much later, when the definitive muscles are laid down, they have different characteristics. One question is what's going on with some of the head muscles which are derived not from a somite but from the pre-caudal plate probably, although their origin is slightly controversial.

Balling: In *undulated*, Grüneberg (1954) described a secondary effect on the muscle pattern. Could that be because the muscles might change according to where the bones develop? Is the muscular system dependent on the pattern of the skeletal system or does it have an intrinsic pattern?

Beddington: In chick, with respect to musculature, it doesn't matter where you take a somite from (with the exception of the most anterior somite) or where you put it, it will make a muscle appropriate for its new location. There is less of an imprint of axial identity with respect to types of muscle.

Balling: Less or none?

Beddington: Maybe none, except the first somite. Other somites from anywhere along the axis will make perfectly normal limb musculature.

References

Bader D, Montgomery M, Litvin J, Bisaka J, Goldhammer D, Emerson C 1991 Commitment and differentiation of avian cardiac progenitor cells. J Cell Biochem Suppl 15C:157

Benezra R, Davis RL, Lockshon D, Turner DL, Weintraub H 1990 The protein Id: a negative regulator of helix-loop-helix DNA binding proteins. Cell 61:49–59

Charles de la Brousse F, Emerson CP 1990 Localized expression of a myogenic regulatory gene, qmfl, in the somite dermatome of avian embryos. Genes & Dev 4:567–581

Choi J, Costa ML, Mermelstein CS, Chagas C, Holtzer S, Holtzer H 1990 MyoD converts primary dermal fibroblasts, chondroblasts, smooth muscle, and retinal pigmented epithelial cells in to striated mononucleated myoblasts and multinucleated myotubules. Proc Natl Acad Sci USA 87:7988–7992

Grüneberg H 1954 Genetical studies on the skeleton of the mouse. XII The development of undulated. J Genet 52:441–455

Hopwood ND, Gurdon JB 1990 Activation of muscle genes without myogenesis by ectopic expression of MyoD in frog embryo cells. Nature (Lond) 347:197–200

Lyons GE, Mühlebach S, Moser A et al 1991 Developmental regulation of creatine kinase gene expression by myogenic factors in embryonic mouse and chick skeletal muscle. Development 113:1017–1030

Montarras D, Chelly J, Bober E, Arnold H, Ott M-O, Gros F, Pinset C 1991 Developmental patterns in the expression of *myf5*, *myoD*, myogenin, and *MRF4* during myogenesis. New Biologist 3:1–10

Schäfer BW, Blakely BT, Darlington GJ, Blau HM 1990 Effect of cell history on response to the helix-loop-helix family of myogenic regulators. Nature (Lond) 344:454–458

Shani M, Emerson C, Magal Y, Dekel I, Faerman A, Pearson-White S 1991 The consequences of a constitutive MyoD1 expression in ES cells and mouse embryos. J Cell Biochem Suppl 15C:31

Solursh M, Meier S 1986 The distribution of somite-derived myogenic cells during early development of the wing bud. In: Bellairs R, Ede DA, Lash J (eds) Somites in developing embryos. Plenum Publishing Corporation, New York p 277–287

Thayer MJ, Weintraub H 1990 Activation and repression of myogenesis in somatic cell hybrids: evidence for trans-negative regulation of MyoD in primary fibroblasts. Cell 63:23–32

Weintraub H, Tapscott SJ, Davis RL et al 1989 Activation of muscle-specific genes in pigment, nerve, fat, liver, and fibroblast cell lines by forced expression of MyoD. Proc Natl Acad Sci USA 86:5434–5438

Development of the skeletal system

R. Balling*, C. F. Lau°, S. Dietrich, J. Wallin* and P. Gruss

Department of Molecular Biology, Max-Planck Institute of Biophysical Chemistry, D-3400 Göttingen, Germany

Abstract. The analysis of the development of the skeletal system has been greatly facilitated by the availability of a large number of mouse mutants with skeletal defects. Whereas for many of these mutants a description of the main phenotypic abnormalities is known, molecular insight into the ontogeny of the skeletal system is limited. One of the few skeletal mutants for which the molecular basis is known is *undulated*. These mice have a defect in the differentiation of the sclerotome and *Pax-1*, a mouse paired-box containing gene, has been identified as a candidate gene for this mutation. A molecular analysis of three independent *undulated* alleles revealed that in each case the *Pax-1* gene is affected. One of the alleles could be classified as a null allele, in which the *Pax-1* gene is deleted. A phenotypic analysis shows that *Pax-1* is required for proper differentiation of intervertebral discs and vertebral bodies.

1992 Postimplantation development in the mouse. Wiley, Chichester (Ciba Foundation Symposium 165) p 132–143

The development of the skeleton is a complex process that requires the coordinate regulation of many molecular and cellular events. From the fossil records we know more about the skeleton of ancestral organisms than about any other body structure (Hall 1975), but it is not only from an evolutionary perspective that the skeleton is receiving so much interest.

The axial skeleton is composed of metameric units—the vertebral bodies and intervertebral discs. The mechanisms that govern the establishment of this metameric pattern are largely unknown, as are the events that control the determination of region-specific modifications of the individual skeletal components along the body axis. The axial skeleton is uniquely suited to serve as a model system to study pattern formation in vertebrates.

Our current concept of how pattern formation is achieved during embryogenesis of vertebrates has been strongly influenced by the insights into how the body plan of insects, particularly *Drosophila melanogaster*, becomes established. The principal mechanisms of insect segmentation and vertebrate

---

Current address: *Department of Developmental Biology, Max-Planck Institute for Immunobiology, D-7800 Freiburg, Germany and °Development Center for Biotechnology, Taipei, Taiwan.

segmentation might be more similar than originally expected. Although the early segmentation of *Drosophila* larvae and the segmentation of vertebrate mesoderm into somites are not homologous, they share important features. An initially homogeneous axis becomes divided into segments, which then become further modified. *Drosophila* cuticle patterns and the vertebrate skeleton can be analysed easily and deviations from a normal pattern detected. In *Drosophila*, extensive mutagenesis screens produced a large number of segmentation mutants, forming the basis of a genetic and molecular analysis of segmentation (Nüsslein-Volhard & Wieschaus 1980). In the mouse, a saturation mutagenesis of this kind is not feasible at present and other strategies have to be employed. However, since mice have been bred systematically for almost 100 years, many skeletal mutants already exist (Lyon & Searle 1989). Most of these mutants arose spontaneously. Skeletal abnormalities are easily recognized in the living animal and often do not interfere with viability and reproductive performance. It is for the same reason that so many coat colour mutations of the mouse are available and are now being characterized at the cellular and molecular level. Closer phenotypic and embryological analysis of the various mouse skeletal mutants allows us to group them into different classes (Table 1, Grüneberg 1963), which reflect the various steps of skeletal development.

In this paper we will give a brief outline of the development of the axial skeleton and then focus on the molecular and phenotypic analysis of a mouse mutation (*undulated*) that specifically affects the differentiation of the sclerotome.

Formation of the primitive streak and the notochord

The development of the vertebrate skeleton is intimately connected with the establishment of the primary body axis. At about 6.5 days of mouse development, epithelial cells from the primitive ectoderm begin to detach and, by migration through the primitive streak, form the mesoderm. With the establishment of the primitive streak the decision about the future anteroposterior axis of the embryo has been made. With the exception of craniofacial structures, the skeleton is derived entirely from the mesoderm; the major decisions that determine the fate of mesoderm cells are made during or shortly after gastrulation. Immediately after gastrulation two essential components of the future skeleton can be recognized: the notochord and the paraxial mesoderm. The notochord arises concomitantly with the regression of the primitive streak. The importance of the notochord during embryogenesis cannot be overestimated: it not only reflects the anteroposterior orientation of the embryonic axis, but also acts as an organizer for adjacent embryonic structures (Watterson et al 1954, van Straaten et al 1988, Placzek et al 1990, Yamada et al 1991). In vertebrates the notochord is present only in the early stages of embryogenesis. In primitive chordates, e.g. amphioxus or cyclostomata,

TABLE 1 Classes of skeletal mutants

Tissue/process affected	Mutant
Notochord/primitive streak	*Brachyury* (*T*) *Danforth short tail* (*Sd*) *Pintail* (*Pt*)
Segmentation	*pudgy* (*pu*) *rhachiterata* (*rh*) *Rib fusions* (*Rf*) *amputated* (*am*)
Sclerotome formation	*undulated* (*un*) *tail-kinks* (*tk*)
General, systemic abnormalities	*cartilage matrix deficiency* (*cmd*) *congenital hydrocephalus* (*ch*) *osteopetrosis* (*op*)

the notochord persists throughout life and is the only skeletal-like structure, no vertebrae develop. In vertebrates the notochord is essential for the induction of the floor plate of the spinal cord (van Straaten et al 1988, Placzek et al 1990, Yamada et al 1991). Excision and transplantation experiments show that the notochord is also important for proper segmentation of paraxial mesoderm and differentiation of somite-derived cells into cartilage (Nicolet 1970, 1971, Cooper 1965).

From paraxial mesoderm to somites

Paraxial mesoderm, the tissue from which somitogenesis will be initiated, is formed flanking the notochord. Blocks of cells bud off from the paraxial mesoderm sequentially in an anteroposterior direction to form the somites. This is the first sign of segmentation in the mouse embryo, starting around Day 8. The molecular mechanisms responsible for subdividing paraxial mesoderm into metameric blocks of tissue are unknown. On the basis of heat shock and cell lineage tracing studies, Stern et al have proposed a role of the cell cycle in the segmentation of mesoderm (Primmett et al 1989, Stern 1990). They suggest the process is cell autonomous and that the somite progenitor cells contain an internal clock which is responsible for forming segments at a specific time.

From somites to sclerotomes

After the somites have formed, they differentiate into myotome, dermatome and sclerotome. Dermatome and myotome form from the more dorsal part of the somites giving rise to the connective tissue of the skin and the muscles of the body wall and limbs, respectively. The sclerotomes form from the ventromedial part of the somites by de-epithelialization. Sclerotome cells then move towards and around the notochord. Condensations lateral to the notochord are observed, dividing the sclerotome into rostral and caudal halves. Neural crest cells migrate specifically through the rostral half of the sclerotome, giving rise to the spinal ganglia of the peripheral nervous system (Rickmann et al 1985).

From sclerotome to prevertebra

Sclerotome-derived cells reach the area surrounding the notochord forming the perichordal tube. At this time no overt segmentation is apparent (Verbout 1985). The perichordal tube then differentiates into regions of higher cell density alternating with regions of lower cell density. This process begins at the anterior end of the vertebral column and progresses posteriorly. The regions of higher cell density form the intervertebral discs, whereas the regions of lower cell density are apparently the anlagen of the vertebral bodies (Claudio D. Stern, personal communication). The spatial and cell lineage relationships between somitic cells and sclerotome-derived structures have been a major issue of controversy for more than a century. Originally, Remak (1855) proposed that the sclerotome undergoes a resegmentation process, in which the anterior half of one somite fuses with the posterior half of an adjacent somite to form a vertebral body. This would shift the original periodicity by half a segment, allowing the attachment of the muscular system derived from one somite to two adjacent vertebrae. Preliminary observations by Claudio D. Stern and co-workers (personal communication) suggest that this concept might not hold true and that one somite will give rise to one vertebra.

The embryological and cellular events described above have been studied mainly in the chick. The major reason for this is the large size and easy manipulation of the chick embryo, allowing transplantation and excision experiments. For a genetic analysis the chick is less suitable and the mouse is considered the model of choice. As already mentioned, many mouse mutants exist in which the development of the skeleton is altered. A molecular understanding of these mutants is limited and the affected genes are known in only a few cases, e.g. *undulated*.

undulated, a sclerotome differentiation mutant

undulated was originally described by M. Wright (1947) as a mutation affecting the spine and the tail. The detailed pathology of the developing and adult *undulated* mice was studied by Grüneberg (1954, 1963). It is a recessive mutation located on chromosome 2 between the *agouti* and the *β2-microglobulin* loci. Adult homozygous mice are viable and fertile and can be recognized by a shortened, kinky tail (Fig. 1). Kyphosis (excessive forward curvature of the spine)

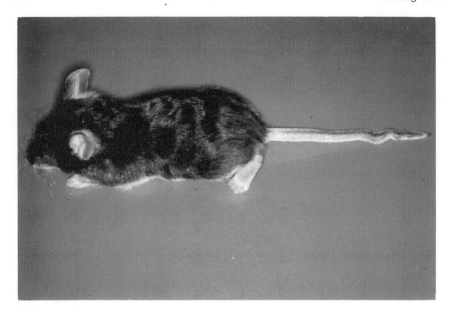

FIG. 1. A homozygous *undulated* mouse. Note the kinky tail, which results from abnormalities in intervertebral disc development.

and scoliosis (lateral curvature of the spine) are common, particularly in the lumbar region. Embryologically, *undulated* has been classified as a sclerotome differentiation defect. The condensation of sclerotome cells along the perichordal tube into alternating zones of higher and lower cell densities seems to be affected. The condensation of the zones with a higher cell density is apparently delayed. One of the major histological observations in *undulated* embryos is that the anlagen of the intervertebral discs are larger and the posterior ends of the vertebral bodies are smaller than normal. The deviation from normal embryos can first be found on Day 11 of development. Abnormalities are seen along the entire body axis. The appendicular skeleton of *undulated* mice is not affected, with the exception of the acromion of the scapula, which can be reduced or absent. Although *undulated* has been studied in detail at the phenotypic level, the molecular and cellular mechanisms that lead to this phenotype are not understood.

Pax-1 and *undulated*

New interest in the *undulated* mutation has recently been triggered by the observation that a mouse gene, *Pax-1*, maps to the same chromosomal location as *undulated* (Balling et al 1988). *Pax-1* is a paired box-containing gene that has been cloned through the sequence similarity in a conserved motif, the paired box, between *Drosophila* segmentation and segment polarity genes and vertebrate genes (Bopp et al 1989, Deutsch et al 1988). *Pax-1* is expressed during mouse development starting between Days 8 and 9 of gestation. Expression can first be found in the developing somites and is later restricted to sclerotome-derived cells in the condensed regions of the perichordal tube. As mentioned above, these cells are thought to give rise to the intervertebral discs, the structures that are most affected in *undulated* mice.

The *Pax-1* gene encodes a protein of 361 amino acids. Gel shift assays established that *Pax-1* is a DNA-binding protein and a core DNA-binding motif was defined. Evidence was obtained that this gene product can act as a transcriptional activator (Chalepakis et al 1991). A genetic and molecular analysis of *undulated* revealed that the *Pax-1* gene is mutated in *undulated* mice (Balling et al 1988). At position 15 of the paired box there is a G-A point mutation, leading to a substitution of serine for glycine in the *Pax-1* protein. This mutation affects the DNA-binding properties of *Pax-1* protein such that both the affinity and the specificity of the DNA binding are altered *in vitro* (Chalepakis et al 1991, Treisman et al 1991).

un, *un^{ex}* and *Un^s*—an allelic series of *undulated*

The chromosome localization of *undulated* and *Pax-1* in the same region on chromosome 2, the phenotype of *undulated* mice where the affected tissues are those that express *Pax-1*, and a mutation in the most conserved region of *Pax-1* constitute strong circumstantial evidence that *undulated* and *Pax-1* are the same gene. Final proof, however, requires experiments in which either the *undulated* mutation is rescued by the introduction of a normal *Pax-1* gene into transgenic mice, or the introduction of the Gly-Ser mutation (by homologous recombination) results in the *undulated* phenotype. While rescue experiments are in progress, we thought to obtain additional support that *undulated* is caused by a mutated *Pax-1* gene by trying to identify and analyse other alleles of *undulated*.

It was possible to obtain two additional alleles of *undulated*. *undulated-extensive* (*un^{ex}*) is a recessive mutation first described by M. Wallace at the University of Cambridge (Wallace 1985) and phenotypically very similar to *undulated*. As its name implies, the manifestation of the vertebral column abnormalities is more severe in *un^{ex}* than in *un*. *Undulated-short tail* (*Un^S*) has been described by Blandova & Egorov (1975) as a semidominant mutation. Heterozygous mice have a short kinky tail; in the homozygous condition *Un^S* is a lethal mutation, but the phenotype and the time of death have not been described yet.

Genetic evidence has been obtained that *un*, *un^{ex}* and *Un^S* are allelic, therefore the *Pax-1* gene would be predicted to be affected in all three mutants. We carried out a molecular analysis of *Pax-1* in *un^{ex}* and *Un^S* and found evidence that the gene is affected in the two new mutants. Less *Pax-1* RNA is found in homozygous un^{ex}/un^{ex} embryos; the basis for this defect is still unknown. In Un^S mice the *Pax-1* gene is deleted, classifying this mutation as a null allele. This deletion allowed a molecular identification of homozygous Un^S embryos. Surprisingly, Un^S/Un^S mice were found up to birth, although none survived postnatally. Preliminary phenotypic analysis shows that in these mice formation of the vertebral column is severely affected. Whereas in *un* and un^{ex} the abnormalities are found mainly in the intervertebral discs, Un^S/Un^S mice completely lack vertebral bodies and intervertebral discs in the lumbar region. The neural arches, however, are almost normal.

The role of *Pax-1* in vertebral column formation

Although we now have the appropriate tools at hand, the precise function of *Pax-1* during formation of the vertebral column is still unknown. A comparative phenotypic analysis of the different *undulated* alleles indicates that both components of the metameric series of the body axis, the vertebral bodies and the intervertebral discs, require *Pax-1* for their proper development, although *Pax-1* is only expressed in the intervertebral disc region. There are at least two possible explanations: intervertebral disc and vertebral body cells could be connected through a cell lineage relationship, i.e. cells in the vertebral bodies may be directly derived from cells in the intervertebral discs. However, so far no evidence for such a relationship between these two structures has been obtained. Alternatively, the two structures may be required for each other's maintenance or there could be inductive effects between the two adjacent structures during the course of differentiation. Unfortunately, a precise cell lineage map of somite-derived cells is lacking, which complicates the interpretation of our results.

Pax-1 is a DNA-binding protein and very likely a transcriptional regulator. One of the major directions of research will be the identification of genes that regulate *Pax-1* transcriptionally and post-transcriptionally. Equally important will be the identification of genes regulated by *Pax-1*. The characterization of a DNA target sequence to which the *Pax-1* protein can bind (Chalepakis et al 1991) will greatly facilitate this search.

One of the most important structures for the development of the vertebral column is the notochord. It is assumed that induction through the notochord is essential for the correct differentiation of sclerotome-derived cells into cartilage. Preliminary analysis of other mouse paired box-containing genes expressed in the developing spinal cord shows that transplantation or excision of the notochord has drastic effects on their expression (M. Goulding, personal communication). Therefore it is conceivable that *Pax* genes are essential for interpretation of inductive signal(s) from the notochord leading to the establishment of proper positional information and/or cellular differentiation.

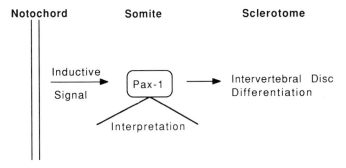

FIG. 2.   Possible role of *Pax-1* in translating a notochord signal into a differentiation programme for intervertebral discs.

In the developing sclerotome it might be *Pax-1* that receives an input from the notochord and translates it into a differentiation programme for the intervertebral disc (Fig. 2). It is obvious that the production, transmission and interpretation of a notochord–somite induction process would require a number of different genes and gene products. Mutations in some of these can be expected to give rise to very similar phenotypes, i.e. kinky tails. On the other hand, many of these genes will have functions outside the vertebral column, making the prediction of a phenotype very difficult. We are optimistic that genetic, molecular and embryological analyses of existing and arising skeletal mutants will prove them to be a treasure-trove for understanding of the development of the vertebral column.

Acknowledgements

We thank Merve Olowson for outstanding technical assistance and Drs M. Wallace and A. M. Malashenko for supplying the *un^ex* and *Un^S* mice, respectively. This work was supported by the Max Planck Gesellschaft.

References

Balling R, Deutsch U, Gruss P 1988 *undulated*, a mutation affecting the development of the mouse skeleton, has a point mutation in the paired-box of *Pax-1*. Cell 55:531–535
Blandova ZR, Egorov IU 1975 *Sut* allelic with *un*. Mouse News Lett 52:43
Bopp LD, Jamet E, Baumgartner S, Frigerio G, Noll M 1989 Conservation of a large protein domain in the segmentation gene *paired* and in functionally related genes of *Drosophila*. Cell 47:1033–1040
Chalepakis G, Fritsch R, Fisckenscher H, Deutsch U, Goulding M, Gruss P 1991 The molecular basis of the *undulated/Pax-1* mutation. Cell 66:873–884
Cooper GW 1965 Induction of somite chondrogenesis by cartilage and notochord: a correlation between induction activity and specific stages of cytodifferentiation. Dev Biol 12:185–212

Deutsch U, Dressler GR, Gruss P 1988 *Pax-1*, a member of a paired box homologous murine gene family, is expressed in segmented structures during development. Cell 53:617–625

Grüneberg H 1954 Genetical studies on the skeleton of the mouse. XII. The development of *undulated*. J Genet 52:441–455

Grüneberg H 1963 The pathology of development. Blackwell Scientific Publications, Oxford

Hall BK 1975 Evolutionary consequences of skeletal differentiation. Am Zool 15:329–350

Lyon MF, Searle AG 1989 Genetic variants and strains of the laboratory mouse, 2nd edn. Oxford University Press, Oxford, New York

Nicolet G 1970 Is the presumptive notochord responsible for somite genesis in the chick? J Embryol Exp Morphol 24:467–478

Nicolet G 1971 The young notochord can induce somite genesis by means of diffusible substances in the chick. Experientia (Basel) 27:938–939

Nüsslein-Volhard C, Wieschhaus E 1980 Mutations affecting segment number and polarity in *Drosophila*. Nature (Lond) 287:795–801

Placzek M, Tessier-Lavigne M, Jessell T, Dodd J 1990 Mesodermal control of neural cell identity: floor plate induction by the notochord. Science (Wash DC) 250:985–988

Primmett DRN, Norris WE, Carlson GJ, Keynes RJ, Stern CD 1989 Periodic segmental anomalies induced in heat-shock in the chick embryo are due to interference with the cell cycle. Development 105:119–130

Remak R 1855 Untersuchungen über die Entwicklung der Wirbelthiere. Reimer, Berlin

Rickmann M, Fawcett JW, Keynes RJ 1985 The migration of neural crest cells and the growth of motor axons through the rostral half of the chick somite. J Embryol Exp Morphol 90:437–455

Stern CD 1990 Two distinct mechanisms of segmentation? Semin Dev Biol 1:109–116

Treisman J, Harris E, Desplan C 1991 The paired box encodes a second DNA-binding domain in the paired homeo domain protein. Genes & Dev 5:594–604

Wallace M 1985 An inherited agent of mutation with chromosome damage in wild mice. J Hered 76:217–278

Watterson RL, Fowler I, Fowler BJ 1954 The role of the neural tube and notochord in development of the axial skeleton of the chick. Am J Anat 95:337–400

Wright ME 1947 *undulated*: a new genetic factor in *Mus musculus* affecting the spine and the tail. Heredity 1:137–141

van Straaten HWM, Hekking JWM, Wirtz-Hoessels EL, Thors F, Drukker J 1988 Effect of the notochord on the differentiation of a floor plate area in the neural tube of the chick embryo. Anat Embryol 177:317–324

Verbout AJ 1985 The development of the vertebral column. Adv Anat Cell Biol. Springer-Verlag, Berlin vol 90

Yamada T, Placzek M, Tanaka H, Dodd J, Jessell TM 1991 Control of cell pattern in the developing nervous system: polarizing activity of the floor plate and notochord. Cell 64:635–647

DISCUSSION

Joyner: Have you tried to make compound heterozygotes of the different *undulated* alleles?

Balling: Yes. I crossed *un/un* and *un*[ex]*/un*[ex] mice with *Un*[s]*/* + mice and analysed the phenotypes of the *Un*[s]*/un* and *Un*[s]*/un*[ex] compound heterozygotes.

Each has a phenotype very similar to that of the deletion homozygote (Un^s/Un^s). This supports the assumption that the phenotype of Un^s/Un^s is a consequence of the loss of *Pax-1* function and not a result of deletion of other genes flanking the *Pax-1* locus. However, because the size of the deletion is unknown, the clearest evidence on whether Un^s/Un^s is only a *Pax-1* null allele will come from comparing the phenotype of Un^s/Un^s with mutants in which the *Pax-1* gene has been inactivated by homologous recombination. Kenji Imai in Peter Gruss' lab has knocked out *Pax-1* in ES cells but no germline chimeras have been produced yet.

Joyner: Are you suggesting that Un^s is a null allele of *undulated* and that therefore one wild-type copy of *un* is not sufficient for normal development?

Balling: I do think that Un^s is a null allele of *undulated* because the *Pax-1* gene is apparently completely deleted in these mutants. The phenotype of the Un^s/un compound heterozygote suggests that one copy of the mutated *Pax-1* (*un*) gene is not sufficient.

Joyner: Are the *un* and un^{ex} mutations leaky alleles with a low level of gene activity?

Balling: From the comparison of the phenotypes of *un/un* and un^{ex}/un^{ex} with those of Un^s/Un^s, Un^s/un and Un^s/un^{ex}, the *un* and un^{ex} mutations are very likely leaky alleles and the mutants probable hypomorphs. Molecular evidence supports this: the DNA-binding affinity of the *Pax-1* protein containing the *un* mutation for a core DNA-binding motif is reduced and the specificity is altered (Chalepakis et al 1991); also, in the un^{ex}/un^{ex} embryos we still find some *Pax-1* RNA. However, the *un* mutation might also be a dominant negative mutation. On some genetic backgrounds, the *un* heterozygote does show a slight phenotype (Grüneberg 1954, 1963).

Joyner: What about the compound heterozygote, un/un^{ex}? Does it look like either the *un/un* or the un^{ex}/un^{ex}?

Balling: It looks like the un^{ex}/un^{ex} homozygote.

Hastie: Can you say any more about the rescue experiments that are in progress?

Balling: We have established 11 transgenic lines that harbour a 45 kb *Pax-1* cosmid. We don't have any results yet.

Krumlauf: Have you looked at the pattern of expression of *Pax-1* in the chick? Some of these other sites of expression in the cranial regions may not be the same between species and that might be a help.

Balling: The chick *Pax-1* gene has not been cloned; we have just made the primers for PCR. Paula Timmons from Peter Rigby's lab has done preliminary *in situ* hybridization on chick embryos using the mouse *Pax-1* gene. The expression patterns look identical, as far as has been analysed.

Krumlauf: Are there any hints at all of rostrocaudal or temporal changes in expression in the mouse that would help you think about the lumbar phenotype? Have you looked carefully at the timing?

Balling: I can't see any differences in the expression of *Pax-1* along the anteroposterior axis that would explain why there should be a specific phenotype in the lumbar region. *Pax-1* is expressed in the intervertebral disc anlagen at apparently similar levels along the entire vertebral column. The expression in the skull hasn't been looked at carefully yet.

Buckingham: What happens to expression of the *Pax-1* gene in notochordal mutants—those resulting from mutated *T* gene alleles, for example?

Balling: We haven't looked yet. We are going to look in the T^c mutants that Bernhard Herrmann described (this volume).

We will also look at the expression of *Pax-1* in *Danforth short tail* (*Sd*) mice. The phenotype of *Sd* is very interesting: as in $Un^s/+$ the dense axis is missing and there is a peculiar articulation between the atlas and axis (Grüneberg 1953).

Another genetic interaction has been found by Rick Woychik in Oak Ridge. He made a transgenic insertional mutant that might turn to be very interesting. The phenotype is very similar to *undulated*. When he crossed it with *undulated*, the double heterozygotes displayed the *undulated* phenotype, although each mutation alone is recessive. The insertional mutant is not allelic with *undulated* because it maps to a different chromosome. This phenomenon is known in yeast; it is called non-allelic non-complementation. The genetic interaction might reflect a molecular interaction between the *Pax-1* protein and other proteins, for example in a multimeric complex. It will be important to see whether the other *un* alleles also show non-complementation with this insertional mutant.

Beddington: Could the new mutation be a *Pax-1*-related gene?

Balling: It could, but there is no evidence for this.

McLaren: There are at least two reported examples of non-allelic interactions affecting tail development, namely *Brachyury* and *vestigial tail* (Michie 1956) and *curly tail* and trisomy 16 (Crolla et al 1990).

Copp: The craniocaudal variations seen in your mutants are very interesting. There is a whole constellation of defects in humans broadly classified as neural tube defects. The most severe result from failures of neural tube closure, for which there are a number of mouse models that we and others are studying (Copp et al 1990). These mainly affect the head, a region that is especially susceptible to disturbance of neural tube closure, and the lumbar spine where the posterior neuropore closes.

There is a second group of milder conditions that are poorly understood. These include cervical occipital encephalocoels. Your system shows that the cervical region is an area that's disturbable. There is a third, even more mild and very common condition, spina bifida occulta, which is almost invariably lumbar, again paralleling your mutants. Both the encephalocoels and the spina bifida occulta are presumably primarily skeletal problems. So we are beginning to see how disturbances of axial development of varying severity and at different craniocaudal levels can be correlated between humans and the animal models.

Is there a known human disease that could be caused by a mutation in the human *Pax-1* gene?

Balling: The phenotype of the original *undulated* mutation cannot be easily matched with any of the many known human diseases with vertebral column defects. Similarities between the phenotype of the more severe *un* alleles and of the compound heterozygotes and the phenotype of the Jarcho-Levin syndrome, also known as spondylocostal dysplasia, have been pointed out to me. We are currently analysing DNA from patients with this syndrome for mutations in the *Pax-1* gene. The human *Pax-1* gene has been mapped to chromosome 20q by Ingo Hansmann and co-workers (personal communication). Unfortunately, the chromosomal location of the Jarcho-Levin syndrome is not known yet.

McLaren: There is another sclerotome mutant, *tail-kinks*: do you know anything about its phenotype?

Balling: tail-kinks was also described by Hans Grüneberg. It is a recessive, homozygous-viable and fully fertile mutant. The mutation is located on chromosome 9. The defect seems to be that the differentiation of the sclerotome into anterior and posterior halves is delayed. This is reminiscent of the *undulated* phenotype, in which the differentiation of the mid-axial condensations (intervertebral disc regions) is apparently affected. *tail-kinks* might be regarded as a segment polarity mutant. We have started to make a genetic and physical map of this locus with the long-term goal of cloning the *tail-kinks* gene.

References

Chalepakis G, Fritsch R, Fisckenscher H, Deutsch U, Goulding M, Gruss P 1991 The molecular basis of the *undulated/Pax-1* mutation. Cell 66:873–884
Copp AJ, Brook FA, Estibeiro JP, Shum ASW, Cockcroft DL 1990 The embryonic development of mammalian neural tube defects. Prog Neurobiol 35:363–403
Crolla JA, Lakeman SK, Seller MJ 1990 The induction of tail malformations in trisomy 16 mouse fetuses heterozygous for the *curly tail* recessive gene. Genet Res 55:27–32
Grüneberg H 1953 Genetical studies on the skeleton of the mouse. VI Danforth's Short-Tail. J Genet 51:317–326
Grüneberg H 1954 Genetical studies on the skeleton of the mouse. XII The development of undulated. J Genet 52:441–455
Grüneberg H 1955 Genetical studies on the skeleton of the mouse. XVI Tail-kinks. J Genet 53:536–550
Grüneberg H 1963 The pathology of development. Blackwell Scientific Publications, Oxford
Herrmann BG 1992 Action of the *Brachyury* gene in mouse embryogenesis. In: Postimplantation development in the mouse. Wiley, Chichester (Ciba Found Symp 165) p 78–91
Michie D 1956 Genetical studies with 'vestigial tail' mice. IV. The interaction of vestigial with Brachyury. J Genet 54:49–53

Development of the left-right axis

Nigel A. Brown, Afshan McCarthy and Jeong Seo

MRC Experimental Embryology and Teratology Unit, Saint George's Hospital Medical School, Cranmer Terrace, London SW17 0RE, UK

Abstract. Left–right is not an axis in the conventional sense but rather two mirror-image proximodistal axes, upon which a quantal piece of positional information (leftness or rightness) is superimposed for laterally asymmetric organ development. We are attempting to establish the stages at which left–right is specified and determined, but this is complicated by the apparent loss of normal handed development in embryos that are cultured from pre-neural plate stages. Experiments suggest that left–right is determined by the first somite stage. The loss of normal left–right development in early cultures is probably not due to removal of some maternal signal, even though embryos do develop *in vivo* with their axes in a specific orientation relative to the uterus. The fact that there are two random embryonic axis orientations, 180° opposed to one another, and that the axes of the two uterine horns are mirror-images of each other make it unlikely that the uterus could impart a sense of left–right to the embryo. The right ovary produces more eggs than the left one; this is reversed in *iv/iv situs inversus* mice. Analysis of *iv/iv* mice shows a correlation of left–right abnormalities with sex and close relationships between the abnormal left–right development of some organs, for example the heart and spleen, that have no obvious developmental connection.

1992 Postimplantation development in the mouse. Wiley, Chichester (Ciba Foundation Symposium 165) p 144–161

It is clear that the early mammalian embryo possesses a mechanism to distinguish its left from right and to specify different patterns of morphogenesis in these two halves of the basic body plan. This is first apparent with the handed asymmetry in the looping of the cardiac tube, and shortly after with embryo turning at the mid-neurulation stage. The heart, lungs, liver, spleen and the whole of the gut and associated structures develop with handed asymmetry.

The left–right axis in early mammalian development is different from the anteroposterior and dorsoventral axes in two important respects. First, it is not a single axis in the conventional sense, but is two mirror-image proximodistal (or lateromedial) axes (Fig. 1). For those structures that are bilaterally symmetrical (limbs, the skeleton, sense organs, etc), there is no requirement for any distinct left–right information. Left and right limbs develop as different non-superimposable shapes because the arrangements of the anteroposterior, dorsoventral and lateromedial axes on the left and right are mirror images of one another (Fig. 1). In the case of laterally asymmetric organs, left and right

144

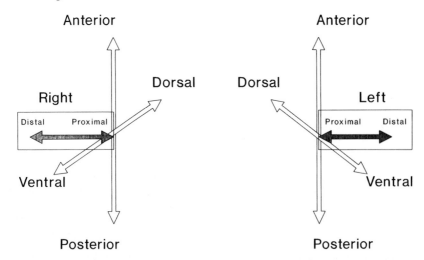

FIG. 1.   Proposed axes of positional information on left and right sides of the body plan. The three axes are mirror images on the two sides, which generates the pattern of bilaterally symmetrical structures such as the limbs. For laterally asymmetric structures, some additional information—'rightness' or 'leftness' or perhaps both—must be superimposed on this arrangement.

positional information must be superimposed upon lateromedial axial information. Thus, left–right information is a quantal, not a graded property. The left and right sides of the body may be considered as analogous to two developmental segments or compartments. We have begun to investigate whether they meet the usual developmental criteria for compartments (see below). The second major difference is that left and right have no meaning in the absence of anterior and posterior, and dorsal and ventral. Therefore, left–right must be specified after the other two axes have been established, and in relation to them.

   We have proposed a model for left–right specification, which has two fundamental assumptions: that left–right positional information is intrinsic to the embryo; and that this information is derived from a molecule (the 'F' molecule) which is, itself, handed and which is held in a specific orientation in the embryo, relative to the anteroposterior and dorsoventral axes (Brown & Wolpert 1990). The model has three components: *conversion*, in which the molecular asymmetry is converted to the cellular level; *random generation of asymmetry*, which can be biased by *conversion* to produce handed asymmetry; and *interpretation* in which individual organs use left–right information. We have reviewed the supporting evidence (Brown & Wolpert 1990, Brown et al 1991), in particular the phenotype of *iv/iv* mutant mice which develop with random, rather than handed, asymmetry (Layton 1976).

Here, we describe our recent observations on the abnormal specification of the left–right axis in cultured embryos; a re-evaluation of embryonic axis orientation in the uterus; a study of organ *situs* relationships in *iv/iv* mice; and preliminary investigations of left and right as compartments.

(Re)specification of left–right in cultured embryos

We have described a series of experiments designed to establish the period during which the left–right axis can be respecified (inverted) by chemical treatment of the rat conceptus (Brown et al 1991). A consistent observation was the inability to induce inversion, or at least inversion of heart looping and embryo turning, once the conceptus had passed the foregut pocket stage. This suggests that the left–right axis is determined by this stage. Treatment with an α_1-adrenergic agonist (Fujinaga & Baden 1991) at the notochordal process stage, before the neural groove was visible, was also without effect. However, other treatments, which were generally more toxic than the adrenergic agonists, did induce inversion at the notochordal process stage. Perhaps the most interesting observation in these studies was the significant incidence of inversion in control, untreated conceptuses. Spontaneous inversion *in utero* is rare. We do not have an accurate figure, but it is certainly less than 1% in our Wistar rats. We have now extended the analysis of inversion in cultured rat conceptuses.

Table 1 shows the incidence of inversion for rat conceptuses explanted on Day 9 or 10 and cultured to the 25-somite stage (72 to 48 hours of culture). These were cultured using our standard methods (see Brown et al 1989) and were not disturbed during culture, except to re-gas bottles for two minutes at about 12-hourly intervals. A small number of conceptuses explanted at all stages failed to develop; these are not reported. The remaining conceptuses grew normally. Development in culture from before the notochordal process stage consistently interfered with left–right specification. Up to half the conceptuses developed with inverted looping and/or turning (Table 1). At later stages a smaller, but still significant, proportion of conceptuses developed with abnormalities of laterality.

There are several possible reasons for the defective left–right development in control cultures. The simplest explanation is the lack of a maternal factor. It has been proposed that the uterus provides positional cues at, and immediately after, implantation (Smith 1980, 1985). We have argued (Brown & Wolpert 1990) against a role of the maternal environment in left–right specification, on the basis that mouse embryos can develop normally in culture from cleavage stages to mid-organogenesis (Hsu 1982). However, few embryos have been cultured over this period and the incidence of inversion is unknown. We found no influence of uterine environment on the development of asymmetry after reciprocal blastocyst transfers between *iv/iv* and +/+ mice (Brown et al 1990). Although this proves that the *iv* phenotype is not caused by an abnormal uterine

TABLE 1 The effects of handling[a] cultured rat embryos on the development of left–right asymmetry

Explant stage	Non-handled					Handled				
	Number of embryos	situs inversus	Inverted heart	Inverted turning	% Left/ right abnormal	Number of embryos	situs inversus	Inverted heart	Inverted turning	% Left/ right abnormal
Amniotic folds	56	9	7	5	38	ND				
Closed amnion	130	29	19	17	50	ND				
Notochordal process	8	0	1	0	12.5	48	3	2	0	10.4
Neural groove	62	0	2	0	3.2	348	23	14	4	11.8
Neural folds	90	1	2	1	4.4	273	14	13	4	11.4
Foregut pocket	48	0	1	0	2.1	145	6	1	2	6.2

Embryos were explanted on Day 9 (amniotic folds and closed amnion stages) or Day 10, then cultured for 40 to 72 hours to the approximate 25-somite stage when the direction of heart looping and embryonic turning were recorded. *situs inversus* denotes inversion of both heart looping and embryo turning. [a]The handled embryos were controls for short-term treatments. 2–4 hours after the start of culture they were removed from their culture bottles, rinsed in minimal essential medium (MEM), then returned. ND, not done.

environment, it does not rule out a role of the uterus in left–right specification. We have repeated some of Smith's observations and re-evaluated her proposal (see next section).

An alternative explanation for aberrant left–right development in culture has been suggested by M. Fujinaga (personal communication). He has observed that the external shape of conceptuses in culture is significantly more spherical than that *in utero*, possibly because of the removal of Reichert's membrane and all tissue external to it. He suggests that this shape change may be sufficient to disturb the development of left–right asymmetry.

A second set of control cultured embryos is also described in Table 1. These were controls for short-term treatments and so were removed from culture, rinsed in medium, then returned to the culture bottles for the remaining 40–44 hours. In most cases the rinsing was done four hours after the beginning of culture. This procedure alone was sufficient to increase the frequency of left–right inversions, at some stages to 12%. This shows that unspecific (overtly innocuous) treatments can disrupt left–right information and suggests that the handed aspect of left–right specification (*conversion* in our model) is a 'weak', labile process.

It is worth stressing that these cultured embryos were apparently normal in all respects other than the handedness of their asymmetry. That they always developed asymmetrically, i.e. turned to one side and their hearts looped to one side, suggests that the process which ensures that the two sides are different (*random generation* in our model) is a 'stronger' mechanism. We have observed rat and mouse embryos developing without turning, or with a ventral heart loop, but these have been very rare in our studies.

Embryonic axis orientation in the uterus

The orientations of the embryonic anteroposterior, dorsoventral and left–right axes in the mouse uterus have previously been examined by Snell & Stevens (1986) and by Smith (1980, 1985). They showed specific orientations relative to the uterus, and Smith suggested that uterine cues could provide the embryo with a sense of left and right. We have now confirmed many of their observations in the rat, and in the *iv/iv* mouse, but cannot agree that left–right positional information could be provided by the uterus.

We examined axis orientation in fresh uteri dissected on the morning of Day 10, when most embryos were at the neural groove stage. Uteri were removed from the mother with the right ovary attached and the left ovary detached. The uterus was stretched out, the mesometrial side conveniently sticking to paper, with the right ovary to the operator's left so the lateral, or 'outside', walls of both horns were facing the operator. The uterine wall was cut along its length at the antimesometrial apex. Then each implantation site was bisected in this same plane, along the long axis of the uterus. Reichert's membrane and trophoblast were split, and the orientation of the embryo was observed *in situ*.

There were three major findings. First, there were more implantation sites and more viable embryos in the right horn than the left (Table 2). This confirms earlier observations (McLaren 1963, Smith 1985) in the mouse, and suggests a subtle difference between left and right in ovarian function. Whether this is due to a primary asymmetry in ovarian development or a secondary effect of another asymmetry, such as in the vasculature or early growth rate, is not known.

A re-evaluation of observations on the progeny of *iv/iv* mothers examined on Day 18 of pregnancy (Brown et al 1989) shows that the *situs solitus* mothers (24 mice) show the bias in favour of implantation sites in the right uterine horn (134 compared to 89 in the left horn). In the mothers showing *situs inversus*, this bias was reversed: in 28 pregnant females there were 119 implantation sites in the right horn and 149 in the left horn.

Embryos were observed to be in one of two specific orientations in the uterus. The long axis of the ovoid conceptus is aligned parallel to the mesometrial–anti-mesometrial uterine axis, naturally with the placenta (ectoplacental cone) toward the mesometrial surface. The anteroposterior embryonic axis is aligned perpendicular to this, but in two orientations opposed by 180°, so that either the anterior or posterior end of the embryo faces the outside wall of the uterus. We term these two orientations 'anterior-out' and 'posterior-out'. There was a slight deviation of a few degrees from this orientation in some embryos, but in no case by as much as 22.5°. This embryonic orientation is the same as described by Smith (1985).

The distribution of conceptuses in the two orientations was analysed for each uterine horn and according to the position of the implantation site along the horn. Overall, the distribution is close to random, with half in each orientation and no great difference between the two horns (Table 2). We found no correlation with position along the horn (data not shown). This is largely in agreement with Smith (1985). There is a suggestion from our results that there are fewer conceptuses in the left horn with the posterior-out orientation than any other combination. This is statistically significant, but we await the results of a second, independent, study currently underway before concluding that this is biologically significant. Smith also reported that one orientation was not distributed equally between horns, but this is not equivalent to our results, as we will discuss.

TABLE 2   Orientation of the embryonic axis relative to the uterus at the neural fold stage in the rat

	Right horn	*Left horn*	*Total*
Anterior-out	57	55	112
Posterior-out	53	42	95
Total number of embryos	110	97	207
Number of implantation sites	118	103	221

The orientations anterior-out and posterior-out are described in the text. Results from 18 Wistar rat litters.

That there is any relationship at all between the embryonic and uterine axes is remarkable; that there are two alternative relationships is perplexing. There are obviously two possible mechanisms that could produce any relationship between these axes. Either the anteroposterior axis is already specified in the blastocyst at the time of implantation and this causes some surface polarization of the conceptus to stabilize implantation in a particular orientation or at least one axis of the conceptus is induced by the uterine environment. We favour the former. Clearly, at the time of implantation the embryonic–abembryonic axis of the conceptus (which is equivalent to the embryonic dorsoventral axis) is determined, with the inner cell mass at the embryonic pole. There are well established differences between the mural and polar trophectoderm cells, which presumably account for attachment of the conceptus initially at the abembryonic pole.

Smith suggests that an attached conceptus rotates about the embryonic-abembryonic axis, to align its putative anteroposterior axis across the uterus, and that the abembryonic pole moves simultaneously to point at the mesometrial surface. She cites the observation by Hsu (1981) of counter-clockwise rotation of mouse blastocysts in culture, although it is not known if this is an *in vitro* artifact; it has not been confirmed by others. Smith further suggests that attachment, largely at random, to either side-wall of the uterus, followed by these movements, would bring the conceptus into a consistent relationship with the axes of the uterus, such that the embryo could derive left–right positional information. Here we disagree.

Smith's proposal is based on an analysis in which each uterine horn was considered to have a left and a right wall. This is not the case. The axes of the uterus are mesometrial–antimesometrial, ovarian–cervical and lateromedial (or inside-outside). The uterus is a bilaterally symmetrical organ, so the axes of each horn are mirror images of one another, just like the limbs, for example. There is no known intrinsic 'leftness' or 'rightness' about the two horns, nor about the two side-walls of each horn. Since the uterine horns are mirror images (but conceptuses in each are not) and conceptuses implant in two opposed orientations, we cannot think of any mechanism by which the embryro could derive a consistent sense of left and right from the uterus. Indeed, each of the four combinations of 'inward' and 'outward' orientation in left and right horns has a different set of relationships between the anteroposterior, left–right, ovarian–cervical and lateromedial axes. This returns us to the conclusion that the development of left–right positional information must be intrinsic to the embryo, but leaves the question of why embryos lose this sense of left–right in culture unanswered.

Organ relationships and sex ratios in *iv/iv* mice

In a previous study of Day 18 *iv/iv* fetuses (Brown et al 1989) we found an excess of female fetuses. There were 140 females and 104 males, a sex ratio of 1.3:1. The frequency of *situs inversus* and *situs solitus* was 50:50 in both sexes (67:68 females; 51:52 males), but there were five females with discordant heart and stomach *situs* and there was only one such male. The deficiency of males may be because they were more severely affected and did not survive to term. It is also possible that the skewing of the sex ratio is unrelated to the *iv/iv* genotype. This study was performed with the SI/Col strain which is a unique inbred strain carrying the *iv* mutation. There is no equivalent strain with the wild-type alleles at the *iv* locus.

We have now examined a further 275 SI/Col *iv/iv* mice with the aim of establishing the relationships of *situs* and morphology between different organs (Seo et al 1992). Some of the observed sex ratios are shown in Table 3. Isomerism and all morphological abnormalities were up to four times more common in females. As before, we cannot tell if this is because of loss of affected males.

This apparent interaction of left–right asymmetry with sex is interesting in the light of suggestions that differential growth rates may be involved in sex determination (Mittwoch 1989). We have suggested that differential growth of left and right halves may be sufficient to establish left–right asymmetry. It is intriguing that in human true hermaphrodites ovarian tissue is preferentially located on the left, testicular tissue on the right (Simpson 1978). Mittwoch has reported that in normal development the right gonad is more advanced than the left. We are planning further studies of the interaction of *iv/iv* and sex using mice of different genetic backgrounds.

TABLE 3   Sex ratios for organ situs and morphology in *iv/iv* mice

Organ		Female (%)	Male (%)	Ratio female/male
Atria	usual	62 (41.9)	62 (48.8)	0.86
	mirror-image	75 (50.7)	63 (49.6)	1.02
	isomeric	11 (7.4)	2 (1.6)	[a]4.72
Spleen—abnormal		66 (64.6)	34 (26.8)	[b]1.67
Liver—abnormal		63 (43.6)	34 (26.8)	[a]1.59
Portal vein—ventral		49 (37.1)	24 (18.9)	[b]1.75
Systemic vein—abnormal		29 (19.6)	20 (15.7)	1.24
Lungs	usual	63 (42.6)	62 (48.8)	0.87
	mirror-image	75 (50.7)	63 (49.6)	1.02
	isomeric	10 (6.7)	2 (1.6)	[a]4.29
Heart—abnormal		17 (11.5)	4 (3.1)	[b]3.64

148 female and 127 male SI/Col *iv/iv* mice (fetuses, neonates and adults) were examined. The ratio is the proportion of females with the given morphology compared to the proportion of males showing the same morphology.
[a]$P<0.05$ ($\chi^2>3.84$, df = 1).
[b]$P<0.01$ ($\chi^2>6.63$, df = 1).

Although it is clear that lateral asymmetry in *iv/iv* mice is, overall, random (50% are normal and 50% have *situs inversus*), there are many abnormalities caused by the mutation, presumably as a consequence of the loss of handed information. Because the organs within an individual animal need not be of the same *situs*, we have argued that some aspects of left–right specification (perhaps *random generation* in our model) must be tissue specific. However, the exact relationships of *situs* among organs in *iv/iv* mice have not been well studied. We have examined many relationships (Seo et al 1992), but summarize just one here.

The relationship between atrial *situs* and splenic morphology is shown in Table 4. (The location of the spleen is not reported in the Table.) We have not observed true multiple spleens in the mouse, unlike in humans, but bilobed and fissured spleens appear to be the mouse equivalent. Two points are clear from these observations. First, there are many more cases of abnormal spleens than isomeric atria. Second, all cases of atrial isomerism had abnormal spleens, and right isomerism was associated with absent spleen, while left was associated with bilobed and fissured spleens. This is interesting because there is no obvious connection of tissue type between these two organs. It appears that isomerism may be a tissue-independent phenomenon, at least for these organs.

Left and right compartments?

If our model of left–right development is correct and the whole of one half of the embryo acquires some stable property as a result of *conversion*, and our representation of left–right handedness as a quantal process is accurate, then we should expect the left and right halves of the embryo to behave as compartments, at least for a limited period. The lineage studies of Lawson & Pedersen (this volume) at early/mid-primitive streak stages show that daughter cells often cross

TABLE 4   The relation between the atrial arrangement and splenic morphology in *iv/iv* mice

	Spleen				
	Normal	*Fissured*	*Bilobed*	*Elongated*	*Absent*
Atria					
usual	84 (67.7%)[a]	16	24	0	0
mirror-imaged	91 (65.9%)[a]	5	42	0	0
Right isomerism	0	0	0	1	3
Left isomerism	0	0	8	1	0
All	175	21	74	2	3
	63.6%	7.6%	26.9%	0.7%	1.1%

[a]Percentage of normal splenic morphology in each type of atrial arrangement.

the midline of the embryo. However, there is little information on later stages. We have begun to label small groups of cells just lateral to the midline of late streak to early neural plate stage embryos with the fluorescent dye DiI. So far, in 10 embryos labelled in ectoderm or mesoderm and examined at the 12–14 somite stage, all labelled areas are restricted to the injected side. There is a great deal more systematic work to be done.

Conclusions

In our model of left–right development we proposed two separate processes, one (*random generation*) that ensures the two sides are different, and another (*conversion*) that imparts handedness to the process. It appears that handedness is weak and labile since a variety of unspecific treatments, or even explantation and culture *in vitro*, can disrupt it. In contrast, it is difficult to prevent embryos from developing asymmetrically, suggesting that *random generation* is a robust process. In normal development, the left–right axis appears to be determined by the first-somite stage, but the timing of earlier events remains obscure. The specific orientations of embryonic axes within the uterus are intriguing because they suggest that the anteroposterior axis is already specified at implantation of the late blastocyst. However, the uterus probably cannot impart a sense of left–right handedness to the embryo.

The phenotypes of *iv/iv* mutant mice suggest two areas for investigation. The skewed sex ratios imply a relationship between the development of sex and that of asymmetry. A study of differential growth rates would be of interest. The relationships between the asymmetries of various organs suggest a role for the developing vasculature, or perhaps fields of *random generation* of asymmetry.

References

Brown NA, Wolpert L 1990 The development of handedness in left/right asymmetry. Development 109:1–9

Brown NA, Hoyle CI, McCarthy A, Wolpert L 1989 The development of asymmetry: the sidedness of drug-induced limb abnormalities is reversed in *situs inversus* mice. Development 107:637–642

Brown NA, McCarthy A, Wolpert L 1990 The development of handedness in aggregation chimeras of situs inversus mutant and wild-type embryos. Development 109:949–954

Brown NA, McCarthy A, Wolpert L 1991 Development of handed body asymmetry in mammals. In: Biological asymmetry and handedness. Wiley, Chichester (Ciba Found Symp 162) p 182–201

Fujinaga M, Baden JM 1991 Evidence for an adrenergic mechanism in the control of body asymmetry. Dev Biol 143:203–205

Hsu Y-C 1991 Time-lapse cinematography of mouse embryo development from blastocyst to early somite stage. In: Glasser SR, Bullock DW (eds) Cellular and molecular aspects of implantation. Plenum Press, New York p 383–392

Hsu Y-C 1982 Development of mouse embryos *in vitro*: preimplantation to limb bud stage. Science (Wash DC) 218:66–68

Lawson K, Pedersen RA 1992 Clonal analysis of cell fate during gastrulation and early neurulation in the mouse. In: Postimplantation development in the mouse. Wiley, Chichester (Ciba Found Symp 165) p 3–26

McLaren A 1963 The distribution of eggs and embryos between sides in the mouse. J Endocrinol 27:257–277

Mittwoch U 1989 Sex differentiation in mammals and tempo of growth: probabilities vs. switches. J Theor Biol 137:445–455

Seo J-W, Brown NA, Ho SY, Anderson RH 1992 Abnormal laterality and congenital cardiac anomalies: relationships of visceral and cardiac morphologies in the iv/iv mouse model. Circulation, in press

Simpson JL 1978 True hermaphroditism: etiology and phenotypic considerations. In: Summitt RL, Bergsma D (eds) Sex differentiation and chromosome abnormalities. Alan R. Liss, New York p 9–35

Smith LJ 1980 Embryonic axis orientation in the mouse and its correlation with blastocyst relationships to the uterus. 1. Relationships between 82 hours and 4½ days. J Embryol Exp Morphol 55:257–277

Smith LJ 1985 Embryonic axis orientation in the mouse and its correlation with blastocyst relationships to the uterus. II. Relationships from 4¼ to 9½ days. J Embryol Exp Morphol 89:15–35

Snell GD, Stevens LC 1966 Early embryology. In: Green EL (ed) Biology of the laboratory mouse. McGraw-Hill, New York p 205–245

DISCUSSION

McLaren: What do you see as the advantages of handed lateral asymmetry? One can see that with brain laterality the two sides of the brain do different things, but as far as one knows the two sides of the lung don't do different things. Do you think that all handed asymmetry might stem from brain laterality? Are there other organisms that are random rather than handed in their lateral asymmetry?

Brown: These problems were discussed extensively at the recent Ciba Foundation Symposium on 'Biological asymmetry and handedness'. One explanation for the existence of lateral asymmetry is that we have evolved from an organism which was completely asymmetric, rather than bilaterally symmetrical. The process is called dexiothetism (Jefferies 1991). It is proposed that one of our ancestors lay down onto its right side such that its anteroposterior axis became its left–right axis. From that completely laterally asymmetric organism a new, vertical, plane of bilateral symmetry developed perpendicular to the old one, and we have slowly evolved back towards complete bilateral symmetry, but have retained some bilateral asymmetry. The question is not so much why do we have asymmetry but why we haven't got rid of it.

I agree that there is no apparent advantage in having a handed asymmetry. About one person in 10 000 has complete *situs inversus* and they are functionally normal.

There is no clear relationship between cerebral dominance in humans and body plan asymmetry. For example, there are the same proportions of left- and right-handers in a population with *situs inversus* as in the normal population. It has been suggested that the process for brain lateralization is derived from a duplication of the gene responsible for body asymmetry (McManus 1991).

Smith: I was wondering about flatfish. These originally have a 'normal' dorsoventral axis, then they lie down on one side. Is that an example of what you called dexiothetism?

Brown: Policansky (1982) has described the development of flatfish in which they fall over onto one side and then their eyes migrate around to the top side. The confusing thing about flatfish, as far as our model is concerned, is that there are left and right sublines of the same strain. This is very difficult to account for in our model. Our model fits nicely with the fact that there is no known mammalian mutant in which all offspring have *situs inversus*. Such a mutant would require that the asymmetric 'F' molecule was changed such that it was fixed in exactly the opposite orientation to normal. One can imagine lots of loss-of-function of mutations that would randomize things, but complete reversal seems unlikely.

Saxén: One way to test the lability of left–right specification in early embryos is to transplant tissue, for example lung mesenchyme, from one side to the other. My guess would be that in the lung such reciprocal transplantation wouldn't work, because the branching programme is already set in the epithelial component, as in many other organs. Could it be possible for other visceral organs? If you take heart primordia from a mouse with *situs inversus* and culture them *in vitro*, what would happen to heart looping?

Brown: I don't know. I am not sure that the experiment would be possible with a mouse heart, although it has been claimed that amphibian hearts will loop independently in culture. I would expect the heart to loop randomly in an *iv* mutant embryo in culture anyway, since that's what they do in the intact embryo.

Kaufman: The simplest explanation for the asymmetry of the lungs is that it is an adaptation to the asymmetry of the heart. The heart is initially in the midline, then it is principally on the left side in the vast majority of mammals, so there is consequently less room in the left hand side of the thorax, and only one or two lobes develop on that side. One lobe develops on the left side in the mouse, and two in the human. There is more room on the right side of the thoracic cavity for the development of three lobes in the human, and four lobes in the mouse, though, admittedly, the right accessory lobe is mostly located on the left side of the thorax.

Lawson: It would be very difficult to do the critical experiment with the lungs. It would entail performing the epithelium/mesenchyme separation and recombination before the lung buds have appeared at all or very early. Then you would have to be able to rotate the mesenchyme precisely.

There is an observation about differences between left and right lung epithelia. This is a situation which the embryo would never normally be confronted with but there was a time when we were interested in whether there was a specificity in mesenchyme–epithelial interactions in the production of branching morphogenesis in organs such as the lung and salivary glands. We recombined lung epithelium with salivary mesenchyme at a reasonably advanced level, after lung buds had been formed, and obtained branching morphogenesis. When we did the experiment before secondary bronchial buds had been formed, i.e. when only the primary bronchial rudiment had appeared, there was no morphogenesis in the salivary mesenchyme. When we left those cultures for about three days, then just removed the salivary mesenchyme and put it back, we did get branching morphogenesis in the salivary gland pattern. There is a difference between the ability of right and left lung epithelia to respond to salivary mesenchyme at this very early stage. Left lung epithelium responds significantly more frequently than right epithelium does (Lawson 1983). I have no explanation for this. It seems to be a difference in the epithelium rather than in the mesenchyme.

Jaenisch: Is there evidence that cells can cross the midline during gastrulation or before?

Lawson: After labelling at the pre-streak or early streak stage you often find epiblast clones that spread across the midline. Axial clones do it all the time; slightly paraxial clones do it a bit. It is an even spread towards the streak. Presumably, lateral cells also go through the streak without respect to side. The midline poses no barrier to cell movement at this stage.

Jaenisch: It has been reported that a patient with the dominant disorder Marfan's syndrome was affected unilaterally. Such a condition could arise by an early somatic mutation and the mutated cells could become lateralized and remain confined to one side of the embryo. This might suggest that in this case no extensive crossing of cells to the other side had occurred during gastrulation.

Lawson: In normal mice we do not see early lateralization, but that could be something specific to the mutation. Clones generated at the early streak stage produce descendants that go equally into somites or paraxial mesoderm on both sides.

Kaufman: Have you done the obvious experiment of isolating embryos at say the primitive streak stage and then putting them in your culture system, but keeping embryos from the left uterine horn and those from the right uterine horn completely separate? There is such a high incidence of *situs inversus* in the control situation that it would be of interest to establish whether there is any difference in the incidence of *situs inversus* between these two groups?

Brown: We haven't done the experiment, but I would not anticipate any difference. Embryos develop essentially at random in the two orientations in both right and left horns.

We have examined axis orientation in developing *iv/iv* embryos *in utero*. There was the possibility that the 50% which develop with *situs inversus* would be in one orientation and the other half, that develop with normal situs, would be in the opposite orientation. In fact, there is no relationship between axis orientation and the development of situs.

McLaren: Could the fact that the right ovary in the mouse tends to shed more eggs than the left ovary (McLaren 1963) be due to a very slight difference in blood pressure, since the right ovarian artery comes off the dorsal aorta more anteriorly? Quantitatively, is the difference in blood pressure anything more than microscopic?

Brown: I am not sure whether this would be blood pressure *per se*, but certainly the vascular supply is different.

Tam: Is there also an asymmetry in the rate of maturation of paired organs? For example, does the left lung mature faster than the right?

Brown: There are many organs where the right appears to be more advanced than the left at a particular stage of development in wild-type mice. In *situs inversus*, it is the other way around. In *situs inversus* embryos everything that we have ever looked at is inverted. In normal embryos the limb buds develop slightly asymmetrically and that is reversed in the *iv/iv* mouse.

Tam: So it is not just morphological asymmetry, the whole developmental programme may be different between left and right.

Beddington: Do you know when that asymmetry in growth first starts in mouse or rat development?

Brown: It has been shown that the number of nuclei labelled with [³H]thymidine is different on the left and right sides of 10–12 somite mouse embryos (Miller & Runner 1978). However, the embryos were already turning and thus morphologically asymmetric in those experiments.

Beddington: Do you know if there is asymmetry earlier?

Brown: No, we are looking at that. It would be interesting to pick up such an asymmetry before there is an overt difference between the two sides. It's not surprising to detect asymmetric growth rates once the embryo starts turning, because the two sides are very different sizes.

Beddington: Such an early subtle asymmetry in growth rate is an obvious thing that you may interrupt briefly when you explant the embryos into culture.

Brown: It's possible, although the embryos that develop with the highest incidence of *situs inversus* in the control cultures are the ones which are explanted at the mid-streak stage.

Lawson: The mid-streak stage is the time when they are growing very rapidly, so any upset might be disturbing.

Hogan: Have you tried making chimeras of wild-type and *iv* mice? Do you think the wild-type cells would successfully rescue the phenotype?

Brown: We had a great deal of difficulty making chimeras with *iv/iv* embryos; we ended up with only 21 real chimeras. Two of those had abnormalities of left–right. These were examined at Day 10 when turning and heart looping can be observed. One had an inverted heart but normal turning; and the other was *situs inversus*. Those two were unique in having more than 67% *iv* contribution to both the heart and visceral yolk sac. All other embryos developed normally and many of them had 70–80% contribution of mutant cells to some tissues. The problem is that half of *iv/iv* embryos develop with normal *situs*. We had chimeras that had just more than the lowest detectable amount of chimerism, perhaps 5% wild-type cells by analysis of glucose phosphate isomerase isoforms. They were normal, but that may not be rescue since half the mutant embryos develop with normal *situs* anyway.

Krumlauf: Do strain differences affect the sorts of things you see with the *iv/iv* mice?

Brown: Very little in our hands; we have the mutation on three genetic backgrounds. There are some differences in so-called venous heterotaxias, but other than that the phenotypes are very much the same.

Saxén: In Rudi Balling's talk (this volume), it looked from the pictures as though many of the defects in the *undulated* mice were asymmetric or even unilateral.

Balling: Yes, they are. Some embryos have defects on both sides, but mostly they were on the left side.

Saxén: Do you have any explanation?

Balling: No, I don't. Unilateral defects might indicate that the right and left sides are independently specified. It doesn't explain why there is a preferential bias for one side.

McLaren: Rudi, many mutations affect the vertebral column and the limbs; those are the systems that have been worked on most. Not so many mutations are known that affect the pelvic girdle, pectoral girdles or skull bones. Do you think it is simply a question of what mutants have been observed and therefore what genes have been picked out?

Balling: There are mutants such as *luxate* which show preferential problems with the fore limb buds or the hind limb buds with prevalence for one specific side (Carter 1954). Whether people have selected for axial skeletal mutants as opposed to mutants with defects in the appendicular skeleton, I don't know, but I don't think that is the case.

McLaren: Tail-short (*Ts*) shows a very odd asymmetry. *Ts/+* mice often have a shortening of the left fore leg (mainly the humerus) and/or the right hind leg (mainly the tibia) (Deol 1961). I wonder if both asymmetries would be reversed in *iv/iv* mice.

Balling: When Michael Kessel treats pregnant mice with retinoic acid, some of the effects on the axial skeleton occur unilaterally; for example, homeotic shifts only on the right-hand side. This again indicates that positional information can be affected on just one side. The question is at what stage this occurs.

Lawson: From your work and the normal symmetrical distribution of descendants of early gastrulae, one would expect the critical time to be at late gastrulation or after gastrulation.

Brown: The effects on the somites that occur at relatively late stages, after the embryo has turned and when it is essentially helical, are not so difficult to explain. The side of the embryo on the 'inside' of the helix is actually smaller than the other side, so the somites may be 'cramped'. Several treatments that disrupt somite formation, like valproic acid or heat shock, will affect the left-hand side more than the right-hand side. I think that the explanation is simply a structural constriction.

Balling: It may also depend on the timing. If somitogenesis is dependent on the cell cycle of the progenitor cells, the time at which the treatment is given could be critical.

Solter: Is bilateral asymmetry mandatory? Your model allows that *situs viscerum inversus* is possible, yet left and right differences (handedness) persist, because there is no bilateral symmetrically mirror-image mammal. In humans, there is isomerism of some organs. Is complete bilateral symmetry possible but extremely rare? How would the molecules which provide the signal for asymmetry in your model become isomeric?

Brown: Isomerisms certainly do occur in several organs. Even isomerism in heart looping is possible. Hearts that have looped ventrally in the midline, or even in rare cases dorsally in the midline, have been observed. However, for normal development the heart must develop in an asymmetric way; it's a prerequisite for ventricular formation. Presumably, the reason you don't see complete isomerism of the heart is that it is incompatible with functional circulation.

Isomerism is most likely the result of loss of both left–right specification and, somehow, loss of the randomly generated difference between left and right, be it a gradient or whatever. So, it is not really a mutation to isomerism but rather a loss of the normal processes that dictate an asymmetry between the two sides.

Solter: Can you produce that with the isomeric molecule? Your F molecule would have to lose handedness.

Brown: Isomerism in our model is a loss of function or orientation of the F molecule, not an isomeric molecule.

Solter: There are some very subtle differences between the two sides. For example, in humans the right testicular vein enters the vena cava and the left one enters the renal vein. From the two completely symmetrical organs, symmetrical blood vessels enter a basically symmetrical part of the venous system, nevertheless their paths are quite different. What makes these two veins choose, with great precision, such different pathways? It is probably easier to explain the global bilateral asymmetry than to explain these very subtle differences in handedness.

Smith: How many other chemicals, apart from α_1-adrenergic agonists, affect *situs inversus*?

Brown: Lots, and it appears to be unspecific. Heat shock, retinoic acid, lithium, colcemid, all work, but it may be simply a toxic effect in these cases.

Smith: So why did you choose α_1-adrenergic agonists?

Brown: Of everything we have ever tried, it is the 'cleanest', certainly not a toxic effect. All the other treatments also cause growth retardation and in some cases they cause other abnormalities. That's not the case with the α_1-adrenergic agonists. At least in these culture systems, they do nothing other than randomize the asymmetry.

Smith: How about their mode of action?

Brown: We don't know. One of the α receptors is coupled to the inositol phosphate pathway and we have started some investigation of steps in that cascade. We have examined the actions of calcium ionophores which do not induce randomization, and lithium, which does.

McLaren: Nigel, what about monozygotic twins? Is the degree of mirror-imaging related to how late in development the splitting occurs?

Brown: Presumably, that's the case. There are very few objective data, although there are plenty of anecdotal data on mirror imagery of identical twins: hair whorl patterns, finger prints and so on (Burn 1991). There is some controversy in the field, with people either not believing it's a real phenomenon or being firmly convinced that it is. One explanation of mirror imagery is that these are monozygotic twins that cleaved relatively late. There is no way, with current data, that we can establish whether that's the case or not.

McLaren: There have been quite extensive surveys that include information on zygosity and placentation in twins (e.g. Burn & Corney 1988). There is also a sex ratio bias towards female twin pairs in monozygotic twins that cleave later (James 1980).

Brown: There are certainly lots of data. One thing that is clear from those data is that certain kinds of cardiovascular malformations are significantly more common in monozygotic twins. The idea again is that they cleaved at a stage when there already was some sense of left–right, and this disrupted the normal asymmetry of the heart.

Kaufman: I have the impression that there are probably two types of *situs* in the human: one where there is a definite genetic lesion, and a second group where there were monozygotic twins. It has been suggested that there is a quite high incidence of mirror imaging (up to about 25%) associated with monozygotic twinning (Hamilton & Mossman 1972). However, not infrequently one of a pair of monozygotic twins completely disappears (the so-called 'vanishing' twin) after the diagnosis of twinning has been made by ultrasound scanning. If the normal twin dies and disappears, the other twin would then survive as an isolated *situs inversus*. Some surveys suggest 30–50% morbidity in monozygotic twins, so this may be the explanation for the sporadic incidence of individuals with *situs inversus*.

McLaren: That would be quite difficult to check but it's an interesting hypothesis.

References

Balling R, Lau CF, Dietrich S, Wallin J, Gruss P 1992 Development of the skeletal system. In: Postimplantation development in the mouse. Wiley, Chichester (Ciba Found Symp 165) p 132–143

Burn J 1991 Disturbance of morphological laterality in humans. In: Biological asymmetry and handedness. Wiley, Chichester (Ciba Found Symp 162) p 282–299

Burn J, Corney G 1988 Zygosity determination and the types of twinning. In: MacGillivray I, Campbell DM, Thompson B (eds) Twinning and twins. Wiley, Chichester p 7–25

Carter TC 1954 The genetics of luxate mice. IV Embryology. J Genet 52:1–35

Deol MS 1961 Genetical studies on the skeleton of the mouse. XXVIII. Tail-short. Proc R Soc Lond B Biol Sci 155:78–95

Hamilton WJ, Mossman HW 1972 Hamilton, Boyd and Mossman's human embryology: prenatal development of form and function, 4th edn. Heffer & Sons, Cambridge

James WH 1980 Sex ratio and placentation in twins. Ann Human Biol 7:273–276

Jefferies RPS 1991 Two types of bilateral symmetry in the Metazoa: chordate and bilaterian. In: Biological asymmetry and handedness. Wiley, Chichester (Ciba Found Symp 162) p 94–127

Lawson KA 1983 Stage specificity in the mesenchyme requirement for rodent lung epithelium in vitro: a matter of growth control? J Embryol Exp Morphol 74:183–206

McLaren A 1963 The distribution of eggs and embryos between sides in the mouse. J Endocrinol 27:157–181

McManus IC 1991 The inheritance of left-handedness. In: Biological asymmetry and handedness. Wiley, Chichester (Ciba Found Symp 162) p 251–281

Miller SA, Runner MN 1978 Tissue specificity for incorporation of [^{3}H]thymidine by the 10- to 12-somite mouse embryo: alteration by acute exposure to hydroxyurea. J Embryol Exp Morphol 44:181–189

Policansky D 1982 Flatfish and the inheritance of asymmetries. Behav Brain Sci 5:262–265

The role of *Sry* in mammalian sex determination

Robin Lovell-Badge

Laboratory of Eukaryotic Molecular Genetics, National Institute for Medical Research, The Ridgeway, Mill Hill, London NW7 1AA, UK

Abstract. The testis-determining gene is the Y-linked gene responsible for initiating the developmental pathway leading to testis formation in males. A strategy based on determining the precise chromosomal location of this locus has been used to clone a new gene which has been called *SRY* in humans (*Sry* in mice). A variety of studies now show that this is indeed the testis-determining gene. *Sry* has a spatial and temporal pattern of expression which correlates with the initiation of testis differentiation. The amino acid sequence encoded by the gene suggests that the protein may function as a transcription factor, which fits well with models of sex determination. Some cases of XY sex reversal in humans and mouse have been attributed to mutations in *SRY/Sry*, indicating that it is normally necessary for testis determination. The finding that a genomic fragment carrying *Sry* can cause male development in XX mice has proved that *Sry* is the only gene from the Y chromosome necessary for testis determination.

1992 Postimplantation development in the mouse. Wiley, Chichester (Ciba Foundation Symposium 165) p 162–182

The processes of sex determination and differentiation involve interacting networks or pathways of gene activity which lead to the development of male or female characteristics. These pathways must include a switch mechanism responsible for the decision to become male or female, and a wide variety of mechanisms appear to have been adopted for this purpose during evolution. In many lower organisms such as yeast, the switch is the presence of a particular allele at an active mating type locus (Kushner et al 1979). Mammals have a conceptually similar mechanism in that the two sexes differ in genetic make-up, but in this case it is the presence of the Y chromosome that acts as a dominant male determinant (Jacobs & Strong 1959, Ford et al 1959, Welshons & Russel 1959). This may be contrasted to the situation found in many other species where all the genes responsible for sexual dimorphism are present in both males and females. In some, such as *Drosophila*, it is the X:autosome ratio that is critical for the activation of one or other pathway, and although there is a Y chromosome in males it is required only for fertility (Baker 1989). In others, such as alligators, there are no differences at all, even in chromosome constitution, and the switch is environmental (Ferguson & Joanen 1982).

Sexual dimorphism normally affects the whole animal and there are different ways in which this can be achieved. Thus, in *Drosophila* the sex determination signal operates probably in all somatic cells in a cell-autonomous fashion (Baker 1989). In eutherian mammals, on the other hand, all the secondary sexual characteristics are a result of the action of hormones or factors produced by the developing gonads (in marsupials some characteristics may be determined independently from the gonad, perhaps by the X: autosome ratio) (O et al 1988). It was shown almost 50 years ago, by Jost (1947), that the presence of testes was necessary for the development of male characteristics, because castrated rabbit embryos of either chromosomal sex develop as females. In mammals, therefore, the primary event in sex determination is the differentiation of the indifferent gonads (or genital ridges) into testes rather than ovaries. Female development can be considered the normal or default pathway, and the Y chromosome diverts development into the testicular pathway. The male-determining activity of the Y chromosome has therefore been attributed to a gene or genes termed *TDF*, for testis-determining factor in humans, and *Tdy*, for testis-determining Y gene in mice.

There has been considerable interest in isolating and defining the testis-determining gene, partly because it would allow an understanding of sex determination itself, but also because it may help us to understand other processes in the embryo that involve developmental decisions. For example, *Tdy* could be a member of a family of genes with similar functions in development, and/or it could lead to an understanding of a common type of pathway used in cell differentiation and morphogenesis.

Gonadal differentiation

Before we consider the testis-determining gene itself, it is relevant to discuss the origins of the testis and ovary and the cell types involved. The genital ridge arises as a thickening along the length of the mesonephros, through a process thought to involve epithelial–mesenchymal induction events similar to those seen in kidney development and elsewhere. The genital ridges become more prominent from 10.5 to 11.5 days post coitum (p.c.), as shown in Fig. 1; however, no differences are seen between males and females until about 12.5 days p.c., when the male gonad takes on a characteristic striped appearance. This is due to the differentiation and alignment of Sertoli cells into the testis cords. Other characteristics of early testis differentiation are its rapid growth and the formation of a prominent vasculature. Apart from some changes in overall shape there is little to distinguish the differentiating ovary from the indifferent gonad until follicle cells aggregate around oocytes several days later. This does not mean there are no underlying molecular changes; indeed, there is some evidence

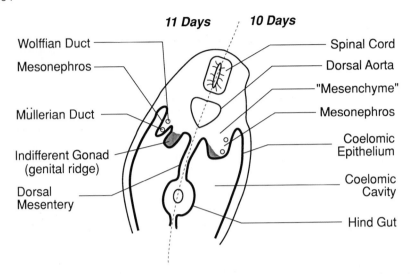

FIG. 1.   Diagram representing transverse sections through the posterior region of 10- and 11-day mouse embryos showing the position of the developing genital ridge. The genital ridge is thought to arise in an induction process involving coelomic epithelium and the mesenchyme of the mesonephros. Subsequent testicular development, first visible at about 12.5 days p.c., is due to the differentiation and alignment of Sertoli cells into epithelial cords.

to suggest that the ovarian pathway rapidly becomes insensitive to *Tdy* action (Burgoyne & Palmer 1991).

There are essentially four cell lineages that make up the developing gonad; these are illustrated in Fig. 2. The origin of the somatic cell types is poorly understood. For example, there is some debate as to whether Sertoli cells arise from coelomic epithelium or directly from mesenchymal cells of the mesonephros, and about the relationship between Sertoli and Leydig cells. On the other hand, the origin of the germ cells is comparatively well understood. These cells are first distinguishable at the posterior end of the primitive streak in an embryo 7.25 days p.c. (Ginsburg et al 1990). Over the following three or four days they migrate along the hind-gut and dorsal mesentery into the developing genital ridge. However, the germ cells are essentially irrelevant to testis determination, as seen in embryos homozygous for mutations affecting the proliferation and migration of germ cells. For example, in embryos with extreme mutations at the *W* locus the genital ridges are devoid of germ cells, yet testes develop normally (McLaren 1985). Germ cells are required for the proper organization and differentiation of the ovary: their absence results in the formation of so-called 'streak gonads'. It is possible that they are also irrelevant to ovarian determination, but this is difficult to ascertain without early markers of follicle cell differentiation.

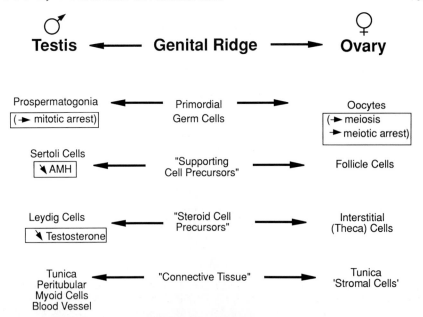

FIG. 2.   Theoretical lineage relationships and cell types making up the testes and ovaries. The female pathway is the normal or default pathway for both somatic and germ cell sex. In the male, Sertoli cells produce anti-Müllerian hormone (AMH) and Leydig cells produce testosterone; these are responsible for the elimination of the anlagen of the female reproductive tract (Müllerian duct derivatives) and development of the male tract (Wolffian duct derivatives), respectively.

Experiments by Burgoyne et al (1988), and more recently by Palmer & Burgoyne (1991), argue that the testis-determining gene acts cell autonomously within a population of cells termed the supporting cell precursors, directing them from the default follicle cell pathway to that of Sertoli cells. This conclusion was based on the finding that XX cells rarely formed Sertoli cells in XX ↔ XY male chimeras. Sertoli cells are then thought to influence the differentiation of the other cell types along the testicular pathway. For example, Leydig cells probably appear a day later, and connective tissue is rapidly organized into the testicular pattern. Importantly, the sex of the germ cells is also determined by their environment. Germ cells enclosed within testis cords enter mitotic arrest, which is characteristic of spermatogenesis, instead of following the default pathway of entering meiosis and meiotic arrest which is characteristic of oogenesis (McLaren 1988).

Subsequent differences between the sexes are due largely to two factors produced by the testes. Sertoli cells make anti-Müllerian hormone (AMH, also known as Müllerian inhibiting substance or MIS), which is responsible for the regression of the Müllerian ducts (Josso & Picard 1986). These would otherwise

166

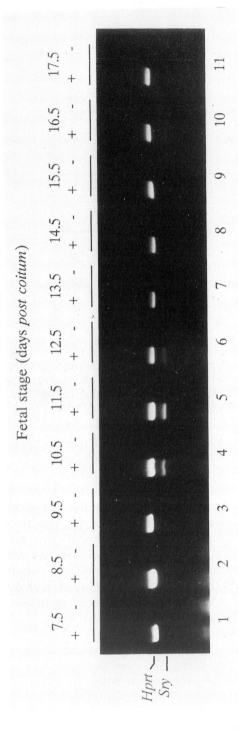

FIG. 3. A. Reverse transcriptase–polymerase chain reaction (RT–PCR) analysis of *Sry* expression during gonadal development demonstrates the presence of *Sry* transcripts for a brief period between 10.5 and 12.5 days of development. RNA was prepared from whole embryos 7.5 and 8.5 days p.c., from posterior portions of embryos 9.5 and 10.5 days p.c., from urogenital ridges of embryos 11.5 and 12.5 days p.c. and from testes of late stages. Primers were included for *Hprt* as a control for quality of RNA in the RT–PCR reactions; + and – refer to the presence or absence of reverse transcriptase as a control for DNA contamination.

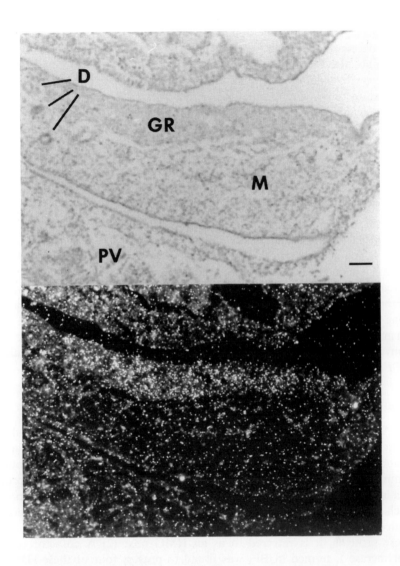

FIG. 3. B. *In situ* hybridization using an antisense RNA probe detects *Sry* transcripts only in the genital ridge of embryos 11.5 days p.c. The signal is weak because of the low abundance of the *Sry* mRNA. *Top*: bright field; *bottom*: dark field view of the same section. GR, genital ridge; M, meso-nephros; D, mesonephric ducts; PV, prevertebrae. Scale bar, 50 µm. For details see Koopman et al (1990).

develop into components of the female reproductive tract (oviducts, uterus and upper vagina). Leydig cells make testosterone, which promotes development of the Wolffian duct system into the male reproductive tract (vas deferens, seminal vesicles, etc) (Grumbach & Ducharme 1960). It can be seen again that the female pathway is the default one, with no special factors required (Wolffian ducts degenerate in the absence of testosterone and Müllerian ducts develop in the absence of AMH).

The problem of sex determination in mammals can therefore largely be reduced to one of finding the trigger that initiates Sertoli cell differentiation in the genital ridge 11.5 days p.c. All subsequent differentiation of male characteristics follows on from the expression of *Tdy*, and female characteristics develop in its absence.

Isolation and properties of *Sry*

The most promising strategy for locating the testis-determining gene was based upon one of deletion mapping the human Y chromosome in a class of XX males that arise from abnormal X:Y interchange at meiosis (Petit et al 1987). Testis-determining activity was found to be localized within approximately 35 kb of Y-unique sequence adjacent to the boundary with the pseudoautosomal region. A search of this region led to the discovery of a conserved sequence that mapped to the Y chromosome of all mammals tested. This sequence formed part of a gene which was named *SRY* (for gene in the <u>s</u>ex determining <u>r</u>egion of the <u>Y</u>) in humans, *Sry* in mice (Sinclair et al 1990, Gubbay et al 1990a). There is now convincing evidence that this gene is genetically and functionally equivalent to *TDF/Tdy*, and I review some of its properties below.

Structure of *SRY/Sry*

DNA sequencing of the human, rabbit and mouse *SRY/Sry* genes revealed the presence of a conserved open reading frame. Within this the region of greatest homology corresponded to a 79 amino acid domain similar to one found in a number of known or suspected DNA-binding proteins. These other proteins fall into essentially three classes. (1) There are a number of non-histone proteins such as *hmg-1* which have essentially no sequence specificity in their interaction with DNA (Einck & Bustin 1985). Because the domain was first recognized in these proteins, it is referred to as an HMG box. (2) A cofactor of RNA polymerase I, termed hUBF, was found to possess four of these HMG box motifs, but its association with its target ribosomal RNA genes shows some degree of sequence specificity (Jantzen et al 1990). (3) An increasingly large class of more *bona fide* transcription factors, including gene products such as TCF-1 and LEF-1 (van de Wetering et al 1991, Travis et al 1991), is being described. These appear to have a much higher affinity for their DNA targets and show greater sequence specificity than the other HMG box-containing proteins. Although the similarity between the HMG boxes of *SRY* and these other proteins is not high (it varies between about 40% and 50%), the HMG box of *SRY* resembles most strongly that of the third class. This suggests that it is a sequence-specific DNA-binding protein and probably a transcription factor. Preliminary studies with *SRY* protein made in *Escherichia coli* show that it will bind to DNA *in vitro* in a sequence-specific manner (Harley et al 1992).

While screening for cDNA clones corresponding to the mouse *Sry* transcript, we obtained a number of clones from an 8.5 days p.c. library that were from genes other than *Sry* (Gubbay et al 1990a). On sequencing, it was found that these clones were isolated as a result of their possession of an HMG box showing extensive homology (greater than 77%) to that found in *Sry*. Three of these genes have been studied in some detail and show dynamic patterns of expression associated mainly with the developing central nervous system (J. Collignon & R. Lovell-Badge, unpublished). Their similarity to *Sry* suggests that they may also be involved in specifying cell fate; if so, this would bear out one of the original reasons for wanting to identify *Tdy*.

Expression of Sry

Northern analyses have revealed *Sry* transcripts of about 1.3 kb in adult mouse testis (Koopman et al 1990) and similar sized transcripts from *SRY* in human testis (Sinclair et al 1990), but failed to detect anything in differen-tiating genital ridges or embryonic testes from mice. Using the much more sensitive reverse transcriptase–polymerase chain reaction (RT–PCR) technique, it was possible to demonstrate that *Sry* transcripts were indeed present for a brief period, just before overt testis differentiation (Koopman et al 1990, and Fig. 3A). Other PCR experiments and *in situ* hybridization studies (Fig. 3B) provide evidence that the gene is expressed only by cells within the genital ridge. Furthermore, *Sry* expression can be seen in genital ridges isolated from 11.5 days p.c. embryos homozygous for W^e, an extreme mutant allele at the *W* locus, which completely lack germ cells. This shows that the gene is expressed by a somatic cell type within the genital ridge, which is an essential condition for the gene to be *Tdy*. The pattern of expression of *Sry* in the embryo is therefore entirely consistent with a role in testis determination. The expression seen in adult testis appears to be confined to postmeiotic germ cells (probably round spermatid stages); for example, it is absent in testes of XX *Sxr* males which are devoid of germ cells, and comes on at about 20 days after birth (Koopman et al 1990, and P. Koopman, unpublished results). However, it is conceivable that the gene has no specific function in spermatogenesis, because germ cells in which *Sry* was deleted, within a male chimera, were able to give rise to offspring (Lovell-Badge & Robertson 1990, Gubbay et al 1990b).

Mutation studies

Proof that *SRY* is normally necessary for testis determination came from studying the gene sequence in *SRY*-positive XY female humans. It seems that about 10–15% of such females possess mutations specifically within the DNA-binding domain of SRY. Some of these are point mutations leading to amino acid substitutions or frame shifts; others involve small deletions. Most of these are *de novo* mutations, because they are not present in the *SRY* gene from the father of the XY female. However, several XY females have now been found whose father and/or brothers have the same amino acid substitution (Berta et al 1990, Jäger et al 1990, and J. R. Hawkins et al, unpublished data). These are unlikely simply to be variants, as similar substitutions have not been found in many control XY males, and it is suspected that they are conditional mutations dependent on some aspect of genetic background. Study of such cases may help towards the identification of autosomal or X-linked genes involved in testis determination.

The remaining 85% or so of cases of *SRY*-positive XY females could be due to mutations affecting the structure of the gene product outside the HMG box, to mutations affecting the regulation of the gene, or to mutations in genes elsewhere in the testis-determining pathway.

In the mouse, there is no evidence yet of specific point mutations affecting *Sry* structure. However, *Sry* appears to be the only gene deleted in a line of XY females that had previously been shown by genetic means to be mutant in *Tdy* (Lovell-Badge & Robertson 1990, Gubbay et al 1990b). Recent results suggest that this deletion is of only about 11 kb around the *Sry* gene (J. Gubbay et al, unpublished).

Transgenic studies

Final proof that *Sry* is the only gene from the Y chromosome necessary for testis determination has come from transgenic experiments (Koopman et al 1991). A 14 kb genomic fragment carrying the mouse *Sry* gene was injected into fertilized mouse eggs which were then reimplanted and allowed to develop *in utero* for about 14 days. Initially, embryos were sexed by gonad morphology; then their chromosomal sex was determined by looking for sex chromatin, indicative of two X chromosomes, in amnion cells. Two embryos were found that were anomalous in that testes were developing despite the fact that they were chromosomally female. Southern blot analysis confirmed that they lacked a Y chromosome, but possessed copies of *Sry* as a transgene. Detailed sequence analysis of the 14 kb genomic fragment and cross-hybridization studies with human DNA failed to reveal the presence of any other genes within the injected DNA, so it can be concluded that *Sry* alone is able to initiate male development in an otherwise chromosomally female background.

TABLE 1 **Summary of data from the analysis of mouse fragment 741 transgenic embryos**

No. of embryos	Sex chromatin	Sry	Zfy	Deduced karyotype	Transgenic	Phenotypic sex
63	+	−	27-/40 ND	XX	−	♀
27	−	+	+	XY	ND	♂
58	−	ND	ND	XY	ND	♂
2	−	−	−	XO	−	♀
6	+	+	−	**XX**	+[a]	♀
2	+	+	−	**XX**	+	♂

A total of eight chromosomally female transgenic mouse embryos were identified at about 14 days after injection of the 14 kb genomic DNA fragment, clone 741 (see Fig. 4). Two of these showed testicular development and are therefore classified as XX males. A further six were normal females. Of these, four[a] were mosaic, as they had less than one copy of the transgene per cell. (See Koopman et al 1991 for details.) *Zfy*, Y-linked gene sequences (assay for presence of Y chromosome); ND, not determined.

To determine the frequency with which this 14 kb DNA fragment containing *Sry* was able to produce sex reversal, we examined all the female embryos for the presence of the transgene. Six embryos were found that possessed *Sry* sequences (Table 1). There are two simple reasons to explain why the transgene may have failed to reverse the sex of the embryo in these cases. It is known from chimera and mosaic studies that more than about 25% of the somatic cells of the genital ridge have to carry *Tdy* for the gonad to develop as a testis. Four of the six transgenic females had less than one copy per cell of *Sry*, suggesting that they were in fact mosaic, a common finding for transgenics; so perhaps the *Sry* gene was present in too few cells in these embryos. Alternatively, it is known that transgenes often show very different levels of expression from one line to another, probably owing to the site of integration. Normally, this is inconsequential to the experiment because any expression is sufficient to determine, for example, tissue specificity. However, in the case of the *Sry* transgenics, it is function that is being assayed, and the level of expression may be vital for this.

Some of the injected embryos were allowed to develop to term and a number of adult transgenic mice were obtained (Fig. 4). One of these, m33.13, was chromosomally female, but had a normal external male phenotype (Fig. 5). This animal exhibited apparently normal male reproductive behaviour, but was sterile, as would be expected for an XX male. On internal examination he was found to have a normal male reproductive tract (Fig. 6), with no signs of hermaphroditism. This indicates that Sertoli cells and Leydig cells must have been functionally normal during embryonic development of m33.13, at least in terms of AMH and testosterone production. However, his testes were considerably smaller than those of control XY littermates. Histological

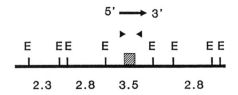

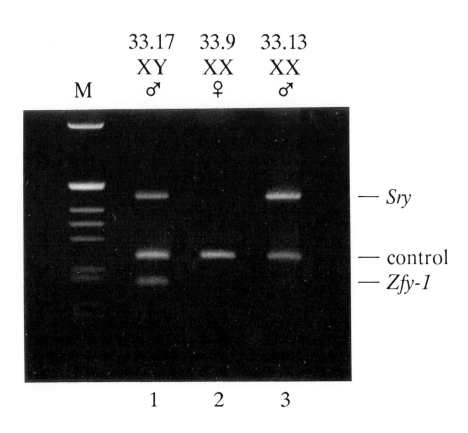

FIG. 4.   Identification of mice transgenic for the mouse *Sry* gene. The top panel shows the 14 kb genomic DNA fragment (f741) used for pronuclear injection. The shaded box represents the conserved DNA-binding domain of *Sry*. E, *Eco*R1 restriction sites (fragment sizes are indicated in kb below); ▶◀, position of primers used for PCR analysis. The lower panel shows PCR analysis of DNA from three littermates obtained from pronuclear injection with f741. An *Sry*-specific band is seen on the ethidium-stained agarose gel with either the endogenous gene or the transgene. Primers to *Zfy-1* were included to reveal the presence or absence of the Y chromosome and primers to *MyoD* were included as a control for DNA quality. M, marker bands. This analysis reveals animal m33.13 to be an XX male transgenic mouse. For details see Koopman et al (1991).

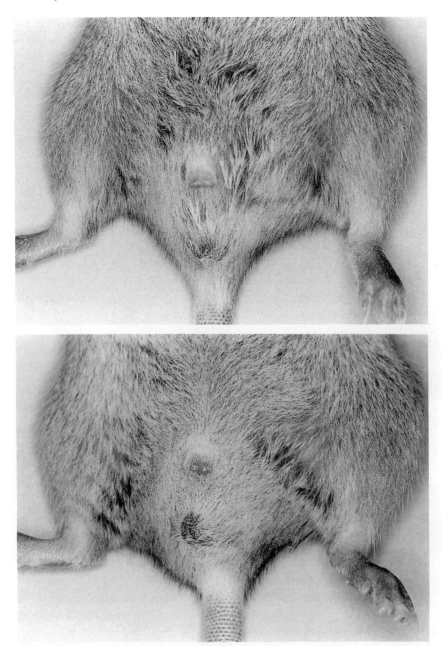

FIG. 5.   External genitalia of XX male m33.13 (*upper panel*) and a control, non-transgenic XY male littermate (*lower panel*). No abnormalities of male development could be seen.

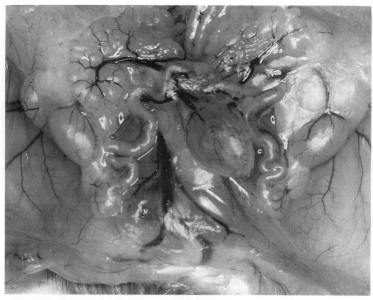

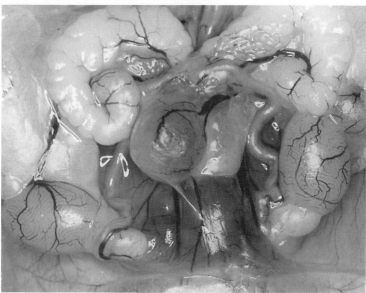

FIG. 6.   Reproductive tract and testes of XX male m33.13 (*upper panel*) and a control, non-transgenic XY male littermate (*lower panel*). The only significant difference seen is the small size of the testes in the sex-reversed male. The latter had mated during the preceding night which probably accounts for the different appearance of the seminal vesicles compared to the control.

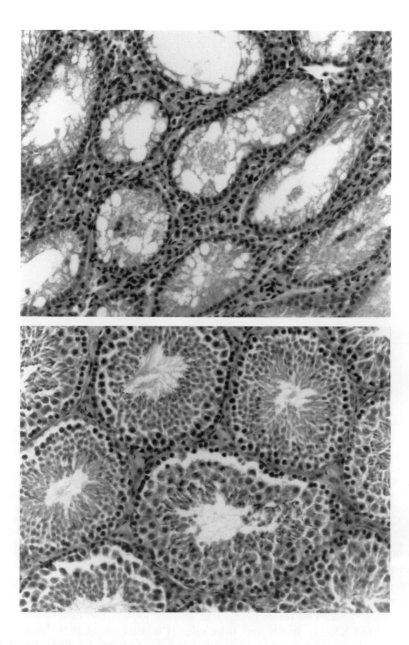

FIG. 7.   Section through testes of XX male m33.13 (*upper panel*) and the control XY male (*lower panel*) reveal the complete lack of spermatogenesis in the former. Other cell types and testicular organization appear normal.

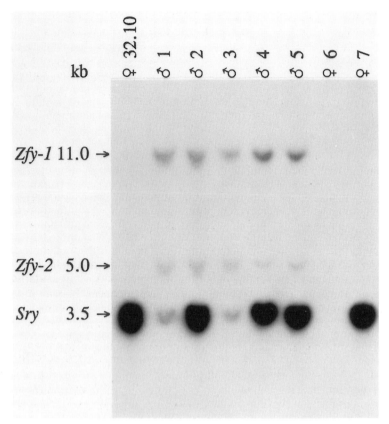

FIG. 8. Southern blot analysis of DNA from founder XX female m32.10 and some of her offspring, showing transmission of the transgene to both male and female offspring. The transgenic daughter cannot be mosaic, suggesting that her failure to undergo sex reversal is due to insufficient or inappropriate expression of the transgene. *Sry* and *Zfy-1* probes were combined, the former revealing endogenous *Sry* (single copy) and the transgene (10–20 copies); the latter reveals the presence or absence of a Y chromosome. See Koopman et al (1991) for details.

examination of the testes revealed an essentially normal structure and the presence of all somatic cell types, but with a complete lack of spermatogenesis (Fig. 7). This is due to several factors, most notably the possession of two X chromosomes and the absence of Y chromosomal genes involved in spermatogenesis. The testes of m33.13 in fact looked identical to those of sex-reversed XX *Sxr* or XX *Sxr'* male mice (Cattanach et al 1971, Sutcliffe & Burgoyne 1989). This result proves that *Sry* is functionally equivalent to *Tdy*, and indicates that all other genes required for the development of the male phenotype reside on chromosomes other than the Y.

Two adult female mice were also obtained that carried multiple copies of *Sry* as a transgene. One of these has transmitted the transgene to her offspring, in which it has also failed to cause sex reversal (Fig. 8). Because the offspring cannot be mosaic for the transgene, it seems likely in this case that it is not being expressed appropriately. Preliminary results suggest that the transgene is expressed at a lower level than normal. Careful measurements of the number of transcripts over the time course of development will be required before any conclusion can be reached.

Summary and conclusions

With the isolation and characterization of *Sry*, a gene with a critical role in development has been defined. It can be described as a master regulatory gene in that it has a pivotal role in the sex determination pathway. However, much has to be done to find out how the gene fits into such a pathway. First, its precise pattern of expression suggests that it itself must be regulated by tissue-specific transcription factors, or by a unique combination of such factors, present within the supporting cell precursor lineage. These factors are clearly not sex specific. Secondly, there are likely to be gene products that interact with the *Sry* protein in order to generate the specificity of its interaction with target genes. Lastly, there are the downstream target genes themselves that need to be defined. *Sry* appears to act only for a very brief period just before testis cord formation. It is therefore not required for any long-term maintenance of gene activity; however, it must initiate (or repress) the expression of some critical genes. The gene for AMH has long been thought to be a reasonable target for regulation by *Tdy*. In the embryo it is expressed uniquely by Sertoli cells, mRNA being first detectable at about 12.5 days p.c. (Münsterberg & Lovell-Badge 1991). This is perhaps a little late, because it means that AMH expression does not overlap with *Sry* expression at all, so there could be an intermediate. Also, AMH is expressed by follicle cells in the postnatal ovary, where *Sry* cannot be required to trigger its expression. Determination of the nature of the interaction of *Sry* with DNA, and the definition of target genes by rigorous methods, will clearly be required before the role of *Sry* in sex determination can be understood at the molecular level.

References

Baker BS 1989 Sex in flies: the splice of life. Nature (Lond) 340:521–524
Berta P, Hawkins JR, Sinclair AH et al 1990 Genetic evidence equating SRY and the male sex determining gene. Nature (Lond) 348:248–250
Burgoyne PS, Palmer SJ 1991 The genetics of XY sex reversal in the mouse and other animals. Semin Dev Biol 2:277–284

Burgoyne PS, Buehr M, Koopman P, Rossant J, McLaren A 1988 Cell-autonomous action of the testis-determining gene: Sertoli cells are exclusively XY in XX ↔ XY chimaeric mouse testes. Development 102:443–450

Cattanach BM, Pollard CE, Hawkes SG 1971 Sex reversed mice: XX and XO males. Cytogenetics 10:318–337

Einck L, Bustin M 1985 The intracellular distribution and function of the high mobility group chromosomal proteins. Exp Cell Res 156:295–310

Ferguson MWJ, Joanen T 1982 Temperature of egg incubation determines sex in *Alligator mississippiensis*. Nature (Lond) 296:850–853

Ford CE, Jones KW, Polani PE, De Almeida JC, Briggs JH 1959 A sex-chromosome anomaly in a case of gonadal dysgenesis (Turner's syndrome). Lancet 1:711–713

Ginsburg M, Snow M, McLaren A 1990 Primordial germ cells in the mouse embryo during gastrulation. Development 110:521–528

Grumbach MM, Ducharme JR 1960 The effects of androgens on fetal sexual development: androgen-induced female pseudo-hermaphroditism. Fert Steril 11:157–180

Gubbay J, Collignon J, Koopman P et al 1990a A gene mapping to the sex-determining region of the mouse Y chromosome is a member of a novel family of embryonically expressed genes. Nature (Lond) 346:245–250

Gubbay J, Koopman P, Collignon J, Burgoyne P, Lovell-Badge R 1990b Normal structure and expression of *Zfy* genes in XY female mice mutant in *Tdy*. Development 109:647–653

Harley VR, Jackson DI, Hextall PJ et al 1992 Sequence-specific DNA binding by the human sex determining gene product, *SRY*, is abolished by mutation in some XY females. Science (Wash DC) 255:453–456

Jacobs PA, Strong JA 1959 A case of human intersexuality having a possible XXY sex-determining mechanism. Nature (Lond) 183:302–303

Jäger RJ, Anvret M, Hall K, Scherer G 1990 A human XY female with a frame shift mutation in *SRY*, a candidate testis determining gene. Nature (Lond) 348:452–454

Jantzen H-M, Admon A, Bell SP, Tjian R 1990 Nucleolar transcription factor hUBF contains a DNA-binding motif with homology to HMG proteins. Nature (Lond) 344:830–836

Josso N, Picard JY 1986 Anti-Mullerian hormone. Physiol Rev 66:1038–1090

Jost A 1947 Recherches sur la différenciation sexuelle de l'embryon de lapin. Arch Anat Microsc Morphol Exp 36:271–315

Koopman P, Münsterberg A, Capel B, Vivian N, Lovell-Badge R 1990 Expression of a candidate sex-determining gene during mouse testis differentiation. Nature (Lond) 348:450–452

Koopman P, Gubbay J, Vivian N, Goodfellow P, Lovell-Badge R 1991 Male development of chromosomally female mice transgenic for *Sry*. Nature (Lond) 351:117–121

Kushner PJ, Blair LC, Herskowitz I 1979 Control of yeast cell types by mobile genes: a test. Proc Natl Acad Sci USA 76:5264–5268

Lovell-Badge RH, Robertson EJ 1990 XY female mice resulting from a heritable mutation in the murine primary testis determining gene, *Tdy*. Development 109:635–646

McLaren A 1985 Relating of germ cell sex to gonadal differentiation. In: Halvorsen HO, Monroy A (eds) The origin and evolution of sex. Alan R. Liss, New York, p 289–300

McLaren A 1988 Somatic and germ-cell sex in mammals. Philos Trans R Soc Lond B Biol Sci 322:3–9

Münsterberg A, Lovell-Badge R 1991 Expression of the mouse anti-mullerian hormone gene suggests a role in both male and female sexual differentiation. Development 113:613–624

O W-S, Short RV, Renfree MB, Shaw G 1988 Primary genetic control of somatic sexual differentiation in a mammal. Nature (Lond) 331:716–717

Palmer J, Burgoyne PS 1991 *In situ* analysis of fetal, prepuberal and adult XX ↔ XY chimaeric mouse testes: Sertoli cells are predominantly, but not exclusively, XY. Development 112:265–268

Petit C, de la Chapelle A, Levilliers J, Castillo S, Noel B, Weissenbach J 1987 An abnormal terminal X–Y interchange accounts for most but not all cases of human XX maleness. Cell 49:595–602

Sinclair AH, Berta AP, Palmer MS et al 1990 A gene from the human sex-determining region encodes a protein with homology to a conserved DNA-binding motif. Nature (Lond) 346:240–244

Sutcliffe MJ, Burgoyne PS 1989 Analysis of the testes of H-Y negative XOSxrb mice suggests that the spermatogenesis gene (*Spy*) acts during the differentiation of the A spermatogonia. Development 107:373–380

Travis A, Amsterdam A, Belanger C, Grosschedl R 1991 LEF-1, a gene encoding a lymphoid-specific protein with an HMG domain, regulates T-cell receptor alpha enhancer function. Genes Dev 5:880–894

van de Wetering M, Oosterwegel M, Dooijes D, Clevers H 1991 Identification and cloning of TCF-1, a T lymphocyte-specific transcription factor containing a sequence-specific HMG box. EMBO (Eur Mol Biol Organ) J 10:123–132

Welshons WJ, Russell LB 1959 The Y chromosome as the bearer of male determining factors in the mouse. Proc Natl Acad Sci USA 45:560–566

DISCUSSION

Hogan: What is the evidence that entry into meiosis is the default pathway of germ cells?

Lovell-Badge: Some germ cells fail to enter the genital ridge during migration. If, for example, they end up in the adrenal in an XY embryo, the germ cells will enter meiosis. It is only when they are enclosed in testis cords that they don't.

McLaren: If very early genital ridges (10.5 days p.c.) are removed from the embryos and maintained *in vitro*, the germ cells will enter meiosis within a few days, whether they are XX or XY in chromosome constitution (McLaren & Buehr 1990).

Robertson: Robin, have you looked at spatial and temporal patterns of expression of the other autosomal *Sry*-related genes? Are these genes conserved in humans and other vertebrates?

Lovell-Badge: The autosomal genes with a similar DNA-binding domain are conserved in humans. They are also conserved in birds, *Xenopus* and reptiles. There are at least six genes in this family, plus *Sry*. We've done preliminary expression studies on three: we haven't seen any expression of those three in genital ridge. They are expressed in other sites fairly early, particularly in the developing central nervous system.

Robertson: What's the degree of sequence conservation outside the HMG box?

Lovell-Badge: Between *Sry* and the other genes there is none. Among those other genes it varies. There is a family of genes we are calling the *Sox* family: *Sox-1* and *Sox-3* are fairly similar outside the DNA-binding domain; they are different from *Sox-2*; *Sox-4* looks very different again.

Goodfellow: The other transcription factors are a long way away from the *Sry* family. There is only 40% homology between *TCF-1*, *LEF-1* and autosomal *Sry*, whereas among the *Sry* family there is 70% homology or more.

Jaenisch: Is there any reason to assume there is a gene upstream of *Sry*?

Lovell-Badge: I don't think there is any evidence to suggest that there has to be a specific upstream gene, except that we have shown that *Sry* is precisely regulated. There are clearly autosomal and/or X-linked genes that are involved in sex determination, but we can't say whether these are upstream or downstream. For example, studies on wood lemmings have shown that there is an X-linked gene involved in sex determination. Wood lemmings have two types of X chromosome: there are normal XY males and X*Y females. The X* chromosome always gives females, so even an X*YY individual will be female. There is some rearrangement on the X* chromosome which may be affecting a particular gene.

Goodfellow: In humans, XX males without Y sequence could be explained by gain-of-function mutations in genes downstream of *SRY*. Of the XY females, 15% or more are mutant in *SRY*. There were some patients in America who were considered to have pure gonadal dysgenesis by their clinical criteria. Among those cases, the proportion went up to three out of five instead of the one out of 10 we were seeing previously. So if you check very carefully that there is no evidence of testicular structure, you may be able to identify those with mutations in *SRY*. The XY females who are not mutant in *SRY* are obviously candidates for carrying mutations either upstream or downstream in the sex-determining pathway.

Recently, we re-discovered something that was first published in 1980 by Bernstein et al. The original patients were XY females with an interstitial duplication of part of Xp. Dr T. Ogato, in my lab, has been studying a patient where a partial duplication of Xp is transferred onto the Y chromosome. In this patient *SRY* is perfectly normal. Our hypothesis is that there's a gene, probably downstream of *SRY*, which is affected by gene dosage, because the only difference between someone who is XXY and someone who is XY with a duplicated Xp is that in the XXY you would expect an inactivated gene to be switched off, whereas it wouldn't be switched off in the duplication of Xp. That's interesting, because it might just be a remnant of linking sex determination with X inactivation and dosage compensation.

Hastie: From the XX males who don't have any Y sequences: if that is a downstream effect, either there is only one essential downstream gene or there is ectopic expression of something else which can then do the job of the gene on the Y chromosome. Is that reasonable?

Lovell-Badge: Yes, that's reasonable.

McLaren: In the mouse, it seems that there must be at least one essential gene downstream of *Sry* in the sex determination pathway. Sertoli cells in XX/XY chimeras develop almost entirely from the XY component, unlike other cell types in the testis and elsewhere (Burgoyne et al 1988). This indicates that *Sry* is expressed in Sertoli cell precursors cell autonomously, as would be expected of a gene encoding a DNA-binding protein. A few XX cells are, however, included in the Sertoli cell population (Palmer & Burgoyne 1991), suggesting the existence of a downstream gene switched on by the *Sry* gene product that has a paracrine recruitment effect into the Sertoli cell lineage. Such a gene would be a candidate for what Nick is suggesting.

Lovell-Badge: We also know from freemartin effects that AMH can have a profound effect in terms of conferring male characteristics on an XX individual. This could be a direct effect on follicle cells or supporting cell precursors, or an indirect effect through the elimination of germ cells. Again, you can end up with XX Sertoli cells through the action of AMH.

Goodfellow: It's worth emphasizing the apparent temporal gap between *Sry* expression and the gene for AMH being switched on.

Lovell-Badge: We can't tell at the moment whether there is a direct interaction between *Sry* and AMH. Using PCR we see *Sry* mRNA rise and then fall. *In situ* hybridization shows AMH mRNA rising later and staying at a high level. PCR might show a low level of AMH at the earlier time.

Goodfellow: Or the protein might persist.

McLaren: It seems unlikely that AMH would ever produce full sex reversal. It produces some degree of masculinization in freemartins, but they are masculinized females rather than males.

Hastie: People forget that when you look at time courses, the time it takes for the protein to reach saturation depends on the half-life of the molecule. You can be misled sometimes by comparing when the peak concentrations of mRNAs occur.

Buckingham: Robin, have you tried the rabbit gene in the transgenic experiments?

Lovell-Badge: We have one XX transgenic with the rabbit gene, which is female. So far, the human gene has also been unable to produce an XX male mouse.

Hastie: The failure of the human sequence to work in the transgenic mice is perhaps not surprising. Eva Eicher showed years ago that the Y chromosome of a different mouse species could not function to give a normal testis on an inbred background (Eicher et al 1982).

Lovell-Badge: That story is about the *Mus poschiavinus* Y chromosome, which on a C57BL/6 background failed to give testes in all cases. There were some XY males, some XY hermaphrodites and some XY females. Recent studies by Steve Palmer and Paul Burgoyne suggest that it's a timing problem. With the

M. poschiavinus Y chromosome, testicular cord formation is delayed relative to that with a normal *M. musculus* Y chromosome by about 14 hours. So we believe that it is a combination of a late-acting testis-determining gene and a genetic background which provides an early ovarian signal (if such a thing exists), that leads to XY females in this case.

Buckingham: Have you compared the exonic/intronic structure of the mouse *Sry* gene with that of the human *SRY* gene? In most genes that kind of structure is conserved between humans and mice.

Lovell-Badge: For *Sry/SRY* the structure is completely different. The only argument we have is that as this gene is on the Y chromosome, there are few constraints to prevent it diverging.

Goodfellow: The thing which looks like an intron is pretty garbagey! It looks as though something has inserted into the gene. The rabbit and human *SRY* genes are closely related but the mouse *Sry* gene is more distantly related; we don't know why this is so. Rodents appear to have very rapid evolution of their sex chromosomes. For some reason, they seem to be under fewer constraints than other mammals when it comes to the chromosomal basis of sex determination. It is a weak argument, but that's what you observe.

McLaren: Marsupials have a very minute Y chromosome. Is their Y chromosome sex determining, and does it contain an *Sry* homologue?

Lovell-Badge: The evidence seems pretty good that the Y chromosome in marsupials is sex determining; but, it's not the whole story. Some male and female characteristics are present before gonadal differentiation, and this could be explained if the X:A ratio had an effect on, for example, the development of pouch versus scrotum but, unfortunately, very few aneuploids have been looked at. However, the Y chromosome is necessary for testis determination. There appears to be a Y-linked *Sry* homologue but it has not yet been cloned.

References

Bernstein R, Jenkins T, Dawson B et al 1980 Female phenotype and multiple abnormalities in sibs with a Y chromosome and partial X chromosome duplication: H-Y antigen and Xg blood group findings. J Med Genet 17:291–300

Burgoyne PS, Buehr M, Koopman P, Rossant J, McLaren A 1988 Cell-autonomous action of the testis-determining gene: Sertoli cells are exclusively XY in XX ↔ XY chimaeric mouse testes. Development 102:443–450

Eicher EM, Washburn LL, Whitney JB III, Morrow KE 1982 *Mus poschiavinus* Y chromosome in the C57BL/6J murine genome causes sex reversal. Science (Wash DC) 217:535–537

McLaren A, Buehr M 1990 Development of mouse germ cells in cultures of fetal gonads. Cell Differ & Dev 31:185–195

Palmer J, Burgoyne PS 1991 *In situ* analysis of fetal, prepuberal and adult XX ↔ XY chimaeric mouse testes: Sertoli cells are predominantly, but not exclusively, XY. Development 112:265–268

Epithelial-mesenchymal interactions in murine organogenesis

Lauri Saxén* and Irma Thesleff†

*Department of Pathology, University of Helsinki, Haartmaninkatu 3, SF-00290 Helsinki and †Department of Pedodontics and Orthodontics, University of Helsinki, Mannerheimintie 172, SF-00300 Helsinki, Finland

Abstract. Reciprocal, sequential interactions between embryonic epithelia and their mesenchymal stroma guide the cytodifferentiation and organization of both components. These morphogenetic interactions and their consequences are examined in two model systems *in vitro*: the mouse metanephric blastema and the tooth rudiment. Experimental approaches include dissection and recombination of the interacting tissues, localization of molecular changes by immunohistology and *in situ* hybridization. An early response of the mesenchyme is increased proliferation of cells in the vicinity of the epithelial inductor and their subsequent aggregation (condensation). In the kidney model disruption of this aggregation or prevention of assembly of the programmed cells results in impaired cytodifferentiation. If the cells are allowed to reaggregate, a phenotype is expressed not unlike that seen in normal *in vivo* conditions. Our present interest is focused on the early metabolic events associated with the condensation phenomenon. The cell surface proteoglycan syndecan and the matrix glycoprotein tenascin are expressed in the condensed mesenchyme and may mediate cell–matrix interactions. The expression patterns of certain growth factors suggest functions in signal transduction.

1992 Postimplantation development in the mouse. Wiley, Chichester (Ciba Foundation Symposium 165) p 183–198

Epithelia derived from all three germ layers coat the outer surface and line the inner cavities of vertebrates and show a remarkable repertoire of functionally and morphologically specialized cell types. Diversification and organization of epithelial cells are associated with differentiation of their mesenchymal (stromal) counterpart, and both contribute to the ultimate development of organs. The prerequisite for spatially and temporally synchronized development of the two cell lineages is an informative communication, a continuous dialogue, between them. This has been experimentally demonstrated in a great variety of tissues and organs, e.g. in the skin and its appendages (Sengel 1990), in many glandular organs including the lung, the pancreas, and the salivary and mammary glands (Bernfield et al 1984, Kratochwil 1987), and in the intestinal and genitourinary systems (Mizuno & Yasuki 1990, Cunha et al 1991).

Thus far, isolation and characterization of signal transmitters in epithelial–mesenchymal interactions have not been successful; exposure of the responding tissues to known molecules involved in embryogenesis has yielded inconclusive results. This is not unexpected because the outcome of inductive interactions—monitored as new functional or structural phenotypes and organized tissues—is undoubtedly a consequence of a cascade of such events. Each interaction leads to a new, often covert change in the responding cells and is immediately followed (or even overlapped) by other post-inductory changes. Causal linkage of these, only partially known events of determination and differentiation is an ultimate though distant goal of research in this field and should lead to the identification of the so far hypothetical signal molecules.

In this paper we will explore the early post-inductory events in our model systems and speculate about their causal relationships, though we are fully aware of the complexity of the metabolic network triggered during any morphogenetic cell interaction. On the other hand, the availability of new and sensitive methods to examine small tissue fragments and their metabolism means that the classic problems can now be revisited.

Two model systems of organogenesis

The metanephric kidney initially develops from two cell lineages, the ureter epithelium and the mesenchymal blastema. A reciprocal interaction between these components is a prerequisite for the regular branching of the ureter and for the epithelial conversion of the mesenchyme (Saxén 1987, Fig. 1A). The teeth likewise originate from two cell lineages: the neural crest-derived jaw mesenchyme and the presumptive dental epithelium which interact in a reciprocal and sequential manner (Thesleff & Hurmerinta 1981, Fig. 1B). Experimental separation of the two components followed by recombination to form homo- and heterotypic tissues allows various modifications of the interactive processes to be studied *in vitro*. A useful tool is the trans-filter technique where a porous filter is placed between the two tissues. These can then be examined separately at different post-inductory stages and kinetic studies can be performed (Grobstein 1956, Saxén & Lehtonen 1978, Thesleff & Hurmerinta 1981).

Directive and permissive inductions

It has become customary to distinguish between directive (instructive) and permissive inductions. 'When an embryonic cell possesses more than one developmental option, the choice between them is affected by extracellular factors, which thus exert a true directive action on differentiation. A permissive action, on the other hand, refers to a step of development, in which the cell has become committed to a certain pathway, but still requires an exogenous stimulus to express its phenotype' (Saxén 1977). As cells and tissues undergo

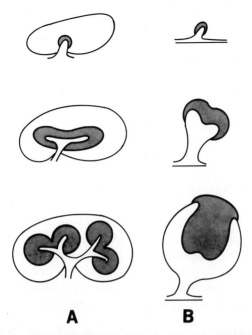

FIG. 1. A scheme of the two model systems investigated. (A) The mouse metanephric kidney on embryonic Days 11 to 13. (B) The mouse tooth rudiment on embryonic Days 11 to 16. The condensing mesenchyme is shaded; the epithelial 'inductor' is white.

a series of directive and permissive interactions, their competence, the ability to react to inductive signals, is constantly changing; when a new phenotype is expressed, cells lose their capacity to respond to certain determinative stimuli but retain (or gain) a state where new signals can be received and meaningfully interpreted.

In the two interactive systems explored by us, both directive and permissive interactions can be shown. The metanephric mesenchyme responds to a great variety of heterotypic tissues by tubule formation (Saxén 1987). Yet, only the predetermined blastema cells respond properly; all other embryonic mesenchymes so far tested are unresponsive. This interaction thus represents a permissive type. A directive mode of action can be demonstrated in the tooth model. In the tooth rudiment, the odontogenic potential has been shown to reside primarily in the epithelium. It subsequently becomes transferred to the mesenchymal component which thereafter acts as a directive inductor (Mina & Kollar 1987). The following progressive differentiation of the tooth, including sequential expression of new proteins, suggests further inductive processes (Thesleff et al 1990).

Transmission of signals between epithelia and mesenchyme

Until the molecules transmitting inductive signals are characterized, their localization and transmission in developing tissues can be followed only indirectly, by monitoring the response. The trans-filter technique has proven useful in such studies; it can be employed to distinguish between major modes of transmission, i.e. through intimate cell-to-cell apposition or via long-range diffusion of the signal substances.

The induction of kidney tubules belongs to the former category, as interposed filters block induction, if they prevent cytoplasmic processes from making contact between the interactants (Lehtonen 1976). Quantitative evaluation of the response through filters that allow such contacts shows that induction is a function of the extent and duration of intimate association of the cells (Saxén & Lehtonen 1978). The trans-filter technique has further allowed us to explore the kinetics of the interactive events in the kidney. In this *in vitro* system the first mesenchymal cells become irreversibly determined after 12 hours of trans-filter contact, and the induction is completed some 12 to 16 hours later (Saxén & Lehtonen 1978). Interruption of the interaction at that stage no longer alters the subsequent differentiation or segregation of the induced tubules (Ekblom et al 1981).

None of the experiments suggesting the importance of close cell–cell apposition has demonstrated an actual transfer of molecules from the inductor to the responding cell (e.g. via specialized membrane structures). As pointed out by Gurdon (1987) and others, a permissive epithelial–mesenchymal interaction could be implemented by relatively simple, non-informative compounds of the extracellular matrix. Such modulating effects become, however, less likely in instructive inductions that lead to gene activation in the target cells (Takeda et al 1990).

There is now substantial evidence for growth factors mediating inductive signals during primary mesoderm formation in *Xenopus laevis* (New & Smith 1990), particularly factors that belong to either the TGF-β superfamily or the fibroblast growth factor family. In *Drosophila* embryos also, a member of the TGF-β family mediates inductive interactions across germ layers (Panganiban et al 1990). Direct evidence for the role of growth factors in the transfer of signals during epithelial–mesenchymal interactions regulating organ development in vertebrate embyros is still lacking. However, the amount of circumstantial evidence is constantly increasing. TGF-β has been implicated in the development of several organs, the evidence deriving mainly from *in situ* hybridization analyses and partly from the localization of the protein by immunohistological methods (Thompson et al 1989, Pelton et al 1990, Potts et al 1991).

The pattern of TGF-β1 expression during early tooth development is suggestive of a role in signal transduction. Our *in situ* hybridization analysis has shown that TGF-β expression appears in a restricted area at the tip of the invaginating

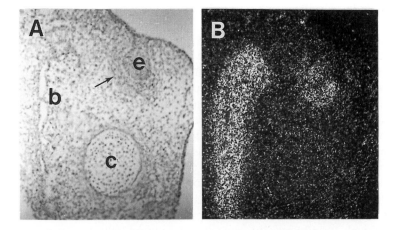

FIG. 2.   Distribution of TGF-β1 transcripts in bud-stage tooth germs shown by *in situ* hybridization. In the transverse section of the lower jaw, the transcripts are presented both in the dental epithelium (e) and in the condensed mesenchyme (arrow). Intense expression is evident in the developing bone (b), but not in Meckel's cartilage (c). (A) Bright field; (B) dark field illumination of the same section.

dental epithelium during the bud stage (Vaahtokari et al 1991). This expression is then rapidly extended in the epithelium as well as into the condensing mesenchymal cells around the epithelial bud (Fig. 2). It is an interesting possibility that the mesenchymal expression results from an autoinductive process, i.e. that epithelium-derived TGF-β induces its own expression in the mesenchyme and acts as a true morphogen. The assumption that the target tissue of epithelial TGF-β is the mesenchyme is supported by immunohistochemical localization studies which have revealed extracellular TGF-β in the mesenchymal component of many developing organs (Thompson et al 1989).

The known biological activities of TGF-β are well suited to a role in inducing developmental changes in the mesenchymal tissue. TGF-β enhances the accumulation of extracellular matrix by several mechanisms and increases mesenchymal cell proliferation, probably via induction of other growth factors such as platelet-derived growth factor (Moses et al 1990). In particular, TGF-β regulates the expression of both the matrix glycoprotein tenascin and the matrix receptor syndecan, which accumulate in the condensed dental mesenchyme (see below).

Hence, our studies as well as those from other laboratories support the view that epithelial TGF-β may act as a signal molecule during morphogenetic tissue interactions. It may exert both paracrine and autocrine effects on the organ-specific mesenchymal cells and its effect may be modulated and/or mediated by the extracellular matrix.

Aggregation of cells during organogenesis

In both the model systems, like in many other steps during early embryogenesis and subsequent organogenesis, an early detectable consequence of an induction is condensation or aggregation of the induced mesenchymal cells. Cells committed to the same developmental pathway recognize each other and form a distinct colony within which they show 'homotypic interactions'.

An early aggregation and its significance for further cytodifferentiation can be demonstrated in the kidney model by the following experiment: the responding metanephric mesenchyme is induced through a filter for 24 hours to allow complete induction, whereafter the cells are separated from the inductor and dissociated into single-cell suspensions. Development of these programmed cells then proceeds either in monolayer cultures that prevent aggregate formation or in hanging-drop cultures that allow the dispersed cells to form three-dimensional assemblies. The capacity of induced cells to form pretubular aggregates in the hanging drop cultures indicates that they have acquired two new characteristics during the 24 hour induction: they have developed mechanisms by which they recognize similarly programmed cells and adhere to them. These induced features are essential for further cytodifferentiation, as shown by comparisons between the aggregated cells and their monolayer controls: monitored by several criteria, the aggregated cells regularly manifest a differentiated phenotype, whereas these properties were severely confused in the dispersed cells (Saxén et al 1988 and unpublished results).

The molecules that mediate this early post-inductory aggregation are not known, but some candidates can be tested. The extracellular matrix of the metanephric mesenchyme does change after induction; fibronectin and interstitial collagens are lost and epithelial matrix proteins such as type IV collagen, the A-chain of laminin and uvomorulin are expressed (Ekblom 1989). Most of these changes seem, however, to be associated with later stages of tubulogenesis, especially with polarization of the aggregated cells. Syndecan is a heparan sulphate-proteoglycan which spans the cell surface and acts as a matrix receptor (Bernfield & Sanderson 1990). During early kidney development, both syndecan core protein and its mRNA can be demonstrated in the condensed mesenchyme and pretubular aggregates (Vainio et al 1989a and unpublished). Syndecan is also expressed in and around the experimental reaggregates (Fig. 3). Interestingly, when mouse and rat mesenchymes are dissociated and mixed after induction, our antibody detects syndecan only in mouse-derived aggregates. Hence, we have a species-specific compound that has adhesive properties associated with aggregation.

In the developing tooth also, the condensation of mesenchymal cells around the epithelial bud is accompanied by intense expression of syndecan (Thesleff et al 1988, Fig. 4A). We have examined whether syndecan mediates a cell–matrix interaction and is involved in the compartmentalization of the dental

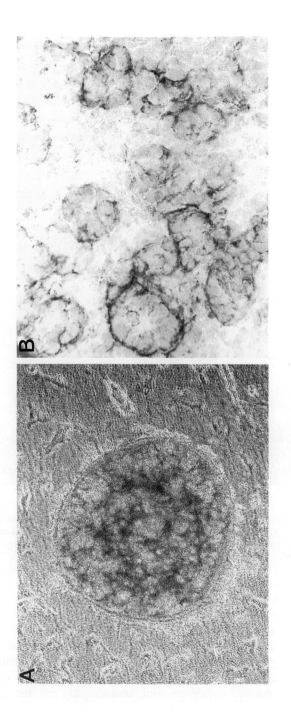

FIG. 3.   An aggregate of nephrogenic mesenchymal cells induced through a filter for 24 hours, then dissociated and cultivated in a hanging drop. (A) A living culture on Day 5 of cultivation showing the abundant pretubular aggregates. (B) A section of an aggregate on Day 3 treated with anti-syndecan antibody and stained with the peroxidase technique. The aggregates and some aggregating single cells are surrounded by the proteoglycan syndecan, which is not detectable in uninduced stromal cells.

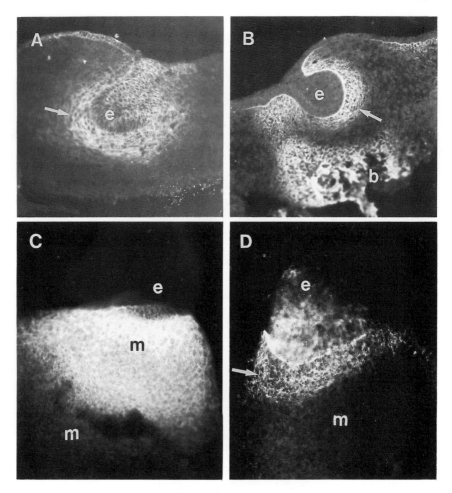

FIG. 4.   Immunofluorescent localization of the cell surface proteoglycan syndecan (A,C) and the extracellular matrix glycoprotein tenascin (B,D) in bud-stage tooth germs *in vivo* (A,B) and in cultured recombinants of presumptive dental epithelium and mesenchyme (C,D). Accumulation of syndecan and tenascin is apparent *in vivo* in the condensed dental mesenchyme (A,B, arrow) and in the experimental recombinants in the mesenchymal tissue which is in contact with the epithelium (D, arrow). e, epithelium; m, mesenchyme; b, bone. From Thesleff et al (1990).

mesenchymal cells. One potential ligand in the extracellular matrix for syndecan is tenascin (cytotactin), a glycoprotein that has been specifically associated with organ-specific mesenchyme in several organ rudiments, including hair follicles, mammary glands and teeth (Chiquet-Ehrismann et al 1986). During the bud stage of tooth development, syndecan and tenascin are co-distributed

(Fig. 4A,B), whereas other components of the extracellular matrix, such as fibronectin, are not specifically confined to the condensed mesenchyme. We have shown by experimental tissue recombination studies that the expression of both syndecan and tenascin is induced by the dental epithelium (Vainio et al 1989b, Fig. 4C,D). Furthermore, solid-phase binding assays have indicated that syndecan isolated from dental mesenchyme binds tenascin (Salmivirta et al 1991). Hence there are several pieces of evidence which support the hypothesis that syndecan and tenascin are involved in the condensation of early organ-specific mesenchymal cells in the developing teeth, and, more specifically, that these two molecules mediate a cell–matrix type of interaction which promotes mesenchymal cell aggregation.

The aggregation process in kidney and teeth is apparently a consequence of several post-inductory events. The surface properties of the cells change to create new recognition and adhesion properties, and the expression of extracellular matrix molecules is altered, which influences cell–matrix interactions. Enhanced cell proliferation as a response to induction has been directly measured in the kidney system (Saxén et al 1983). Localization of dividing cells by immunohistochemical analysis of incorporation of 5-bromo-2′-deoxyuridine indicates that the condensation or aggregation of mesenchymal cells is accompanied by rapid cell proliferation both in the kidney and tooth (Vaahtokari et al 1991 and unpublished results).

Concluding remarks

It is apparent that epithelial–mesenchymal interactions trigger a cascade of regulatory events, and that the advancing specialization of the cells involves programmed sequential gene expression. Some of the very early molecular changes were described above, and our previous studies have pointed out other developmentally regulated changes during subsequent kidney and tooth morphogenesis (Saxén 1987, Thesleff et al 1990). It is conceivable that the tissue interactions regulate the expression of transcription factors which give rise to tissue-specific gene expression. Potential candidates for transcription factors regulating early kidney and tooth development are the zinc finger protein-encoding genes, the Wilms' tumour gene *WT2-1* expressed in the condensed metanephric mesenchyme (Pritchard-Jones et al 1990 and unpublished results), and the early growth response gene *Egr-1* expressed in the condensed dental mesenchyme (McMahon et al 1990).

References

Bernfield M, Sanderson RD 1990 Syndecan, a morphogenetically regulated cell surface proteoglycan that binds extracellular matrix and growth factors. Philos Trans R Soc Lond A Math Phys Sci 327:171–186

Bernfield MR, Banerjee SD, Koda JE, Rapraeger AC 1984 Remodeling of the basement membrane as a mechanism of morphogenetic tissue interactions. In: Trelstad RL (ed) The role of extracellular matrix in development. Alan R Liss, New York p 545–596

Chiquet-Ehrismann R, Mackie EJ, Piarson CA, Sakakura T 1986 Tenascin: an extracellular matrix protein involved in tissue interactions during fetal development and organogenesis. Cell 47:131–139

Cunha GR, Young P, Higgins SJ, Cooke PS 1991 Neonatal seminal vesicle mesenchyme induces a new morphological and functional phenotype in the epithelia of adult ureter and ductus deferens. Development 11:145–158

Ekblom P 1989 Developmentally regulated conversion of mesenchyme to epithelium. FASEB (Fed Am Soc Exp Biol) J 3:2141–2150

Ekblom P, Miettinen A, Virtanen I, Wahlström T, Dawnay A, Saxén L 1981 In vitro segregation of the metanephric nephron. Dev Biol 84:88–95

Grobstein C 1956 Trans-filter induction of tubules in mouse metanephrogenic mesenchyme. Exp Cell Res 10:422–440

Gurdon JB 1987 Embryonic induction—molecular prospects. Development 99:285–306

Kratochwil K 1987 Tissue combination and organ culture studies in the development of the embryonic mammary gland. In: Gwatkin RBL (ed) Developmental biology: a comprehensive synthesis. Plenum Press, New York p 315–334

Lehtonen E 1976 Transmission of signals in embryonic induction. Med Biol (Helsinki) 56:372–379

McMahon AP, Champion JE, McMahon JA, Sukhatme VP 1990 Developmental expression of the putative transcription factor *Egr-1* suggests that *Egr-1* and *c-fos* are coregulated in some tissues. Development 108:281–287

Mina M, Kollar EJ 1987 The induction of odontogenesis in nondental mesenchyme combined with early murine mandibular arch epithelium. Arch Oral Biol 32:123–127

Mizuno T, Yasuki S 1990 Susceptibility of epithelia to directive influences of mesenchymes during organogenesis: uncoupling of morphogenesis and cytodifferentiation. Cell Differ Dev 31:151–159

Moses HL, Yang EY, Pietenpol JA 1990 TGF-β stimulation and inhibition of cell proliferation: new mechanistic insights. Cell 63:245–247

New HV, Smith JC 1990 Inductive interactions in early amphibian development. Curr Opin Cell Biol 2:969–974

Panganiban GEF, Reuter R, Scott MP, Hoffmann FM 1990 A drosophila growth factor homolog, *decapentaplegic*, regulates homeotic gene expression within and across germ layers during midgut morphogenesis. Development 110:1041–1050

Pelton RW, Dickinson ME, Moses HL, Hogan BLM 1990 *In situ* hybridization analysis of TGFβ3 RNA expression during mouse development: comparative studies with TGFβ1 and β2. Development 110:609–620

Potts JD, Dagle JM, Walder JA, Weeks DL, Runyan RB 1991 Epithelial mesenchymal transformation of embryonic cardiac endothelial cells is inhibited by a modified antisense oligodeoxynucleotide to transforming growth factor β-3. Dev Biol 134:392–401

Pritchard-Jones K, Fleming S, Davidson D et al 1990 The candidate Wilms' tumour gene is involved in genitourinary development. Nature (Lond) 346:194–197

Salmivirta M, Elenius K, Vainio S et al 1991 Syndecan from embryonic tooth mesenchyme binds tenascin. J Biol Chem 266:7733–7740

Saxén L 1977 Directive versus permissive induction: a working hypothesis. In: Lash JW, Burger MM (eds) Cell and tissue interactions. Raven Press, New York p 1–9

Saxén L 1987 Organogenesis of the kidney. Cambridge University Press, Cambridge

Saxén L, Lehtonen E 1978 Transfilter induction of kidney tubules as a function of the extent and duration of intercellular contacts. J Embryol Exp Morphol 47:97–109

Saxén L, Salonen J, Ekblom P, Nordling S 1983 DNA synthesis and cell generation cycle during determination and differentiation of the metanephric mesenchyme. Dev Biol 98:130–138

Saxén L, Vainio S, Jalkanen M, Lehtonen E 1988 Intercellular adhesion and induction of epithelialization in the metanephric mesenchyme. In: Eguchi G, Okada TS, Saxén L (eds) Regulatory mechanisms in developmental processes. Elsevier Scientific Publishers, Amsterdam p 111–118

Sengel P 1990 Pattern formation in skin development. Int J Dev Biol 34:33–50

Takeda H, Suematsu N, Mizuno T 1990 Transcription of prostatic steroid binding protein (PSBP) gene is induced by epithelial–mesenchymal interaction. Development 110:273–282

Thesleff I, Hurmerinta K 1981 Tissue interactions in tooth development. Differentiation 18:75–88

Thesleff I, Jalkanen M, Vainio S, Bernfield M 1988 Cell surface proteoglycan expression correlates with epithelial–mesenchymal interactions during tooth morphogenesis. Dev Biol 129:565–572

Thesleff I, Vaahtokari A, Vainio S 1990 Molecular changes during determination and differentiation of the dental mesenchymal cell lineage. J Biol Buccale 18:179–188

Thompson NL, Flanders KC, Smith JM, Ellingsworth LR 1989 Expression of transforming growth factor-β1 in specific cells and tissues of adult and neonatal mice. J Cell Biol 108:661–669

Vaahtokari A, Vainio S, Thesleff I 1991 Associations between transforming growth factor β1 RNA expression and epithelial–mesenchymal interactions during tooth morphogenesis. Development 113:985–994

Vainio S, Jalkanen M, Lehtonen E, Bernfield M, Saxén L 1989a Epithelial–mesenchymal interactions regulate the stage-specific expression of a cell surface proteoglycan, syndecan, in the developing kidney. Dev Biol 134:382–391

Vainio S, Jalkanen M, Thesleff I 1989b Syndecan and tenascin expression is induced by epithelial–mesenchymal interactions in embryonic tooth mesenchyme. J Cell Biol 108:1945–1954

DISCUSSION

Hogan: The hanging drop culture seems to be a very nice system for testing whether conditioned medium from spinal cord or from a neuronal cell line or other cell lines could induce tubular formation. Could you use it to purify the factor that is being contributed by such cells?

Saxén: That's a very good suggestion but it doesn't work. No conditioned media that have been tested are effective in this system; only living cells seem to be able to induce.

Hogan: Does the inductive process require contact between the two types of cells? Can you get the induction with membrane extracts?

Saxén: We can prepare a membrane fraction by letting a spinal cord send its processes through a filter. Then one can scrape off the inducer and the membranes remain. Those membranes do not induce. No one has ever succeeded in inducing with any cell-free material or dead cells.

Hogan: How few cells do you need to elicit this response? If a small group of cells passes a signal through to the other side of the filter, does the signal then spread throughout the blastema? Or do all the cells in the blastema need to be touching the spinal cord cells? I can't see that being possible in the culture system, so the signal must be able to spread. Is it therefore auto-induced?

Saxén: It does spread but not by an assimilatory induction. Cells which have been triggered then migrate; by that mechanism the effect is spread about 100 μm. We have measured this using chromosomal markers and using chick and quail cells. You are right that very few cells touch the spinal cord cells. In the kinetic experiments where we remove the inducer, if we leave just a few groups of cells, a tubule forms at that site on the opposite side of the filter.

Hogan: But you do eventually get tubules throughout the blastema?

Saxén: No. If we make a hole so that the inducer can pass through at only one place, the maximum distance at which a tubule will form is about 100 μm.

McLaren: Does it depend on how long the inducing tissue remains in place? Does it take time for the inducing signal to get to 100 μm?

Saxén: By the time tubules have formed, in a day or so, some of them are 100 μm from the hole.

Hogan: But has every cell that forms a part of the tubule made contact with inducing cells?

Saxén: So I believe. The number of tubules induced is a function of the time of contact.

Hogan: But it should be limited by the surface area of the filter which is in contact with the spinal cells underneath. You could leave it forever and you still wouldn't get more than a certain number of tubules forming.

Solter: No, because some cells make contact and then move away, so that new cells get in contact with the spinal cord cells.

Saxén: That's how I believe it happens. We have made time-lapse movies that show those cells moving randomly, but the amplitude of that movement is not more than 100 μm. The number of tubules is also a function of the number of holes in the filter.

Tam: In one experiment, the induced mesenchyme was disaggregated and then cultured in a hanging drop. The cells apparently re-aggregated and formed kidney tubules. The question is whether those cells that eventually form tubules are only those that were previously in contact with the inducer.

Saxén: There is no direct evidence, but if we prevent contact between the blastemal cells and the spinal cord cells by the use of filters with pores small enough to prevent cell contact, we don't get any induction. I don't know how you could trace the cells that had been in contact specifically.

Hogan: Does it matter which part of the spinal cord you take? So much is now known about gene expression in the spinal cord.

Saxén: It's the dorsal part.

Hogan: The roof plate? Well, what about *Wnt* gene products as potential inducers?

McMahon: There are *Wnt* genes that are expressed in the ureter and also in the spinal cord.

Saxén: Spinal cord is not the only tissue that will induce tubule formation. I can give you a long list of tissues which will cause the induction, for example embryonic salivary gland mesenchyme, developing bone, submandibular epithelium, neural teratoma. So far, no common denominator has been found.

Balling: What is responsible for this induction *in vivo*? Can you disrupt the induction using filters?

Hogan: It is the ureter bud *in vivo* that's doing it. Presumably, the spinal cord is substituting *in vitro* for the ureter bud. One should be able to consult a databank of gene expression patterns in the mouse and see what is expressed in ureter bud and in the dorsal root of the spinal cord.

Balling: There are two *Pax* genes expressed in developing kidney: *Pax-2* and *Pax-8*.

Hogan: But they are not membrane associated. Lauri (Saxén) wants a membrane-associated molecule that is only very locally secreted. You need tissue culture cells transfected with every known *Wnt* gene put on the other side of the filter.

Balling: You still want to know where in this hierarchy the *Pax* genes are located. They might be influenced by or influence extracellular matrix proteins or growth factors.

Joyner: Pax might be a downstream gene acting in the mesenchyme that responds to *Wnt* or the inducing factor and then turns on the differentiation programme.

Krumlauf: In the hanging drop culture, if you mixed uninduced and induced cells, you could address whether the induction mechanism works homeogenetically. Have you tried mixing experiments like that?

Saxén: Yes, the induction is not transmitted from an induced cell to an uninduced one.

Hastie: What happens in terms of proliferation during the whole process from condensation to formation of various parts of the nephron? Do you know anything about cell proliferation and how the rate of that changes?

Saxén: I don't know when the mitotic activity ceases. We have measured the uptake of [^{3}H]thymidine as a function of time. Induction is completed within 24 hours and maximum incorporation of [^{3}H]thymidine also occurs at 24 hours. The increase in [^{3}H]thymidine incorporation starts at 12 hours, which is when the first cells become irreversibly induced.

Using 5-bromo-2'-deoxyuridine, we can show that around the ureter bud there is a zone of accumulation of dividing cells. So cell proliferation is stimulated at the time of early condensation.

Copp: Merton Bernfield has also shown that syndecan binds growth factors (Bernfield & Sanderson 1990). Is syndecan acting as a cell adhesion molecule or as a reservoir of TGF-β?

Saxén: It could do both.

Thesleff: There is recent evidence that syndecan-like molecules at the cell surface may be necessary for the introduction of fibroblast growth factor to its receptor (Yayon et al 1991). There are certainly many interactions between growth factors and these proteoglycans.

Saxén: There are too many ways to explain the action of syndecan. That's why inhibition experiments would be useful.

Lawson: I understand that you have an antibody to syndecan. If you do your hanging drop culture after induction in the presence of antibody does that prevent activation?

Saxén: No, it doesn't. But the antibody we have is a monoclonal one. My colleagues in Stanford think that we need a polyclonal for inhibition.

Hastie: I would like to say something about our work on the human kidney that impinges on many of the talks we have heard. It relates a bit to muscle development, minimally to cartilage development, certainly to gonad development and most of all to kidney development. There are mutations that affect kidney development. The obvious ones are those where there are no kidneys; a strict loss-of-structure type mutation where you don't have a kidney i.e. unilateral and bilateral renal agenesis.

The other type of mutant phenotype most of you don't think of and don't include in developmental biology meetings is the embryonic tumours. C. H. Waddington, for example, pointed out the relationship between cancer and development; this is particularly true in paediatric tumours, of which Wilms' tumour is probably the best example. This is a childhood kidney cancer that derives from the metanephric blastemal cells Lauri Saxén described. Sections of these tumours reveal various stages of normal kidney development recapitulated within the tumours; basically, something has gone wrong with nephrogenesis. Within the so-called triphasic tumour, you see tubule development, blastemal stem cells and stroma. The tumours vary in the proportions of these. They are probably clonal tumours that are formed from a single cell which has lost the function of at least one gene on chromosome 11: probably at least 2–3 other genes are involved in this process. These tumours are still able to make these structures in the absence of an inducer one assumes, so these cells are probably ready to go.

In rare cases you see glomerular structures forming in tumours. In a few tumours you see cartilage forming and in 10% of tumours you see skeletal muscle forming as well. Whether this relates to some default pathway in mesenchyme development, I don't know. I would really like to understand why those sorts of structures form in the tumours.

We know about genetics in this system because of the association of a few percent of Wilms' tumour cases with aniridia, where the child has no iris. There seems to be a mouse equivalent of this called the *smalleye* mutation: homozygotes have no eyes at all and no nose. We are working in collaboration with people in the US on a candidate gene for aniridia/*smalleye* and Peter Gruss is studying the same gene.

One in two or one in three people with sporadic aniridia develops a Wilms' tumour; they do this because they have a deletion which takes out the Wilms' tumour gene and an aniridia gene. 50% of the boys with this syndrome suffer from mild to severe pseudohermaphroditism. For example, we have a child with a large deletion in this region who should be a boy, he is XY chromosomally but has female external genitalia. We are interested in whether these genital abnormalities are due to defects in the Wilms' tumour gene itself, because of the intimate relationship between kidney and gonad development. Alternatively, neighbouring genes may be deleted. There is now genetic evidence that when the Wilms' tumour gene itself is mutated or deleted (these are haploinsufficient) gonadal/genital abnormalities may arise.

We and others did lots of genetics and tried to clone the gene. David Housman's laboratory first cloned a candidate Wilms' tumour gene (*WT1*) on chromosome 11p13. It is a zinc finger gene; it has four zinc fingers very like the zinc fingers found in the transcription factor Sp1 and *Krox-20* and *Egr-1*. David generously sent us the *WT1* clone and we wanted to know if its expression during embryogenesis made sense relative to our notions of tumour origins and histology. We predict that the tumours are derived from the mesenchyme-tubule-glomerulus pathway, because the structures we observe are not from the ureter or the stroma. We've found by *in situ* hybridization that the gene is expressed in only a few embryonic tissues; most tissues do not express it at all. It is not expressed in the uncondensed metanephric mesenchyme; it is expressed at quite high levels in the condensed mesenchyme and then the expression increases so in the most specialized layer of the kidney, the glomerulus, the gene is expressed at very high levels. That fits very nicely with what we expect from the gene. It is not expressed in the ureter or in the stroma.

The new nephrons are formed on the outside of the kidney. Expression of the gene is transient: as the nephron is being formed the gene is expressed at high levels and it goes off as mature nephrons are formed. It is a nice transcription factor playing a key role in nephrogenesis.

In early embryogenesis we find only three tissues where the gene is expressed: there are high levels in the condensed metanephric blastema, it is in the mesonephros (primitive kidney) in structures very similar to the glomerulus and it is expressed at very high levels in the genital ridge. We think in the male gonad it's mainly the Sertoli cells, the cells that express the testis-determining factor, which express this gene; we don't know when relative to activation of the testis-determining factor, but there could be an interaction between these genes.

The only other major site of expression of the *WT1* gene during embryogenesis is the mesothelium, the cell layer that goes all the way around the body cavity. The common property of the three main sites of expression is that they are all cells going through a mesenchymal to epithelial transition. We think this gene could play a general role in switching the mesenchyme to an epithelial programme, and we want to test this idea.

We have found two nice mutations in this gene in patients with tumours. The crystal structure of another zinc finger gene protein very similar to this, *Egr-1* (*Krox-24*), is known. The two mutations we have detected are in two of the amino acids predicted to make specific DNA contacts in the finger. We are particularly interested in trying to relate the mutations we see to the pathology of the tumours. We also want to know what the target genes are. Does the *WT1* product activate genes involved in the epithelial programme or does it shut off proliferation or does it do both? With Martin Hooper we are trying to knock out the gene in mice to find out what happens in the homozygotes.

Finally, Jonathan Bard with Lauri Saxén is looking at the gene in the *in vitro* differentiation system. We are using antisense oligonucleotides to try to prevent expression of *WT1* in the system, and we will see what that does to nephrogenesis. We are also trying to put the gene into mesenchymal cells to see if we can switch them to an epithelial programme.

McLaren: Is the gene expressed in female gonads?

Hastie: In females it is expressed in the granulosa cells of the ovary.

Lawson: There are switches from mesenchymal to epithelial cell-type behaviour much earlier in development, for instance, in the formation of somatopleure and splanchopleure, somites and heart myocardium.

Hastie: We have to look at that more closely but preliminary results suggest it is not induced in somitogenesis.

References

Bernfield M, Sanderson RD 1990 Syndecan, a developmentally regulated cell surface proteoglycan that binds extracellular matrix and growth factors. Philos Trans R Soc Lond Biol 327:171–186

Yayon A, Klagsbrun M, Esko JD, Leder P, Ornitz DM 1991 Cell surface, heparin-like molecules are required for binding of basic fibroblast growth factor to its high affinity receptor. Cell 64:841–849

The *Wnt* family of cell signalling molecules in postimplantation development of the mouse

Andrew P. McMahon, Brian J. Gavin, Brian Parr, Allan Bradley* and Jill A. McMahon

*Department of Cell and Developmental Biology, Roche Institute of Molecular Biology, Roche Research Center, Nutley, NJ 07110 and *Institute for Molecular Genetics, Baylor College of Medicine, Houston, TX 77030, USA*

Abstract. The mammalian *Wnt* gene family consists of at least ten members, all of which share a common structure. The N-terminus encodes a putative signal peptide sequence, suggesting that *Wnt* proteins are secreted. A number of absolutely conserved cysteine residues imply that inter- or intramolecular disulphide bonding is important to *Wnt* protein function. *Wnt* RNAs are localized to discrete regions of the postimplantation embryo and fetus, particularly within the developing central nervous system. Studies on *Wnt* gene expression strongly suggest that *Wnt*-mediated signalling is likely to be an important aspect of mouse development. One member of the family, *Wnt-1*, has been studied in some detail. By generating mutant alleles, we have demonstrated that *Wnt-1* regulates regional development of the central nervous system at early somite stages. There is circumstantial evidence that some aspects of the pathway through which *Wnt-1* action is mediated may be evolutionarily conserved. We propose that the *Wnt* family plays a major role in cell–cell interactions in the mouse.

1992 Postimplantation development in the mouse. Wiley, Chichester (Ciba Foundation Symposium 165) p 199–218

After implantation, the mouse embryo undergoes rapid growth during which the early cell lineages of the embryo proper are founded and organized with respect to the principle body axes. The organization, or patterning, of the embryo has recently become a major focus of studies in the mouse, as well as in other vertebrate and invertebrate species. It has long been appreciated that cell-to-cell signalling is an important mechanism by which groups of cells are organized. Recently, a common theme has emerged from studies of organisms as disparate as *Caenorhabditis elegans*, *Drosophila*, *Xenopus* and mice: peptide growth factors and their cognate receptors play pivotal roles in these signalling events.

In the mouse there are three major families of signalling molecules: the fibroblast growth factor (FGF)-related family, the transforming growth factor-β (TGF-β)-related family and the *Wnt* family. A considerable volume of circumstantial evidence and some more direct experimental observations link members of these families to the regulation of mouse development. We will discuss our recent work which explores the roles of the *Wnt* family of putative signalling molecules.

Organization and features of the *Wnt* family

Identification of the first *Wnt* gene, *Wnt-1* (formerly *int-1*), resulted from studies on the molecular aetiology of mammary tumours. In certain strains of mice, pregnancy-dependent tumours arise at a high frequency as a consequence of proviral activation of normal cellular genes following integration of mouse mammary tumour virus (MMTV) at the locus (Nusse 1988). Analysis of a number of clonally derived tumours led to the cloning of *Wnt-1* (Nusse & Varmus 1982). Subsequent studies of the *Wnt-1* locus indicated that proviral activation results in expression of RNA encoding the normal *Wnt-1* protein in the mammary gland (van Ooyen & Nusse 1984), a tissue in which *Wnt-1* is not normally expressed. Results from transgenic mice, in which a naturally occurring MMTV integration at the *Wnt-1* locus was recapitulated, confirmed that *Wnt-1* is indeed a transforming gene (Tsukamoto et al 1988). Analysis of the predicted protein sequence (van Ooyen & Nusse 1984, Fung et al 1985) identified a short stretch of hydrophobic amino acids at the N-terminus, suggesting that *Wnt-1* is a secreted protein. This prediction has been confirmed in several cell culture studies (Papkoff 1989, Papkoff & Schryver 1990, Bradley & Brown 1990). Thus, ectopic expression of *Wnt-1* presumably interferes with some aspect of the cell signalling controls that regulate growth and morphogenesis of the mammary gland.

Since the discovery of *Wnt-1*, the family has expanded dramatically. The first related gene, *Wnt-2* (formerly *irp*), was discovered in a walk on human chromosome 7 about the cystic fibrosis locus (Wainwright et al 1988); murine *Wnt-2* cDNAs were isolated shortly thereafter (McMahon & McMahon 1989). By comparing conserved amino acid stretches within *Wnt-1* and *Wnt-2* from different species, we were able to utilize a strategy based on the polymerase chain reaction (PCR) with degenerate oliconucleotides (Fig. 1) to clone six new family members (Gavin et al 1990). As the cDNA substrate in these experiments was generated from 9.5 day fetal RNA, this approach specifically targets family members which may function at early to mid-somite stages. This approach is applicable to any species and has led to the cloning of multiple *Wnt* members in many other vertebrates (J. A. McMahon & A. P. McMahon, unpublished work 1991) and *Drosophila* (A. P. McMahon, unpublished work 1990). The current tally of murine *Wnt* genes is ten, including a second member, *Wnt-3*, implicated in MMTV-associated mammary tumorigenesis (Roelink et al 1990). In contrast to other superfamilies (cf. *Hox*), there is no evidence for chromosomal linkage of *Wnt* genes. The chromosomal locations of the mouse members are indicated in Table 1.

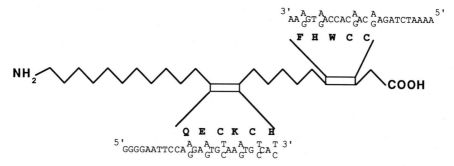

FIG. 1.   Diagram of *Wnt* cDNA cloning strategy. Two pools of degenerate nucleotides were used. The 5′ pool encodes the conserved amino acid sequence QECKCH and the 3′ pool encompasses (on the non-coding strand) the conserved amino acid sequence FHWCC. Successful amplification using the polymerase chain reaction results in a product of approximately 400 bp for most murine *Wnt* genes.

A number of conserved features are evident on alignment of mouse *Wnt* protein sequences (Fig. 2). The N-termini are hydrophobic, suggesting that like *Wnt-1*, other *Wnt* proteins are secreted. A number of absolutely conserved amino acids are scattered throughout the protein, most notably 22 cysteine residues. Thus, it is likely that intra- or intermolecular disulphide bonding is critical to *Wnt* functions, a conjecture supported by experimental studies in *Xenopus* (McMahon & Moon 1989). In addition, there is a single, absolutely conserved, putative N-linked glycosylation site. Whether glycosylation *per se* or some aspect of the primary sequence is responsible for conservation is not

TABLE 1   Chromosomal locations of the mouse *Wnt* genes

Gene	Mouse chromosome
Wnt-1	15
Wnt-2	6
Wnt-3	11
Wnt-3a	11
Wnt-4	4
Wnt-5a	14
Wnt-5b	6
Wnt-6	1
Wnt-7a	6
Wnt-7b	15

For details on chromosome assignments, see Nusse et al (1991).

FIG. 2.   Common features of mammalian *Wnt* proteins. The N-terminus is hydrophobic and probably encodes a signal peptide sequence. Bars indicate positions of cysteine residues. Twenty-two of the cysteine residues present in mammalian *Wnt-1* are absolutely conserved in mouse *Wnt* genes. A single cysteine residue (broken bar) is specific to *Wnt-1* and is found in the *Wnt-1* homologue in all vertebrate and invertebrate species thus far examined. Two additional cysteine residues (arrowheads) are conserved in all mammalian *Wnt* proteins other than *Wnt-1*. A single, putative N-linked glycosylation site (Y) is absolutely conserved in all ten mammalian *Wnt* proteins and all other *Wnt* proteins studied from vertebrate and invertebrate species.

clear. Glycosylation of *Wnt-1* is known to occur during trafficking through the secretory pathway (Brown et al 1987, Papkoff et al 1987), but it is not known if glycosylation is required for *Wnt-1* function.

Before continuing this account, it is necessary to digress somewhat to consider *Wnt* genes in *Drosophila*, since these studies have greatly influenced our models of the action of mammalian *Wnt* genes. There are at least three *Wnt* genes in *Drosophila*, *wingless* (*wg*, Rijsewijk et al 1987) and *Dwnt-2* and *Dwnt-3* (A. P. McMahon unpublished work, R. Nusse, unpublished work). *wg* is the *Drosophila* homologue of *Wnt-1*, an assignment made on the basis of a pattern of cysteine conservation shared by *wg* and *Wnt-1* but not by other mammalian *Wnt* genes (Gavin et al 1990). *wg* is required in embryonic, larval and pupal stages of *Drosophila* development (Baker 1988), but it is the embryonic functions which have been most thoroughly explored. *wg* is a member of the segment polarity class of pattern-regulating genes (Nüsslein-Volhard & Wieschaus 1980), all of whose members are required in each segment for normal patterning. In the absence of any one of those genes, pattern elements are lost. In the case of *wg*, the result is a loss of naked cuticle in the posterior region of the segment that encompasses the region where *wg* is expressed. *wg* is secreted (van den Heuvel et al 1989) and acts on neighbouring cells to regulate the expression of at least two other segment polarity genes, *engrailed* and *armadillo*. *engrailed*, which is a homeobox-containing DNA-binding protein, is activated in adjacent cells in the absence of *wg*, but *wg* is required for the maintenance of *engrailed* expression (DiNardo et al 1988, Martinez-Arias et al 1988). In contrast, *armadillo*, a *Drosophila* relative of the mammalian gap junction-associated protein, plakoglobin (Peifer & Wieschaus 1990), is widely expressed but post-transcriptionally regulated by *wg* (Riggleman et al 1990). Thus, one role of *wg*/*Wnt-1* in *Drosophila* is to regulate segmental patterning in the embryo, at least in part through the downstream regulation of a transcription factor, *engrailed*, and a cytoskeletal protein, *armadillo*. As will be discussed in the next section, *Wnt-1* also regulates embryonic organization in the mouse, specifically within the developing brain. Moreover, there is circumstantial evidence that some aspects of the regulatory pathway by which *wg* acts may be conserved in the mouse.

Wnt-1 is essential for normal patterning of the central nervous system in the mouse

Normal expression of *Wnt-1* in adult mice is restricted to post-meiotic stages of spermatogenesis (Shackleford & Varmus 1987): in the fetus, *Wnt-1* is expressed exclusively in the developing central nervous system (CNS) (Wilkinson et al 1987). While these early studies provided a moderately detailed account of the spatial and temporal expression of mouse *Wnt-1*, we have recently extended these *in situ* hybridization studies to obtain a more precise picture. Moreover, as it is clear that the domains of expression of *Wnt-1* and the mouse homologues of the *Drosophila engrailed* genes, *En-1* and *En-2*, are initially similar (Davis & Joyner 1988), we have included *en* genes in this study.

At the 5-somite stage, *Wnt-1* and *En-1* share similar anterior domains of expression in the presumptive midbrain (Fig. 3). Both genes are expressed broadly within the presumptive midbrain; however, *En-1* RNA extends caudally into the presumptive anterior hindbrain. In addition to this largely overlapping domain of *Wnt-1* and *En-1* expression, *Wnt-1* RNA is present in the dorsal aspects of the caudal hindbrain (Fig. 3), independently of any detectable *En-1* expression. 24 hours later (23- to 27-somite stages) the open neural plate has closed. *Wnt-1* RNA in the presumptive midbrain at this time is localized predominantly to the dorsal midline, although an almost complete circle of expression remains just anterior to the midbrain–hindbrain junction. In contrast, expression of both *En-1* and *En-2* is centred on the midbrain–hindbrain junction, extending anteriorly to the midbrain and posteriorly into the metencephalon. Thus, at 9.5 days, *Wnt-1* expression extends anteriorly to that of the *en* genes at the dorsal midline of the midbrain, while *en* genes are expressed at high levels in the anterior dorsal hindbrain where *Wnt-1* is never expressed. *Wnt-1* continues to be expressed in the dorsal hindbrain and in a continuous streak down the dorsal midline of the spinal cord. This distribution, which is established at 9.5 days, remains essentially intact until at least 14.5 days. Thus, at early somite stages, *Wnt-1* is switched on in a highly localized region-specific pattern. At this time, there is a substantial overlap between the broad domains of *Wnt-1* and *En-1* expression. Later, as *Wnt-1* expression becomes dorsally localized throughout much of the developing CNS, there is little similarity in the *Wnt-1* and *En-1* and *En-2* expression patterns.

The regional distribution and the timing of *Wnt-1* expression in early somite stages suggest that *Wnt-1* may play some fundamental role in development of the midbrain. Moreover, the relationship between early *Wnt-1* and *en* expression domains points to a possible interaction between these genes. To investigate the function of *Wnt-1*, we (McMahon & Bradley 1990) and others (Thomas &

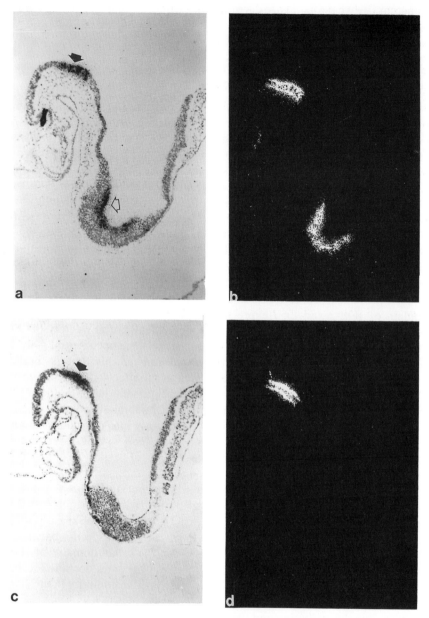

FIG. 3.   Overlapping expression of *Wnt-1* and *En-1* in 8.5 day p.c. mouse embryos.
In situ hybridization of *Wnt-1* (a,b) and *En-1* (c,d) probes to sagittal sections through
embryos at the 5-somite stage. *Wnt-1* and *En-1* share similar anterior domains of
expression in the presumptive midbrain (closed arrow). *Wnt-1* expression is also apparent
more posteriorly in the caudal hindbrain (open arrow).

Capecchi 1990) have used gene disruption in embryonic stem (ES) cells to mutate the *Wnt-1* locus. After homologous recombination in ES cells, a selectable gene was inserted into the normal open reading of the second exon of the *Wnt-1* gene (Fig. 4). Insertion separates the conserved N-terminus in exon 1, which contains the signal peptide sequence, from the remaining cysteine-rich portion of the protein. While it is difficult to prove formally that this generates a null allele, this is almost certainly the case. In the many clonal MMTV-induced tumours in which *Wnt-1* is activated, insertion of *Wnt-1* is never seen within the coding regions of exons or in the introns separating coding exons (van Ooyen & Nusse 1984); thus, it is unlikely that insertion at these positions generates active protein.

Mutated alleles were introduced into the germline and the resultant phenotypes analysed. Mice homozygous for mutated alleles die shortly after birth. Examination of neonates and fetal stages indicates that the detectable effects of loss of *Wnt-1* expression are limited to the CNS, specifically the midbrain and cerebellum. These regions are almost wholly deleted in mutant mice (Fig. 5), while forebrain, caudal hindbrain and spinal cord are apparently normal. The loss of the midbrain region implies an extensive function for *Wnt-1* in this tissue prior to the restriction of *Wnt-1* transcripts to the dorsal line. Indeed, even at 9.5 days p.c. the midbrain is deleted (Fig. 5).

Loss of the cerebellum is more difficult to explain. This tissue is derived from dorsal midbrain, as well as from anterior dorsal hindbrain which never expresses *Wnt-1* (Hallonet et al 1990). However, *en* genes are expressed in the anterior hindbrain; thus, the phenotype resembles more closely the loss of the region expressing *En-1* and *En-2* rather than the smaller domain that expresses both *Wnt-1* and the *en* genes. The lack of an observable phenotype in other *Wnt-1*-expressing areas, such as the caudal hindbrain and spinal cord, could be dismissed simply by proposing that *Wnt-1* has no functional role in this region. Interestingly, expression of one other *Wnt* member, *Wnt-3a*, is reported

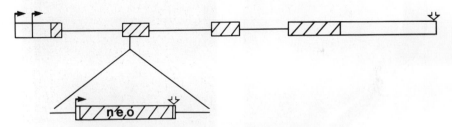

FIG. 4.   Disruption of the *Wnt-1* locus by insertion of a neomycin phosphotransferase (*neo*) expression cassette. The four exons (boxed) of the *Wnt-1* gene are indicated. Insertion of the *neo* gene into the coding (hatched) region of exon 2 results in disruption of the normal *Wnt-1* open reading frame. Transcriptional start sites for *Wnt-1* and the *neo* cassette are indicated by filled arrows and polyadenylation sites by open arrows.

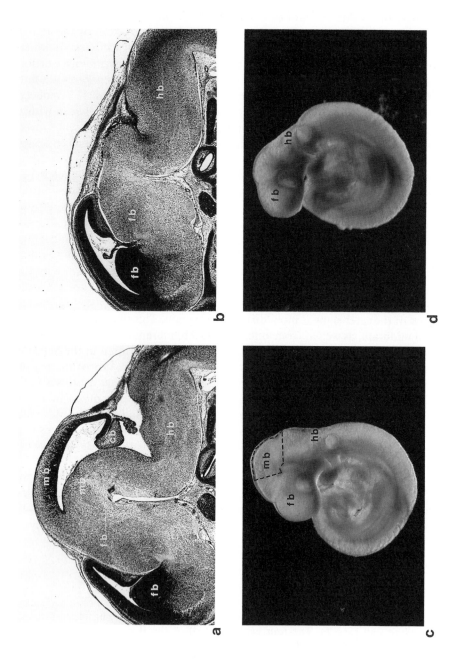

to overlap with *Wnt-1* expression at 10.5 days p.c. (Roelink & Nusse 1991). Thus, *Wnt-1* and *Wnt-3a* may have redundant functions in overlapping regions, an argument which has also been invoked to explain the weak phenotype observed in *En-2* mutants (Joyner et al 1991). However, there is a considerable temporal disparity between the early onset of *Wnt-1* expression in these regions and the first reported expression of *Wnt-3a* two days later (Roelink & Nusse 1991).

These studies on mutant mice have provided a revealing insight into the role of *Wnt-1*, allowing a number of conclusions to be drawn.

1) *Wnt-1* is required before 9.5 days (23 to 27 somites) for development of the midbrain.

2) *Wnt-1* is also required directly or indirectly for development of the cerebellum.

3) *Wnt-1* is apparently not required in the caudal hindbrain and spinal cord.

4) The contiguous domain which is lost in *Wnt-1* mutants is broader than the normal domain of *Wnt-1* expression, but very similar to the wider domain of *en* expression.

5) Thus, as in *Drosophila*, *Wnt-1* is absolutely required at an early stage in the patterning of a structure (the CNS); in the absence of *Wnt-1* that pattern element is lost (midbrain and cerebellum). The concordance between the phenotype and the domains of *Wnt-1* and *en* expression suggests that a *Wnt-1*-mediated signalling pathway may be conserved across very distantly related species. Certainly, the possibility is worthy of further exploration.

Expression of other *Wnt* genes

As indicated earlier, the murine *Wnt* family contains nine members in addition to *Wnt-1*. The expression patterns of several of these have been described: *Wnt-2* (McMahon & McMahon 1989); *Wnt-3* and *Wnt-3a* (Roelink & Nusse 1991); *Wnt-5a* and *5b* (Gavin et al 1990). Unlike *Wnt-1*, these other members show more widespread embryonic, fetal and adult expression: like *Wnt-1*, there are many features of the embryonic and fetal expression which are suggestive of possible roles in pattern regulation. *Wnt-5a* provides some excellent examples.

FIG. 5.   Phenotype of mice homozygous for a disrupted *Wnt-1* allele. Sagittal sections through 14.5 day fetuses (a) heterozygous and (b) homozygous for mutated *Wnt-1* alleles. Both the midbrain region (mb) and the cerebellum (cb) are missing in the homozygous mouse, whereas forebrain (fb) and caudal hindbrain (hb) are apparently normal. Whole embryo view of 9.5 day embryos (c) heterozygous and (d) homozygous for the disrupted *Wnt-1* allele. The midbrain region (boxed) in (c) is missing in the homozygous embryo. Other areas including the hindbrain and spinal cord look normal. Figures reproduced with permission from McMahon & Bradley (1990).

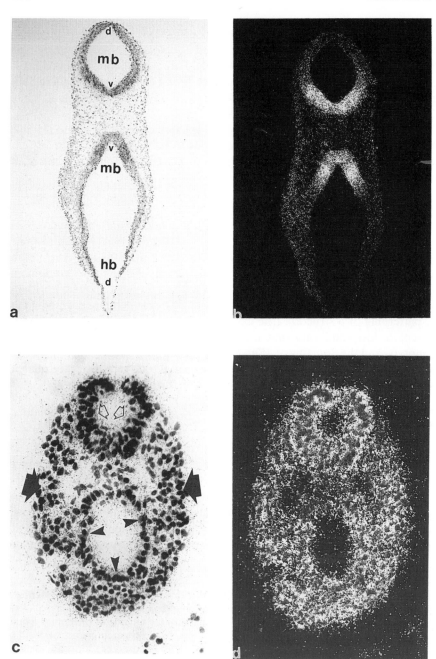

FIG. 6.   (*see legend on opposite page*)

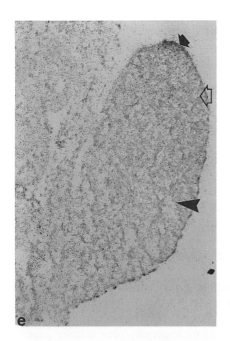
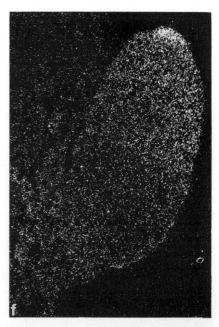

FIG. 6. *In situ* hybridization of a *Wnt-5a* RNA probe to tissue sections through 9.5 day mouse embryos. (a,b) Expression of *Wnt-5a* in a coronal section through the midbrain (mb) and hindbrain (hb). Expression is restricted to the ventral (v) half of the midbrain (d, dorsal). (c,d) Expression of *Wnt-5a* in a transverse section through the pre-somitic region. *Wnt-5a* is widely expressed in neural plate (open arrow), mesoderm (closed arrow) and definitive hindgut endoderm (arrow head). (e,f) Expression of *Wnt-5a* RNA in the developing forelimb bud shows a graded distribution; levels are highest in the apical ectoderm (closed arrow), at intermediate levels in distal mesenchyme (open arrow) and at low levels in proximal mesenchyme (arrowhead). Figures reproduced with permission from Gavin et al (1990).

At 9.5 days p.c., when *Wnt-1* expression in the midbrain is predominantly dorsal, *Wnt-5a* is expressed in the ventral half of the midbrain (Fig. 6a,b). Thus, whereas *Wnt-1* initially functions during open neural plate stages throughout most of the midbrain, *Wnt-5a* may have a slightly later role, specific to patterning the ventral half of this structure.

In gastrulating embryos, *Wnt-5a* expression is seen in posterior regions of the embryo; at 9.5 days p.c. this is not localized to any specific cell type. Thus, neural ectoderm, mesoderm and gut endoderm all express *Wnt-5a* (Fig. 6c,d), but their only obvious relationship is their similar posterior location along the body axis. In the limb, *Wnt-5a* is expressed in both the apical ectoderm and the underlying mesoderm in a graded distribution which correlates with distal–proximal patterning along this axis of the developing limb (Fig. 6e,f).

There are many examples of regional expression of other *Wnt* genes, which space constraints preclude discussing. Clearly, a major objective is to ascribe developmental functions to *Wnt* genes within these regions.

Perspective

There is ample indirect, and in the case of *Wnt-1* some direct, evidence that *Wnt* genes play a substantial role in the developing mouse embryo. Moreover, the highly conserved pattern of expression of *Wnt* genes in various non-mammalian vertebrates (Molven et al 1991, J. A. McMahon & A. P. McMahon, unpublished observations 1991) suggests a widespread conservation of these functions. Since *Wnt* proteins are likely to be secreted, *Wnt* signalling presumably requires a receptor, or more likely a family of receptors, to bind *Wnt* proteins, along with appropriate signal transduction pathways to process the signal and generate appropriate responses. The mammary tumour situation provides a dramatic example of the consequences of inappropriate *Wnt* signalling in an adult tissue.

At present, all details of *Wnt* signalling are a matter for conjecture, a circumstance which will hopefully be resolved in the near future. Moreover, the developmental roles of members other than *Wnt-1* remain to be determined. The generation of new mutant alleles will undoubtedly play a large part in elucidating these roles. One may reasonably suppose that by defining more clearly the action of *Wnt* genes, we will start to unravel some of the molecular aspects of the signalling events that shape mammalian development.

Acknowledgement

We would like to thank Barbara Kerr for preparation of this manuscript.

References

Baker NE 1988 Embryonic and imaginal requirements for *wingless*, a segment polarity gene in *Drosophila*. Dev Biol 125:96–108

Bradley RS, Brown AMC 1990 The proto-oncogene *int*-1 encodes a secreted protein associated with the extracellular matrix. EMBO (Eur Mol Biol Organ) J 9:1569–1575

Brown AMC, Papkoff J, Fung Y-KT, Shackleford GM, Varmus HE 1987 Identification of protein products encoded by proto-oncogene *int*-1. Mol Cell Biol 7:3971–3977

Davis CA, Joyner AL 1988 Expression patterns of the homeo box-containing genes *En-1* and *En-2* and the proto-oncogene *int-1* diverge during mouse development. Genes & Dev 2:1736–1744

DiNardo S, Sher E, Heemskerk-Jongens J, Kassis JA, O'Farrell PH 1988 Two-tiered regulation of spatially patterned engrailed gene expression during *Drosophila* embryogenesis. Nature (Lond) 332:604–609

Fung Y-KT, Shackleford GM, Brown AMC, Sanders GS, Varmus HE 1985 Nucleotide sequence and expression *in vitro* of cDNA derived from mRNA of *int-1*, a provirally activated mouse mammary oncogene. Mol Cell Biol 5:3337–3344

Gavin B, McMahon JA, McMahon AP 1990 Expression of multiple novel *Wnt-1/int-1*-related genes during fetal and adult mouse development. Genes & Dev 4:2319–2332

Hallonet MER, Teillet M-A, Le Douarin NM 1990 A new approach to the development of the cerebellum provided by the quail-chick marker system. Development 108:19–31

Joyner AL, Herrup K, Auerbach BA, Davis CA, Rossant J 1991 Subtle cerebellar phenotype in mice homozygous for a targeted deletion of the *En-2* homeobox. Science (Wash DC) 251:1239–1243

Martinez-Arias A, Baker NE, Ingham PW 1988 Role of segment polarity genes in the definition and maintenance of cell states in the *Drosophila* embryo. Development 103:157–170

McMahon AP, Moon RT 1989 Ectopic expression of the proto-oncogene *int-1* in *Xenopus* embryos leads to duplication of the embryonic axis. Cell 58:1075–1084

McMahon AP, Bradley A 1990 The *Wnt-1* (*int-1*) proto-oncogene is required for development of a large region of the mouse brain. Cell 62:1073–1085

McMahon JA, McMahon AP 1989 Nucleotide sequence, chromosomal localization and developmental expression of the mouse *int-1*-related gene. Development 107:643–651

Molven A, Njolstad PR, Fjose A 1991 Genomic structure and restricted neural expression of the zebrafish *wnt-1* (*int-1*) gene. EMBO (Eur Mol Biol Organ) J 10:799–807

Nusse R 1988 The activation of cellular oncogenes by proviral insertion in murine mammary tumors. In: Lippman ME, Dickson R (eds) Breast cancer: cellular and molecular biology. Kluwer Academic Publishers, Boston p 283–306

Nusse R, Varmus HE 1982 Many tumors induced by mouse mammary tumor virus contain a provirus integrated in the same region of the host chromosome. Cell 31:99–109

Nusse R, Brown A, Papkoff J et al 1991 A new nomenclature for *int-1* and related genes: the *Wnt* gene family. Cell 64:231–232

Nüsslein-Volhard C, Wieschaus E 1980 Mutations affecting segment number and polarity in *Drosophila*. Nature (Lond) 287:795–801

Papkoff J 1989 Inducible overexpression and secretion of *int-1* protein. Mol Cell Biol 9:3377–3384

Papkoff J, Schryver B 1990 Secreted *int-1* protein is associated with the cell surface. Mol Cell Biol 10:2723–2730

Papkoff J, Brown AMC, Varmus HE 1987 The *int-1* proto-oncogene products are glycoproteins that appear to enter the secretory pathway. Mol Cell Biol 7:3978–3984

Peifer M, Wieschaus E 1990 The segment polarity gene *armadillo* encodes a functionally modular protein that is the *Drosophila* homolog of human plakoglobin. Cell 63:1167–1178

Riggleman B, Schedl P, Wieschaus E 1990 Spatial expression of the *Drosophila* segment polarity gene *armadillo* is posttranscriptionally regulated by *wingless*. Cell 63:549–560

Rijsewijk F, Schuermann M, Wagenaar E, Parren P, Weigel D, Nusse R 1987 The *Drosophila* homologue of the mouse mammary oncogene *int-1* is identical to the segment polarity gene *wingless*. Cell 50:649–657

Roelink H, Nusse R 1991 Expression of two members of the *Wnt* family during mouse development—restricted temporal and spatial patterns in the developing neural tube. Genes & Dev 5:381–388

Roelink H, Wagenaar E, Lopes da Silva S, Nusse R 1990 *wnt-3*, a gene activated by proviral insertion in mouse mammary tumors, is homologous to *int-1/Wnt-1* and is normally expressed in mouse embryos and adult brain. Proc Natl Acad Sci USA 87:4519–4523

Shackleford GM, Varmus HE 1987 Expression of the proto-oncogene, *int*-1, is restricted to postmeiotic male germ cells and the neural tube of midgestation embryos. Cell 50:89–95

Thomas KR, Capecchi MR 1990 Targeted disruption of the murine *int-1* proto-oncogene resulting in severe abnormalities in midbrain and cerebellar development. Nature (Lond) 346:847–850

Tsukamoto AS, Grosschedl R, Guzman RC, Parslow T, Varmus HE 1988 Expression of the *int*-1 gene in transgenic mice is associated with mammary gland hyperplasia and adenocarcinomas in male and female mice. Cell 55:619–625

van den Heuvel M, Nusse R, Johnston P, Lawrence PA 1989 Distribution of the *wingless* gene product in *Drosophila* embryos: a protein involved in cell–cell communication. Cell 59:739–749

van Ooyen A, Nusse R 1984 Structure and nucleotide sequence of the putative mammary oncogene *int*-1; proviral insertions leave the protein coding domain intact. Cell 39:233–240

Wainwright BJ, Scambler PJ, Stanier P et al 1988 Isolation of a human gene with protein sequence similarity to human and murine *int*-1 and the *Drosophila* segment polarity gene *wingless*. EMBO (Eur Mol Biol Organ) J 7:1743–1748

Wilkinson DG, Bailes JA, McMahon AP 1987 Expression of the proto-oncogene *int*-1 is restricted to specific neural cells in the developing mouse embryo. Cell 50:79–88

DISCUSSION

Goodfellow: I feel that you are making a large jump in logic. You are not going to convince me that these genes are involved in pattern formation until you show that ectopic expression gives a change in the pattern. The simplest interpretation is that *Wnt* is involved in development and differentiation but in the absence of this growth factor this cell type can't survive. It is T cell growth factor all over again, except that it happens to be in this tissue at the back of the brain. I am not saying it's not important, I just fail to see the logical connection with pattern formation.

McMahon: In *Drosophila* it is quite clear that for several mutants defined as pattern mutants, in the absence of that particular gene product one gets pattern defects caused by the death of cells. It is reasonable to suggest that certain genes that are involved in patterning are involved in switching on developmental programmes to tell cells what they should be doing. If cells don't get the correct information at the correct time, this may compromise their survival in the developing tissue. The net result would be cell death.

One shouldn't view patterning as just the homeotic-type of patterning where cells are always transformed from one fate to another. A loss of a population of cells after mutation of a gene such as *Wnt-1* is consistent with two things: with *Wnt-1* having a role in what people like to think of as patterning; also with the gene encoding a product required for the growth and/or survival of that population of cells.

I thoroughly agree that ectopic expression experiments need to be done. We tried those experiments in the mouse. Unfortunately, there are a lot of problems with this type of approach. One really wants to direct expression of this gene to a rather localized area that's likely to respond in an informative way. That's not easy because the sorts of regulatory sequences that one might want to use are those of genes whose normal regulatory elements have not been defined. We have done some experiments using global promoters, like the cytoplasmic β-actin promoter. The result is a horrible uninterpretable mess. *Wnt* genes are expressed at many different places and, presumably, there is an active receptor pathway at many different places. We really need very specific promoters to do these experiments. For example, the *Hox* gene promoters might allow us to express given genes in closely defined regions of the embryo.

Goodfellow: I am happy with your interpretation; we are having an argument about semantics. In any process where you have an essential gene, if you knock that gene out, the step becomes rate limiting, whether it's right at the beginning or right at the end of the process. To call all the genes required for formation of a tissue, pattern genes, becomes confusing, I think.

McMahon: I think that part of the pattern of the central nervous system is to have a midbrain. I would say that the absence of a midbrain is the loss of part of the pattern.

Goodfellow: I can make such a pattern mutation with a surgical knife!

Hogan: But only because you know the midbrain is there and you know where to cut.

Joyner: Peter, what would you call a gene which had that kind of effect?

Goodfellow: There are lots of examples in haemopoiesis of factors which are needed for maintenance or survival of particular tissues. This is something which I think is going to be generally applicable. So I would call *Wnt* a candidate survival factor for this tissue.

Joyner: If we knocked out the two *engrailed* genes, which are likely transcription factors, and got the same phenotype as Andy sees for *Wnt-1*, what would you call the *engrailed* genes? Would they be patterning genes or genes required for survival?

Goodfellow: Is that what happens?

Buckingham: Andy, do you know how the *engrailed* and *Wnt* genes are interacting? Alex (Joyner) says the *engrailed* gene products are probably transcription factors, do you know if they interact with promoters of the *Wnt* genes, for example?

McMahon: In a *Wnt-1* mutant, *engrailed* comes on normally. So it doesn't require *Wnt-1* for its transcription.

In *Drosophila* there's a reciprocal interaction between the two populations of cells. The cell that expresses *wingless* provides a signal to the *engrailed*-expressing cell which maintains *engrailed* expression. At the same time the *engrailed*-expressing cell sends a signal back to the *wingless*-expressing cell which maintains *wingless* expression. So there's a reciprocal requirement by those two cell types. For the mouse there are no data. I don't want to push the idea that there's a direct interaction between these two genes. At the moment we have coincidental expression; clearly we have to provide more substantial evidence to prove any interaction.

Buckingham: The idea that the engrailed gene product might act as a transcription factor for the *Wnt* gene is not incompatible with your present data.

McMahon: Probably not, we have to define the anterior boundaries of expression of the two genes more closely.

Joyner: It is a different situation in the mouse than in the fly. In *Drosophila wingless* and *engrailed* are expressed in adjacent cells, almost never in the same cells, so clearly all the signalling between the genes has to be through membrane and intracellular proteins. In the mouse we know indirectly that they are expressed in the same cells. I have looked at sections through the midbrain and all the nuclei stain with an anti-*engrailed* protein antibody (Davis et al 1991), therefore some cells must express both *engrailed* and *Wnt-1*.

McLaren: You are referring to the *Wnt* gene products as cell signalling molecules. Do you think the signalling acts through cell membrane contact or by means of a secreted gene product?

McMahon: In Drosophila, where the only real evidence exists, the proteins have been shown by van den Heuvel et al (1989) to exit the *wingless*-expressing cell and to be taken up by the neighbour *engrailed*-expressing cells. So they can pass from one cell to another; however, they probably act over very local distances, 20 μm, something like that.

Solter: How can you explain the interactions in the middle of the segment, which is more than 30 μm away?

McMahon: In the region where there is no overlap between the pattern of expression of *Wnt-1* with that of *engrailed* ventrally? We don't know whether those cells are lost. There is some evidence that part of the ventral midbrain is still present. We have to check, but neuroanatomical experts like Andrew Lumsden are fairly convinced. Using our *Wnt-5a* gene which is expressed in ventral midbrain as a marker, we have evidence that suggests there still is ventral midbrain.

Hogan: Is the floor plate there?

McMahon: As far as we can see, the floor plate is there.

Lawson: Are you looking for receptors for *Wnt-1*?

McMahon: We believe that *Wnt-1* probably operates like any other growth factor that leaves one cell and is taken up by another cell by a receptor-mediated process. To get your hands on the receptors is not easy. Nobody has been able to purify the *Wnt* proteins in any significant quantity; we are all waiting for the *Drosophila* receptor to be cloned. One strong prediction in *Drosophila* is that a mutant receptor should cause a phenotype very similar to the *wingless* phenotype, except that it should be cell autonomous. The signalling molecule gives a non-cell autonomous phenotype. Some candidate genes have been cloned but at the moment none of these genes, as far as I am aware, fits into a normal receptor type model. *dishevelled* is the one most people have put their money on. It's been cloned by Norbert Perriman's group (personal communication).

Copp: From the *in situ* hybridizations of *Wnt-1* in the spinal region, it looks as though cells in the neural plate destined to be neural crest ought to be expressing *Wnt-1*. Nevertheless, in the knock out experiment the neural crest in the spinal region is normal, is that right?

McMahon: That's right. The neural crest in the spinal region seems normal and all the neural crest derivatives that must have come from the midbrain neural crest seem normal too. There are no craniofacial abnormalities in these mutants at all. The trigeminal ganglion that has a large contribution from the midbrain crest is normal.

Copp: How do you explain that?

Lawson: Some mesencephalic neural crest comes from a little further down, from the junction between the neurectoderm and the ectoderm. Some cells come out from what looks like surface ectoderm and under these conditions they might be enough to substitute for the cells with a more dorsal origin. There is very little known quantitatively.

McMahon: Patrick Tam's studies have shown that the mesencephalic neural crest comes out during the 5–7-somite stages.

Tam: The mesencephalic neural crest cells actually leave the neural plate much earlier than that—when the embryos have 1–6 somites (Chan & Tam 1988).

McMahon: That is when we start to see the phenotype, so the neural crest population of cells may escape intact. That's probably the best explanation. We have to look at markers of the neural crest to see what's really happening.

Joyner: Have you ever seen expression of *Wnt* in neural crest migrating away from the midbrain region?

McMahon: No. When we first looked at *Wnt* expression, we thought we saw some expression in neural crest spilling out of the top part of the neural plate. However, if it is expressed in neural crest, it must get down-regulated very quickly. We never see expression of *Wnt-1* in neural crest cells at any distance from their initial site of emigration.

Goodfellow: One explanation for the lack of a phenotype elsewhere is that there is functional redundancy. The alternative explanation is that *Wnt-1* can be expressed anywhere and it doesn't matter because the receptor is not present or you don't have some other component which is needed for functional expression.

McMahon: In terms of functional redundancy, we know that the two most divergent members of the *Wnt* family can produce the same phenotypic outcome in one experimental paradigm. In experiments done in collaboration with Randy

Moon, if we take a *Xenopus* gene *Xwnt-8*, which is as different from *Wnt-1* as you can possibly get in the family, and inject it into *Xenopus* embryos (Christian et al 1991), we get exactly the same phenotypic outcome as with injections of *Wnt-1*.

The mammary tumour situation is another example: three members of the family all give the same sort of mammary tumours. That again suggests they can elicit the same response. So it is quite likely that if two family members are expressed in the same place in normal development, they might be able to substitute functionally for one another.

Goodfellow: It's biochemistry, we all believe in biochemistry, so it's not surprising. But it then begs the question: why bother to express both genes at the same time? Either one is irrelevant or there is something very subtle about expression which we don't yet understand.

McMahon: You could look at it several different ways. You can say that when you duplicate genes, and this family presumably evolved by gene duplication, you don't just duplicate structural parts of the gene, you will duplicate regulatory sequences too. If there is some selective advantage for this animal to have both *Wnt* genes expressed in one cell type in the wild, then it will maintain expression. Laboratory mice don't live in a very selective environment—a small cage in an animal house. Alternatively, and I don't like this explanation, the duplicated regulatory regions mutate in time and acquire new functions but they will also lose the expression in some sites and their expression simply hasn't been lost yet in the roof plate. I suggest that there is a selective advantage in the wild for the mouse to express these two genes at this particular site, the roof plate.

Hogan: If you knock out one *engrailed* gene, you still get a functional mouse (Joyner et al 1991).

Joyner: But that doesn't mean the mutant homozygous mouse would survive in the wild.

McMahon: It has an abnormal cerebellum. If you look at lots of different mice in the wild, you probably find they all have the same type of cerebellum.

Joyner: Yes, they are very similar.

McMahon: Presumably, there is a genetic mechanism to make sure that they have that type of cerebellum. Some laboratory mice are blind but that's probably not very useful in the wild.

Wilkinson: In both of your experimental paradigms showing redundancy, the gene is drastically overexpressed. In *Xenopus*, if you injected more physiological amounts of RNA into early embryos, could you distinguish the activity of different members of the *Wnt* family?

McMahon: Randy Moon finds that *XWnt-8* is more potent than a similar amount of *Wnt-1* RNA injected into *Xenopus* (Christian et al 1991).

Wilkinson: So there could be different receptors, for example, with different affinities.

McMahon: I wouldn't be surprised. There might be subfamilies of receptors: maybe not as many receptors as ligands, but this is the most attractive hypothesis for me.

Jaenisch: One concern about the interpretation of these experiments is that you think you have inactivated exon 2. There is potential for splicing exon 1, which encodes the signal sequence, to exon 3. The resulting truncated protein may be beyond your limit of detection and it could be partially functional in the more posterior parts. Have you directly looked at this?

McMahon: We haven't, because we don't have an antibody that detects the protein *in vivo*. I think it's fairly unlikely for a number of reasons. One is that exon 2, upstream of the insertion, contains a number of the highly conserved cysteine residues. Although we haven't mutated those cysteine residues, we know that a single mutation in another of the highly conserved cysteine residues completely abolishes function of the protein. Irrespective of those cysteine residues, there is quite a high frequency of absolutely conserved amino acids scattered throughout the protein. So I think it's unlikely, but it is a valid concern. It may be more of a concern where you have a modular type of protein: for example, a transcription factor which clearly has different domains. The conservation in the *Wnt* family of proteins is scattered throughout the molecule and there's no evidence of any modular structure.

Goodfellow: Supporting Rudi Jaenisch's model are studies on people with Duchenne muscular dystrophy. Antibody staining has revealed dystrophin in small numbers of fibres even in patients with frameshift, stop codon or deletion mutations. Aberrant splicing which puts something back into frame or removes an exon with a stop codon is one explanation. You never know how much expression you need to provide some phenotypic rescue.

McMahon: Another aspect is that the major transcripts from the disrupted allele seem not to include the 5'-most region. There are two possible explanations: either the endogenous promoter gets hijacked by the tyrosine kinase promoter that's inserted and driving the neomycin gene, or the transcripts that initiate upstream from the *Wnt-1* promoter are unstable. Whatever the mechanism, the transcripts of the mutated site don't include the 5' sequence that encodes the signal peptide.

Goodfellow: You don't actually know the signal peptide is needed for the action of this putative growth factor.

McMahon: We know in *Xenopus* that the signal peptide is absolutely required: if it is deleted, *Wnt-1* is not active. One can delete the N-terminus and replace it with a completely different amino acid sequence constituting another signal peptide sequence that has no homology at all to *Wnt-1* and restore function. The amino acid sequence itself is not necessary but the fact that it encodes a signal peptide is important.

Joyner: The *Xenopus* assay may not reflect what *Wnt-1* normally does in the mouse.

References

Chan WY, Tam PPL 1988 A morphological and experimental study of the mesencephalic neural crest cells in the mouse embryo using wheat germ agglutinin-gold conjugate as the cell marker. Development 102:427–442

Christian JL, McMahon JA, McMahon AP, Moon RT 1991 *Xwnt-8*, a *Xenopus Wnt-1/int-1*-related gene responsive to mesoderm-inducing growth factors, may play a role in ventral mesoderm patterning during embryogenesis. Development 111:1045–1055

Joyner AL, Herrup K, Auerbach BA, Davis CA, Rossant J 1991 Subtle cerebellar phenotype in mice homozygous for a targeted deletion of the En-2 homeobox. Science (Wash DC) 251:1239–1243

van den Heuvel M, Nusse R, Johnston P, Lawrence PA 1989 Distribution of the *wingless* gene product in Drosophila embryos: a protein involved in cell–cell communication. Cell 59:739–749

The TGF-β-related DVR gene family in mammalian development

Karen M. Lyons, C. Michael Jones and Brigid L. M. Hogan

Department of Cell Biology, Vanderbilt University, Nashville, TN 37232, USA

Abstract. The genes that encode the bone morphogenetic proteins and the *Vg*-related proteins are mammalian members of a group of TGF-β-related genes, designated the DVR family, that includes the *decapentaplegic* gene of *Drosophila* and the *Vg1* gene of *Xenopus*. Members of the DVR (*decapentaplegic-Vg-related*) family have been implicated in diverse processes during development, particularly in epithelial–mesenchymal interactions. The results of our *in situ* hybridization studies with postimplantation mouse embryos provide evidence for the involvement of DVR family members, particularly *DVR-2*, *DVR-4* and *DVR-6*, in specific inductive interactions during the development of many organs, including the limb, the whisker follicle and the heart.

1992 Postimplantation development in the mouse. Wiley, Chichester (Ciba Foundation Symposium 165) p 219–234

The morphogenesis of most organ systems in vertebrates proceeds through a series of reciprocal inductive interactions among different cell types. In some cases, the result of the inductive interaction is the realization of a preset developmental programme within the responding cells (permissive induction). In other cases, the responding cells can adopt different fates in response to different inducing signals (instructive induction). In recent years a major effort has been directed towards elucidating the molecular mechanisms that mediate these inductive interactions. An important conclusion is that certain classes of gene products have been used repeatedly during evolution to bring about changes in cell fate in organisms as diverse as fruit flies and mice. For example, closely related families of genes encoding extracellular matrix components, cell adhesion molecules, peptide growth factors and their receptors, and DNA-binding transcription factors have all been implicated in mediating cell–cell interactions during development in vertebrates and invertebrates (for review see Kessel & Gruss 1990, Melton 1991). This paper is concerned with our attempts to define the role of one class of genes, the TGF-β-related DVR (*decapentaplegic-Vg-related*) gene family, in tissue interactions during mammalian development.

TGF-β-related DVR proteins and embryonic induction

Among the peptide growth factors, members of the transforming growth factor-β (TGF-β) family have been shown by genetic and biochemical criteria to play central roles in cell–cell interactions during development in diverse species. The members of this family are secreted, disulphide-bonded dimers which interact with specific receptors on the cell surface. Certain TGF-β family members can be placed into groups on the basis of amino acid sequence comparisons. One group consists of TGF-β1 itself and at least five closely related proteins. The members of this group have so far been detected only in vertebrates.

A second group, now designated DVR (*decapentaplegic-Vg-related*) contains the *Drosophila decapentaplegic* (*dpp*) gene product, the *Xenopus Vg1* gene, the mammalian *Vg-related 1* (*Vgr-1/Bmp-6*) and *Vgr-2* genes, and the genes for the bone morphogenetic proteins (*Bmp-2* to -7) (Gelbart 1989, Weeks & Melton 1988, Lyons et al 1989a, Wozney et al 1988, Celeste et al 1990, G. Thompsen & D. Melton, personal communication) (Table 1). Members of the DVR group have been found in vertebrates and in *Drosophila*. In the latter, two DVR family genes have been identified, *dpp* and *dVgr/60A* (*Drosophila Vg-related*) (J. Doctor & F. M. Hoffmann, and K. Wharton & W, Gelbart, personal communications). The vertebrate *DVR-2* (*Bmp-2/Bmp-2a*) and *DVR-4* (*Bmp-4/Bmp-2b*) genes are more closely related to *dpp* than to other members of the DVR family and may represent the vertebrate cognates of *dpp*. Similarly, the vertebrate *DVR-5* (*Bmp-5*), *DVR-6* (*Bmp-6/Vgr-1*) and *DVR-7*(*Bmp-7/OP-1*) genes are closely related to each other and to the *Drosophila Vgr* (*dVgr/60A*) gene and may thus represent the vertebrate cognates of *dVgr*. This conservation across a wide phylogenetic distance suggests that DVR-type genes play central roles during development.

Various activities have been described for some members of the DVR group that are consistent with this hypothesis. For example, mammalian BMP-2, BMP-3 and BMP-4 are capable of inducing ectopic cartilage and bone formation when implanted subcutaneously in rats, suggesting a role in skeletal development in vertebrates (Wozney et al 1988, Hammonds et al 1991). Genetic analyses have demonstrated that the *dpp* gene product plays multiple roles during development in *Drosophila*; for example, DPP is required for the specification of the dorsoventral axis, since embryos homozygous for recessive null alleles of *dpp* are completely ventralized. The *dpp* gene product is also necessary for normal development along the proximodistal axis in epithelial discs, since certain alleles of *dpp* result in the deletion of distal structures in adult appendages derived from imaginal discs.

The mechanism of DPP action underlying these activities is not yet known. A detailed analysis has recently been made of the role of DPP in the development of the midgut in *Drosophila*, specifically in the inductive interaction between the outer mesodermal layer and the inner epithelial endodermal cells in parasegment 7 (Immerglück et al 1990, Reuter et al 1990, Panganiban et al 1990). Expression of the *dpp* gene in the midgut is normally restricted to the mesoderm of parasegment 7. Using anti-DPP antibodies, it has been shown that the protein first appears in the mesoderm and then diffuses locally across to the endodermal cells, eventually becoming concentrated on their luminal surface. Interaction of DPP with the endodermal cells elicits specific changes in gene expression, such as increased transcription of the homeobox gene, *labial*. The expression of *dpp* itself in the mesoderm is controlled by the action of the homeobox genes, *Ultrabithorax* (*Ubx*) and *abdominal-A* (*abd-A*): in certain *Ubx* and *abd-A* mutants, *dpp* expression in the mesoderm is altered. This in turn, results in changes in the domain of *labial* gene expression in the underlying endoderm and, consequently, a change in the developmental fate of the gut. This model of how DPP acts as an intercellular signalling molecule to mediate inductive mesodermal–endodermal interactions in the *Drosophila* midgut provides an important paradigm for the role of TGF-β-related genes in vertebrate embryos. For example, the TGF-β-related growth factor activin regulates the expression of *Xhox* genes in *Xenopus*, the products of which act as transcription factors activating or repressing sets of genes that specify cell fate (Melton 1991). The regulation of homeobox-containing genes and genes of other transcription factors by polypeptide signalling molecules of the TGF-β family is likely to be a recurring theme in pattern formation during early vertebrate development.

The mouse *labial*-like gene *Hox-2.9* is expressed in a limited domain within the gut epithelium at the level of the forelimb bud in 9.5 days post coitum (p.c.) embryos (Frohman et al 1990, Murphy & Hill 1991). At this stage of development, *DVR-4* is expressed in the associated gut mesenchyme, although transcripts do not appear to be localized to a particular domain (Jones et al 1991). These patterns of expression raise the possibility that in the mouse, as in *Drosophila*, the activation of *labial*-like genes in gut epithelium requires the expression of the *dpp*-like gene *DVR-4* in the overlying gut mesenchyme.

dpp-like genes and pattern formation in mammals

It is reasonable to expect that similar cascades of gene activity involving homeobox-containing genes and peptide growth factors related to TGF-β participate in inductive events during organogenesis in mammals. One prediction of this hypothesis is that TGF-β-related genes are expressed in temporally and spatially restricted patterns that are consistent with a role in pattern formation, and in tissues where development proceeds through a series of epithelial–mesenchymal interactions. A necessary test of this hypothesis is therefore examination of the patterns of expression of members of the TGF-β family

TABLE 1 The DVR (*decapentaplegic-Vg-related*) family of TGF-β-related genes

Mammalian	Xenopus	Drosophila
	DVR-1/Vg1	
DVR-2/Bmp-2/Bmp-2a	DVR-2	
DVR-3/Bmp-3/osteogenin	DVR-3	
DVR-4/Bmp-4/Bmp-2b	DVR-4	dpp/DVR-15
DVR-5/Bmp-5	DVR-5	
DVR-6/Bmp-6/Vgr-1	DVR-6	dVgr/60A
DVR-7/Bmp-7/OP-1	DVR-7	
	DVR-8–14	
Vgr-2 + three others		

Alternative names reported in the literature are indicated (see text for references). Probable cognate genes are listed on the same line, for example, *Drosophila Vgr* (*dVgr/60A*) is closely related to the *DVR-6* gene in mammals and *Xenopus*.

during mammalian development. We have focused on several members of the *dpp* group of TGF-β-related genes: *DVR-2*, *DVR-4* and *DVR-6*. As it turns out, all these genes have cognates in *Drosophila* (Table 1), raising the possibility of some conserved mechanisms of action during development.

We observed expression of each gene by *in situ* hybridization at all stages of development, and in derivatives of all three germ layers (Lyons et al 1989a, b, 1990, Jones et al 1991). *DVR-2*, *DVR-4* and *DVR-6* are expressed in many of the same developing tissues (for example, in heart, limb buds, cartilage and bone, whisker follicles and the central nervous system), but with temporally and spatially distinct patterns, suggesting that the coordinated expression of multiple members of the gene family is required to orchestrate the development of complex organs composed of many cell types (Lyons et al 1990). Furthermore, these genes are often expressed in epithelial and mesenchymal components of tissues where development is known to involve reciprocal epithelial–mesenchymal interactions. In this report, we have focused on the limb bud, the developing whisker follicle and the heart, and we discuss the potential roles that *DVR-2*, *DVR-4* and *DVR-6* may play during the development of these systems.

Expression of *DVR-2* and *DVR-4* in limb buds

Three principal axes must be defined during limb development in vertebrates: the anteroposterior, dorsoventral and proximodistal axes. Two different models have been proposed to explain how these axes are specified: the polar coordinate model and the zone of polarizing activity (ZPA) model (reviewed by Bryant & Muneoka 1986). The polar coordinate model emphasizes the importance of short-range cell–cell interactions and proposes that cells have the general property of intercalation. Thus, when cells from two normally non-adjacent positions contact each other, cell division is stimulated to generate cells with positional values which lie between the non-adjacent cells. This model can account for development along all three axes. The ZPA model emphasizes the importance of long-range interactions and proposes that the ZPA is the source of a diffusible morphogen and cells gain positional value along the anteroposterior axis on the basis of the concentration of morphogen to which they are exposed (Tickle 1980, Smith et al 1989). The ZPA model also proposes that a second morphogenetically active area, the apical ectodermal ridge (AER), is required for development along the proximodistal axis and controls the proliferation and differentiation of the underlying mesenchymal cells within the 'progress zone' (Hinchcliffe & Johnson 1980). As cells leave the progress zone they acquire proximodistal positional information.

Transplantation experiments involving vertebrate limbs clearly demonstrate that inductive interactions between the AER and underlying mesenchymal cells are necessary for development along the proximodistal axis (reviewed in Hinchcliffe & Johnson 1980). Similarly, there is some evidence from transplantation experiments for a role for ventral ectoderm in patterning along the dorsoventral axis (MacCabe et al 1974). These inductive interactions between epithelial and mesenchymal cells can be accounted for by both the polar coordinate and the ZPA models.

As a first step to determine whether the *DVR-2* and *DVR-4* gene products participate in inductive interactions during limb bud development in the mouse, we examined the temporal and spatial patterns of expression of the two genes. At the earliest stage examined, 9.5 days p.c., the forelimb bud is a discrete outgrowth from the body wall. The AER is not yet present in the mouse, but the ventral and distal epithelium is slightly thickened (Fig. 1A). *DVR-2* transcripts are localized in this thickened ventral epithelium (Fig. 1B, Lyons et al 1990). By 10.5 days p.c., the AER is visible as a thickening of the distal tip epidermis (Fig. 1C). At this stage, *DVR-2* transcripts are localized to the AER and are absent from the ventral epithelium (Fig. 1D). *DVR-4* is also expressed in the early mouse limb bud. At 10.5 days p.c., high levels of *DVR-4* transcripts are expressed in the AER (Fig. 1E, F) (Jones et al 1991). From as early as 9.5 days p.c., *DVR-4* RNA is throughout the mesenchyme, with the highest concentrations in the anterior and distal regions.

The patterns of expression of *DVR-2* and *DVR-4* described above are consistent with the idea that these genes are involved in the specification of all three axes of the developing mouse limb. These are several possible ways in which they may affect pattern formation. For example, DVR-2 and DVR-4 homodimers or heterodimers made in the AER may act in an autocrine way to establish and/or maintain the differentiated state of this structure. Alternatively, or in addition, protein made in the epithelium of the AER may diffuse into the

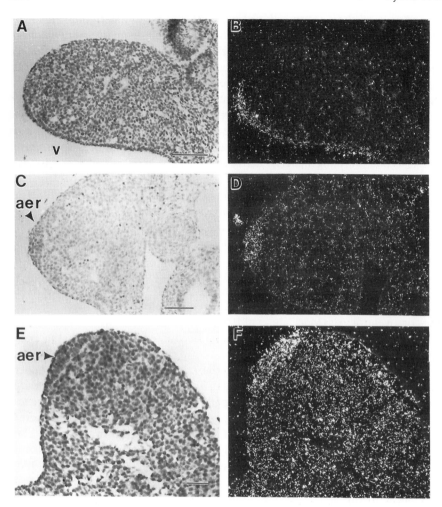

FIG. 1.   Expression of *DVR-2* and *DVR-4* RNA in limb buds. (A) Bright field and (B) dark ground photomicrographs of a section through a forelimb bud 9.5 days p.c. hybridized with the *DVR-2* probe. Hybridization is seen in the thickened ventral and distal epithelium. Bar = 75 μm. (C) Bright field and (D) dark ground photomicrographs of a section through a forelimb bud 10.5 days p.c.. *DVR-2* transcripts are localized to the apical ectodermal ridge (arrow). Bar = 50 μm. (E) Bright field and (F) dark ground photomicrographs of a section through a 10.5 day p.c. limb bud hybridized to the *DVR-4* probe. Bar = 50 μm. d, dorsal; v, ventral; aer, apical ectodermal ridge.

underlying distal mesenchymal cells of the progress zone and play a paracrine role in maintaining their undifferentiated and proliferative state. Finally, DVR-4 in the mesenchyme may act locally on mesenchymal cells; the protein might regulate genes expressed along the anteroposterior axis since the *DVR-4* RNA is localized in the mesoderm in a gradient, with higher levels in the more anterior and distal cells (Jones et al 1991).

A wealth of information is accumulating rapidly about the expression in the developing vertebrate limb of genes encoding nuclear proteins and transcription factors such as homeobox proteins, retinoic acid receptors and the formins, as well as potential intercellular signalling molecules, such as the *Wnt* gene family members (Dollé et al 1989, Nohno et al 1991, Izpisúa-Belmonte et al 1991, Zeller et al 1989). This allows speculation about how DVR-2 and DVR-4 fit into an overall scheme of pattern formation in the developing limb. Both *En-1* and *Wnt-5a* are coexpressed with DVR-4 in the ventral epithelium of the 9.5 day p.c. mouse limb bud (Davis et al 1991, Gavin et al 1990). The homeobox-containing gene, *Hox-7*, and the *Wnt-5a* gene are expressed at high levels in the distal mesenchyme (progress zone) beneath the AER (Robert et al 1989) and are therefore potentially up-regulated in response to DVR proteins secreted by the AER. In addition, it is possible that DVR-4 in the mesenchyme locally down-regulates the expression of the most 5' genes of the *Hox-4* cluster (*Hox-4.5–4.8*), which are expressed in a posterior–distal gradient from the ZPA (Nohno et al 1991, Izpisúa-Belmonte et al 1991), and up-regulates the expression of the cellular retinoic acid binding protein (Dollé et al 1989, Maden et al 1989).

Testing such hypotheses requires information about the distribution of DVR proteins in the limb and their ability to move from the site of synthesis to adjacent tissues. One prediction of our hypothesis is that altering the levels and patterns of expression of DVR-2 and DVR-4 in the developing limb bud, through the implantation of exogenous protein or misexpression in transgenic mice, should result in changes in the patterns of expression of homeobox-containing genes and the morphogenesis of the limb.

Expression of genes of the DVR family in whisker follicle development

The development of whisker follicles provides a clear example of pattern formation in the mouse. Whisker follicles mature in a proximodistal gradient with respect to the maxillary process, and both the number of rows and the arrangement of the follicles within the rows are invariant. Transplantation experiments have demonstrated that the development of the whisker follicle pattern involves reciprocal epithelial–mesenchymal interactions (reviewed by Davidson & Hardy 1952). A localized signal produced in the mesenchyme results in a thickening of the overlying snout ectoderm to produce an epithelial placode. This placode begins to lengthen and migrate downward through the dermis in response to mesenchymal signals from the presumptive dermal papilla. The dermal papilla is required for the continued maintenance of the mature whisker follicle, suggesting that it is the source of a signal that is received by the follicle.

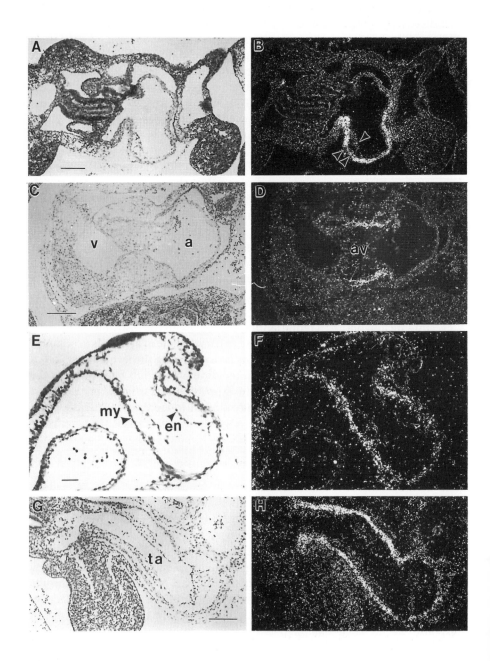

At the earliest stage of whisker follicle development, *DVR-4* transcripts are localized in a punctate manner directly underlying regions of future whisker follicle formation. Once the epithelial placode has formed, *DVR-4* is no longer expressed, but *DVR-2* RNA can be detected within the placode itself. At the epithelial downgrowth stage, *DVR-2* RNA continues to be expressed, but is localized to the basal epithelial cells that are in direct contact with the presumptive dermal papilla. The expression of *DVR-4* in the maxillary mesenchyme raises the possibility that the *DVR-4* gene product is a component of the signalling mechanism involved in the induction of the epithelial placode. Similarly, the expression of *DVR-2* within the epithelial cells in direct contact with the dermal papilla is consistent with a role for the *DVR-2* gene product in mediating epithelial–mesenchymal interactions during the later development and maintenance of the whisker follicle. Unlike *DVR-2* and *DVR-4*, *DVR-6* is not expressed at any stage within the whisker follicle itself, but is expressed at high levels in the surprabasal layers of the interfollicular epidermis, suggesting that the *Vgr-1* gene product plays a role distinct from that of DVR-2 and DVR-4. Our hypothesis that *dpp*-like genes mediate epithelial–mesenchymal interactions during whisker follicle morphogenesis is being tested directly by generating transgenic mice in which these genes are misexpressed under the control of specific keratin promoters normally expressed in different epithelial components of the epidermis and whisker follicle.

Expression of genes of the DVR family in the developing heart

The early embryonic heart consists of two concentric epithelial tubes—an inner endocardial tube and an outer myocardial tube. Beginning at about 9.5–10 days p.c., extensive morphogenesis takes place, resulting in the formation of septa and valves by the localized differentiation and migration of cells from the inner endocardial layer into the extracellular matrix. Fig 2A, B shows that *DVR-2*

FIG. 2.   Expression of *DVR-2* and *DVR-4* RNA in the developing heart. (A) Bright field and (B) dark ground photomicrographs of a parasagittal section through a 9.5 day p.c. embryo hybridized to the *DVR-2* probe. Transcripts are localized to the outer myocardial layer (double arrows) but no hybridization is detected in the inner endocardial layer (single arrow). Bar = 135 µm. (C) Bright field and (D) dark ground photomicrographs of a parasagittal section through the atrioventricular region of a 10.5 day p.c. embryo hybridized to the *DVR-2* probe. *DVR-2* RNA is localized to the myocardial layer of the atrioventricular cushion. Bar = 140 µm. (E) Bright field and (F) dark ground photomicrographs of a section through a 9.0 day p.c. embryo hybridized with the *DVR-4* probe. Hybridization is seen in the myocardial layer of the developing atrioventricular region. Bar = 50 µm. (G) Bright field and (H) dark ground photomicrographs of a parasagittal section through a 10.5 day p.c. embryo hybridized with the *DVR-4* probe, showing expression in the truncus arteriosus. Bar = 100 µm. a, atrium; v, ventricle; en, endocardium; my, myocardium; ta, truncus arteriosus.

RNA is present in the outer myocardial layer at 9.5 days p.c.. By 10.5 days p.c., *DVR-2* transcripts are localized to the myocardial cells of the atrioventricular canal (Fig. 2C, D). Lower levels of hybridization were seen in the dorsal aorta and the myocardium of the truncus arteriosus. Transcripts are no longer detected in the truncus arteriosus at 10.5 days p.c. (Lyons et al 1990). *DVR-4* transcripts first appear in the heart at 9.5–10 days p.c. in the myocardial layer in the region of the atrioventricular canal (Fig. 2D, E) but by 10.5 days p.c. are seen only in the truncus arteriosus (Fig 2E, F) (Jones et al 1991). *DVR-6* is expressed at low levels in the developing heart, in the mesenchyme of the atrioventricular cushions, and the expression is restricted to the region of contact of the dorsal aorta with the truncus arteriosus (Jones et al 1991).

There is evidence that members of the TGF-β family are involved in the morphogenesis of the heart, particularly in the epithelial–mesenchymal transformation that occurs during the formation of the atrioventricular cushions. For example, culture experiments have shown that migration of endocardial cells into a three-dimensional collagen gel is enhanced by TGF-β1 and TGF-β2 (Potts & Runyon 1989).However, TGF-β1 by itself does not cause endocardial cell transformation; an additional signal is required from the myocardium. Our observations that *DVR-2* and *DVR-4* RNA are present within the myocardium raise the possibility that these proteins are necessary for the formation of the atrioventricular cushions. To test this possibility, we are generating transgenic mice in which these genes are misexpressed under the control of the atrial natriuretic factor gene promoter, which drives expression throughout the myocardial layer of the atrium.

Conclusions

Several general conclusions have emerged from our *in situ* hybridization studies. Firstly, the members of the DVR family are commonly expressed in tissues where development proceeds via a series of inductive epithelial–mesenchymal interactions. This observation points to an important role for these genes in the transmission and decoding of positional information during development, but the precise molecular mechanisms of signal transduction are not yet known. However, the role of the *Drosophila dpp* gene product made in the gut mesoderm in the regulation of expression of the homeobox-containing gene *labial* in the underlying endoderm suggests that members of the DVR family in mammals also act upstream and downstream of specific homeobox-containing genes during embryonic inductions. The generation of transgenic mice in which DVR genes are misexpressed during the development of specific organ systems should provide direct evidence for a role in the mediation of epithelial–mesenchymal interactions in mammals. Demonstration of altered patterns of expression of homeobox-containing genes in the resultant transgenic animals may uncover new connections between peptide growth factors and nuclear transcription factors.

Acknowledgements

This work was supported by the NIH. We thank Cheryl Tickle for stimulating discussions about limb development.

References

Bryant SV, Muneoka K 1986 Views of limb development and regeneration. Trends Genet 2:153–159

Celeste AJ, Iannazzi JA, Taylor RC et al 1990 Identification of transforming growth factor β family members present in bone inductive protein purified from bovine bone. Proc Natl Acad Sci USA 87:9843–9847

Davidson P, Hardy MH 1952 The development of mouse vibrissae *in vivo* and *in vitro* J Anat 86:342–356

Davis CA, Holmyard DP, Millen KJ, Joyner AL 1991 Examining pattern formation in mouse, chicken and frog embryos with an *En*-specific antiserum. Development 111:287–298

Dollé P, Ruberte E, Kastner P et al 1989 Differential expression of genes encoding α, β, and γ retinoic acid receptors and CRABP in the developing limbs of the mouse. Nature (Lond) 342:702–705

Frohman MA, Boyle M, Martin GR 1990 Isolation of the mouse *Hox-2-9* gene; analysis of embryonic expression suggests that positional information along the anterior-posterior axis is specified by mesoderm. Development 110:589–607

Gavin BJ, McMahon JA, McMahon AP 1990 Expression of multiple novel *Wnt-1/int-1*-related genes during fetal and adult mouse development. Genes & Dev 4:2319–2332

Gelbart WM 1989 The *decapentaplegic* gene: a TGF-β homologue controlling pattern formation in *Drosophila*. Development (Suppl) 107:65–74

Hammonds RG, Schall R, Dudley A et al 1991 Bone-inducing activity of mature BMP-2b produced from a hybrid Bmp-2a/2b precursor. Mol Endocrinol 5:149–155

Hinchliffe JR, Johnson DR 1980 The development of the vertebrate limb. Clarendon Press, Oxford

Immergluck K, Lawrence PA, Bienz M 1990 Induction across germ layers in *Drosophila* mediated by a genetic cascade. Cell 62:261–268

Izpisúa-Belmonte JC, Tickle C, Dollé P, Wolpert L, Duboule D 1991 Expression of the homeobox Hox-4 genes and the specification of position in chick wing development. Nature (Lond) 350:585–589

Jones CM, Lyons KM, Hogan BLM 1991 Involvement of *Bone Morphogenetic Protein-4* (*BMP-4*) in morphogenesis and neurogenesis in the mouse. Development 111:531–542

Kessel M, Gruss P 1990 Murine developmental control genes. Science (Wash DC) 249:374–379

Lyons KM, Graycar JL, Lee A et al 1989a Vgr-1, a mammalian gene related to *Xenopus* Vg-1 and a new member of the transforming growth factor β gene superfamily. Proc Natl Acad Sci USA 86:4554–4558

Lyons KM, Pelton RW, Hogan BLM 1989b Pattern of expression of murine Vgr-1 and BMP-2a suggest that TGFβ-like genes coordinately regulate aspects of embryonic development. Genes & Dev 3:1657–1668

Lyons KM, Pelton RW, Hogan BLM 1990 Organogenesis and pattern formation in the mouse: RNA distribution patterns suggest a role for *Bone Morphogenetic Protein-2A* (*BMP-2A*). Development 109:833–844

MacCabe JA, Errick J, Saunders JW 1974 Ectodermal control of dorsoventral axis in the leg bud of the chick embryo. Dev Biol 39:69–82

Maden M, Ong DE, Summerbell D, Chytil F, Hirst EA 1989 Cellular retinoic acid–binding protein and the role of retinoic acid in the development of the chick embryo. Dev Biol 135:124–132

Melton DA 1991 Pattern formation during animal development. Science (Wash DC) 252:234–241

Murphy P, Hill RE 1991 Expression of the mouse *labial*-like homeobox-containing genes, *Hox 2.9* and *Hox 1.6* during segmentation of the hindbrain. Development 111:61–74

Nohno T, Noji S, Koyama E et al 1991 Involvement of the Chox-4 chicken homeobox genes in determination of anteroposterior axial polarity during limb development. Cell 64:1197–1205

Panganiban GEF, Reuter R, Scott MP, Hoffmann RM 1990 A *Drosophila* growth factor homolog, *decapentaplegic*, regulates homeotic gene expression within and across germ layers during midgut morphogenesis. Development 110:1041–1050

Potts JD, Runyon RB 1989 Epithelial–mesenchymal cell transformation in the embryonic heart can be mediated, in part, by transforming growth factor β. Dev Biol 134:392–401

Reuter R, Panganiban GEF, Hoffmann FM, Scott M 1990 Homeotic genes regulate the spatial expression of putative growth factors in the visceral mesoderm of *Drosophila* embryos. Development 110:1031–1040

Robert B, Sassoon D, Jacq B, Gehring W, Buckingham M 1989 *Hox-7*, a mouse homeobox gene with a novel pattern of expression during embryogenesis. EMBO (Eur Mol Biol Organ) J 8:91–100

Smith SM, Pang K, Sundin O, Wedden SE, Thaller C, Eichle G 1989 Molecular approaches to vertebrate limb morphogenesis. Development (Suppl) 107:121–131

Tickle C 1980 The polarizing region of limb development. In: Johnson MH (ed) Development in mammals. Elsevier/North-Holland Biochemical Press, Amsterdam vol 4:101–136

Weeks DL, Melton DA 1988 A maternal mRNA localized to the vegetal hemisphere in *Xenopus* eggs codes for a growth factor related to TGF-β. Cell 51:861–867

Wozney JM, Rosen V, Celeste AJ et al 1988 Novel regulators of bone formation: molecular clones and activities. Science (Wash DC) 242:1528–1534

Zeller R, Jackson-Grusby L, Leder P 1989 The limb deformity gene is required for apical ectodermal ridge differentiation and anteroposterior limb pattern formation. Genes & Dev 3:1481–1492

DISCUSSION

McLaren: There are seven exons in *DVR-6* in the mouse and only one exon in that gene in *Drosophila*. Does the homology with *Drosophila* extend across all seven exons?

Hogan: In the C-terminal region the sequence similarity between the *Drosophila* gene, *dVgr*/60A, and mouse *DVR-6* is about 70%. In the propeptide region there are blocks of conserved sequences, but the overall sequence similarity is much less (Wharton et al 1991).

McLaren: Is it possible that exons have simply been added on in the course of evolution?

Hogan: No, the length of the coding sequence is about the same in both species. If anything, introns may have been lost in *Drosophila*. The possible evolution of the DVR gene family has been discussed by Kristi Wharton (Wharton et al 1991).

Goodfellow: One general explanation, when you see something like this, is that there may have been homologous recombination between a retroposon and the original gene: that would account for the loss of introns and would maintain the regulatory sequences.

Hogan: John Doctor in Mike Hoffmann's lab has made an antibody to the *Drosophila dVgr/60A* protein. The protein shows very complex patterns of expression in the mesoderm of the larval mid-gut and in imaginal discs and so forth. Again, it may be one gene with lots of complex regulation in *Drosophila* that in the mouse has been split up into several genes—*DVR-5, DVR-6* and *DVR-7.*

Goodfellow: What is the predicted phenotype of the knock out?

Hogan: For *DVR-6,* I think it may be very similar to the *Wnt-1* knock out.

McMahon: I think that's extremely unlikely because there's no expression of *Vgr-1* in a large dorsal ventral domain at any time and it's that area that we see affected in the *Wnt-1* mutant. The pattern of expression that you describe more closely resembles the late pattern of expression of *Wnt* genes.

Hogan: We haven't looked very carefully at early embryos before the neural tube is closed; we should do.

Another possibility is that *DVR-6* is allelic with *congenital hydrocephalus,* which has defects not only in the brain but also in the meninges, kidney, skin and cartilage (Green 1970). *DVR-6* is expressed in all of these tissues. *DVR-6* maps to mouse chromosome 13, approximately 4 cM proximal to *congenital hydrocephalus* (Dickinson et al 1990). This is the closest known mutation to *DVR-6*; it wasn't that close, but of course they were mapped independently, in different systems.

McMahon: How many of the different members of the DVR family have been mapped?

Hogan: In collaboration with Neal Copeland and Nancy Jenkins' lab, we have mapped murine TGF-β-1–3, *DVR-6, Bmp-1, DVR-2* and two *DVR-4* genes (Dickinson et al 1990). We have pursued some of them. *DVR-4* is on chromosome 14 but is not allelic with *disorganization (Ds).* The *DVR-4* related gene on the X chromosome is not allelic with a cleft palate mutant studied by Philip Avner, Pasteur Institute.

Thesleff: The expression of *DVR-2* in the epithelial placode in the whisker follicle suggests that DVR-2 could be involved in the patterning of the ectoderm.

Hogan: Yes. There is a lot of beautiful experimental evidence saying that the mesenchymal epithelia talk to each other reciprocally during whisker and hair follicle development (Hardy 1992). At the moment these are just correlations; we have to make the transgenics.

Buckingham: You said that *Hox-7* was a potential responder gene to DVR-2 in the developing limb. *Hox-7* is also expressed specifically in the endothelial cushion of the heart.

Hogan: There are quite a few correlations we have noticed between *DVR-4* expression and *Hox-7* expression; they are also both expressed in one of the branchial arches.

Buckingham: *Hox-7* is expressed in the mesenchyme of the branchial arches.

Wilkinson: *DVR-4* is expressed in the central nervous system, including the diencephalon. There is good evidence, from Claudio Stern's lab for example, that the diencephalon is a segmented structure like the hindbrain. Does the expression domain correlate with forebrain neuromeres?

Hogan: We haven't looked at that, it would be very interesting.

Goodfellow: Have you looked in adults in the haemopoietic or lymphocyte lineages?

Hogan: No, we have done hardly anything in adults at all. Erwin Wagner raised this point recently: apparently he has seen expression of *DVR* RNAs in the adult heart. He suggested that this may mean that the genes are not involved in morphogenesis because morphogenesis is no longer taking place in the adult heart. It seems to me that it might be important to maintain a certain level of *DVR* gene expression in an adult organ. Presumably, there is some capacity for regeneration of mesenchymal cells in the heart with use. I can't imagine that there is no turnover of tissue at all in the adult heart. Similarly, in certain disease states ectopic bone and cartilage are formed in the heart. So I wouldn't necessarily rule out that *DVR* genes are playing a role in the adult in tissue turnover and repair.

McMahon: Do you have any results from ectopic expression studies?

Hogan: We have some transgenics which have been made with a cytokeratin gene promoter. The results look promising but it's too early to talk about them.

McLaren: *DVR-6* is expressed in the oocyte. Have you looked in the testis for expression during spermatogenesis?

Hogan: We know that *DVR-6* is expressed in the testis. There are two transcripts for *DVR-6*, but by *in situ* hybridization we couldn't see specific localization in any cell type in the testis. Again, in the embryonic kidney *DVR-6* is expressed at quite high levels as detected by Northern blotting, but from the *in situs* it seems like all the cells are expressing the gene, rather than a particular cell type.

McLaren: You have these chimeras from the knock out experiment but no germline transmission yet. It struck me that there might be a problem in these knock out experiments, in that if absence of *DVR-6* expression in a germ cell was lethal for that germ cell, you could never get germline transmission.

Hogan: It is possible that only making half the normal amount of protein in the testis leads to sterility—that's why we are anxious to find out where it is expressed in the testis.

Robertson: I wouldn't worry too much. That was one of our concerns with the c-*abl* mutation (Schwartzberg et al 1989). c-*abl* is expressed at a high level in late stages of spermatogenesis. Our studies (Schwartzberg et al 1991) and those of Tybulewicz (1991) confirm that absence of c-*abl* protein does not affect fertility in the mutant animals.

Goodfellow: You and Robin Lovell-Badge also showed the same thing for *Sry*.

Balling: John Eppig at the Jackson Lab has evidence that germ cells might in turn influence the somatic follicle cells. Isn't it then dangerous when, for example, Robin (Lovell-Badge) looks at *W* mutants which don't have germ cells and makes conclusions about whether a certain gene product is found in germ cells versus somatic cells? If there is this close interaction and germ cells and somatic cells talk to each other, could that change the gene expression programme in a somatic population? Therefore, when you are looking at gene expression in somatic cells in *W* mice, for example, are you looking at an artifact?

Goodfellow: Of course. But if you read the interpretation of those experiments it says the presence of germ cells is required for expression of this gene. That isn't the same as saying the gene is expressed in the germ cells.

Balling: Is there any evidence that some of those growth factors occur as a matrix-bound form and a non-matrix bound form?

Hogan: Work by Grace Panganiban and Michael Hoffman with the *Drosophila decapentaplegic* antibody has shown very clearly that the *dpp* protein secreted by a Schneider 2 cell line binds to the substrate (Panganiban et al 1990a). It can be extracted with 1% Tween-20 and 500 mM NaCl. They show that mostly it's the C-terminal mature form of the protein that's attached to the dish, but there is some N-terminal propeptide protein as well. There's no evidence I think for alternatively spliced forms, one of which binds to the matrix and another which does not, but that's still a possibility.

They have also reported (Panganiban et al 1990b) that antibody staining shows the *dpp* protein to be retained at the apical surface of the endodermal cells.

Saxén: So it's not internalized?

Hogan: From their antibody staining, I don't think they can tell whether it is in the apical cytoplasm or on the membrane of the apical surface.

References

Dickinson ME, Kobrin MS, Silan CM et al 1990 Chromosomal localization of seven members of the murine TGF-β superfamily suggests close linkage to several morphogenetic mutant loci. Genomics 6:505–520

Green MC 1970 The developmental effects of congenital hydrocephalus (ch) in the mouse. Develop Biol 23:585–608

Hardy MH 1992 The secret life of the hair follicle. Trends Genet 8:55–61

Panganiban GE, Rashka KE, Neitzel MD, Hoffman FM 1990a Biochemical characterization of the Drosophila dpp protein, a member of the transforming growth factor β family of growth factors. Mol Cell Biol 10:2669–2677

Panganiban GEF, Reuter R, Scott MP, Hoffmann FM 1990b A Drosophila growth factor homolog, decapentaplegic, regulates homeotic gene expression within and across germ layers during midgut morphogenesis. Development 110:1041–1050

Schwartzberg PL, Goff SP, Robertson EJ 1989 Germ-line transmission of a *c-abl* mutation produced by targeted gene disruption in ES cells. Science (Wash DC) 246:799–803

Schwartzberg PL, Stall AM, Hardin JD et al 1991 Mice homozygous for the abl[m1] mutation show poor viability and depletion of selected B and T cell populations. Cell 65:1165–1176

Tybulewicz VLT, Crawford CE, Jackson PK, Bronson RT, Mulligan RC 1991 Neonatal lethality and lymphopenia in mice with a homozygous disruption of the c-abl proto-oncogene. Cell 65:1153–1164

Wharton KA, Thomsen GH, Gelbart WM 1991 Drosophila 60A gene, another transforming growth factor-β family member, is closely related to human bone morphogenetic proteins. Proc Natl Acad Sci USA 88:9214–9218

General discussion III

Compiling information on developmental gene expression

McMahon: Something I have discussed with people on a number of occasions is that it would be very useful to have a databank of gene expression. Everybody is accruing all this detailed information from *in situ* hybridization on the sites of gene expression and the data are largely for quite interesting genes, for instance potential transcription factors, growth factors, etc. We need to be able to do our *in situ* hybridizations, find that a gene is expressed at a particular place, then go to a database and find all other genes that are expressed in that site. It would be useful in terms of both markers and targets.

From the work Nick Hastie described (p 196–198), he would like to be able to tap in 'metanephric tubules' and find out everything that was known about gene expression in metanephric tubules. Clearly, he would like to find targets for his gene. The field is growing so fast that unless we do something now we are going to lose a lot of useful information.

Herrmann: The Mammalian Genome, the former Mouse News Letter might be a good forum. It collects any genetic data, why not expression data as well?

Hogan: We really want a computer program.

Evans: It should be an addendum to one of the existing databases, for example the mouse genetics database at Bar Harbor which is now generally available. The people generating those databases could handle this very easily.

Goodfellow: You are describing tissue-specific expression patterns over a very wide period. A blastocyst doesn't look much like a 7 day embryo. If you try and capture all those data as images, you will just eat computer space.

Hogan: No. We don't necessarily want to store the images. One could choose certain times: 7.5, 8.5, 9.5 days.

McMahon: There could be a standard form for people to fill in for the different stages of development (based on Matt Kaufman's wonderful atlas). People could just tick off which gene is expressed where and when and list a reference source, then send the form in.

Krumlauf: The database also needs to incorporate all the enhancer trap and gene trap data that are being accumulated.

Joyner: Janet Rossant and I have been talking about organizing a meeting based around enhancer and gene trap screens. The mouse community has to get together and compile an atlas of gene expression patterns.

Beddington: It would also be important to include data for antibodies, to have the information for proteins as well as transcripts.

Solter: The problem is that when you try to summarize a paper, it is difficult to distinguish whether the gene is definitely not expressed or whether no one has yet looked at the specific place.

McLaren: It would be possible to have a list of tissues where it was known to be expressed and a list of tissues where it was known to be not expressed.

Goodfellow: Is it really worthwhile? Would it be so coarse that the information would never be of any biological use to you?

McLaren: Take the genital ridge, would that not be useful?

Hogan: Lewis Wolpert said that a few years ago he would have given his bicycle for a gene that was specifically expressed in the apical ectoderm ridge of the limb bud. Now, there is almost an embarrassing number of genes known to be expressed there as part of a tissue-restricted pattern. You have to say that either they are all subtly interacting with each other or that they have ended up there and the embryo can't get rid of them. It changes your whole view of the world to know that there are 20 genes expressed there compared with only two.

Goodfellow: I know there are thousands of genes expressed there; you won't find a tissue that expresses fewer than 1000 genes.

Solter: The expression has to be in some sort of specific combination.

Evans: It's a very effective molecular morphology: it describes these tissues in greater detail than we can describe them by histology.

McMahon: Patterns of gene expression clearly define cell populations that are different from one another. If you do see striking correlations, it's up to you to decide whether you want to follow them up.

Buckingham: The important thing is to be able to access the literature easily.

McLaren: Would it also be useful to have a list of cell markers so that if you had some *in vitro* tissue, and you wanted to know whether this was really muscle you had induced, you would know what cell markers to look for?

Buckingham: When Kirstie Lawson was talking about the different cells that she is tracing, she said one of the things that was lacking was markers for these very early cells.

Goodfellow: What are the chances of getting *in situ* hybridization to work on a single cell basis? This is clearly a major technical hang up in the field.

Herrmann: My whole-mount *in situ* hybridization protocol provides cellular resolution. All you have to do is to section the embryo after it has been processed.

Goodfellow: Is that less sensitive than doing radioactive *in situ* hybridization?

Herrmann: Probably more sensitive.

Goodfellow: So in a few years will I never see any more of these radioactive star diagrams?

McMahon: For sections it's very important to be able to compare two things that are very close to one another. I don't think sections will ever fall out of use completely. The *Drosophila* people have found some problems with the digoxy-genin technique in sections, so maybe radiolabelled probes will continue to be used.

Wilkinson: Diethard Tautz seems to have solved those by using frozen sections.

Use of embryonic stem cells to study mutations affecting postimplantation development in the mouse

Elizabeth J. Robertson, Frank L. Conlon, Katrin S. Barth, Frank Costantini and James J. Lee

Department of Genetics & Development, Columbia University College of Physicians & Surgeons, 701 West 168th Street, New York NY 10032, USA

Abstract. The generation and analysis of insertional mutations that perturb early postimplantation development provide a means to identify genes required at this stage of embryogenesis. We have been studying two independently generated insertional mutations termed 413.d and Hβ58 that result in early postimplantation lethality. Each mutation is associated with a distinct phenotype. 413.d mutant embryos become profoundly abnormal around the time of gastrulation: no identifiable embryonic axis or mesodermal structures are formed. Hβ58 mutant embryos proceed further in development, forming a relatively normal anteroposterior axis before developmental arrest occurs. We isolated embryonic stem cell lines homozygous for each of these mutations and assessed their differentiation abilities and developmental potential *in vitro* and after their introduction into wild-type blastocysts. From these studies we conclude that the 413.d mutation acts in a non-cell-autonomous fashion: mutant cells appear capable of participating, in conjunction with wild-type cells, in the formation of derivatives of all three primary cell lineages of the embryo. Hβ58 mutant embryonic stem cells are clearly pluripotent but they appear to be more restricted in their developmental potential, suggesting that the Hβ58 gene product may be required by specific tissues of the embryo.

1992 Postimplantation development in the mouse. Wiley, Chichester (Ciba Foundation Symposium 165) p 237–255

Relatively little is known with respect to specific gene products that may be required before and during early postimplantation development of the mouse embryo. A number of different strategies are being used to identify genes that are involved in early embryogenesis. These include the cloning and characterization of murine homologues of genes known to regulate developmental processes in other species (reviewed by Kessel & Gruss 1990). The study of spontaneous and experimentally induced recessive lethal mutations should also allow the identification of gene products required for the completion of embryonic development (reviewed by Magnuson 1986, Lyon & Searle 1989).

This mutagenesis approach has recently been extended by using a 'gene trap strategy' in embryonic stem (ES) cells, to identify genes that exhibit specific temporal and spatial patterns of expression (Gossler et al 1989). Finally, screening of stage-specific or tissue-specific cDNA libraries has led to the isolation of developmentally regulated genes.

Another approach has been the study of insertional mutations. A number of routes have been used to generate insertional mutations, including microinjection of DNA into the zygote and retroviral infection of embryos and ES cells (reviewed by Gridley et al 1987, Jaenisch 1988). An advantage to studying experimentally induced recessive insertional mutations is that the introduced DNA sequences may facilitate the identification and study of the affected gene(s) (Jaenisch et al 1983, Schnieke et al 1983, Soriano et al 1987, Maas et al 1990, Woychik et al 1990). We have studied two independent recessive mutations, 413.d (Conlon et al 1991) and Hβ58 (Constantini et al 1989, Radice et al 1991), that result in lethality of the early postimplantation embryo.

There are several possible schemes that could account for failure of early postimplantation stage embryos. Developmental arrest could simply reflect the absence of a gene product that is required by all cells. Alternatively, disruption of a gene required for the formation of specialized tissues (block to differentiation) or of a gene necessary for specifying positional information might be lethal for the embryo as a whole. On the basis of gross morphology or molecular characterization alone it is difficult to attribute any specific developmental function to genes identified by mutational analysis. One approach that allows these possibilities to be distinguished is to derive and analyse ES cell lines carrying insertional mutations. Here we analyse the abilities of 413.d and Hβ58 mutant ES cells to differentiate in isolation and their developmental fates following blastocyst injection.

The 413.d recessive lethal mutation

The 413.d recessive lethal mutation was identified during an extensive screen of transgenic mice carrying multiple copies of an MPSVmos-[1]neo retroviral vector (derived from the myoproliferative sarcomavirus) (Robertson et al 1986). The mutation co-segregates with a single proviral insertion site (Conlon et al 1991). The retroviral insertion site was cloned from a genomic library prepared using DNA from heterozygous mice. Unique sequence flanking probes were identified that allow the wild-type and interrupted alleles to be distinguished by Southern analysis (Conlon et al 1991). Interspecific backcross analysis has placed the mutation on chromosome 10 (E. J. Robertson, K. S. Barth, F. L. Conlon, N. Jenkins, unpublished data).

Gross inspection of embryos obtained from heterozygous matings indicates a striking phenotype of mutant embryos at 8.5 days p.c.. Presumed homozygous embryos are typically reduced in size, enclosed in a mass of primary and

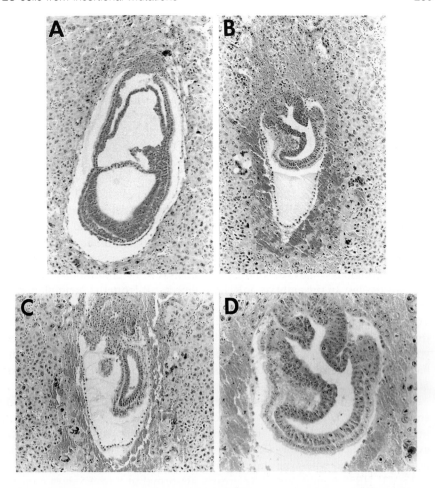

FIG. 1.   Sections of 7.5 day p.c. embryos derived from intercrosses of animals heterozygous for the 413.d provirus insertion. (A) Sagittal section through a wild-type embryo.
(B,C) Sections through presumptive homozygous mutant embryos within the decidual
capsule. Note the absence of a recognizable fetal portion, visceral yolk sac and allantois.
Panels A, B and C are at the same magnification to show the relative size discrepancy
between the wild-type and presumed mutant embryos. (D) A higher magnification of the
section shown in (B) to illustrate the cellular morphology of the disturbed region of the
embryo and the presence of apparently normal trophoblast giant cells and parietal yolk sac.

secondary giant cells and have a well formed parietal yolk sac. The remainder
of the embryo is grossly abnormal and no recognizable fetal structures are present.
By 9.5 days p.c. the abnormal embryos are highly degenerate. Histological analysis
of presumed mutant embryos and a normal litter mate is shown in Fig. 1. There
is no evidence for the formation of mesoderm: morphologically recognizable

visceral yolk sac, allantois or amnion, and a definitive fetal portion do not appear to be present. The region that would normally be occupied by the early somite stage embryo appears to contain randomly organized, highly convoluted tissue layers inserted into the region of the ectoplacental cone. One day earlier, at 7.5 days p.c., a proportion of embryos appear retarded and morphologically abnormal with respect to the spatial arrangement of the tissue layers.

The Hβ58 recessive lethal mutation

The Hβ58 mutation is one of three embryonic lethal mutations identified during a screen of 20 independent transgenic mouse strains (Costantini et al 1989). This mutation results from the insertion of 10–20 copies of a DNA fragment containing the human β-globin gene. Molecular analysis determined the structure of the integration site and the corresponding wild-type allele. A candidate transcription unit of 2.7 kb that is affected by insertion of the transgene has been identified (Radice et al 1991); this potentially encodes a 38 kDa protein. Preliminary analysis has shown the Hβ58 mRNA to be ubiquitously expressed in adult tissues. In the embryo, mRNA is detectable from the preimplantation stage onwards. There are significant quantitative differences in the level of expression between different tissues. At 7.5 days of development Hβ58 transcripts are most strongly expressed in the ectoplacental cone and the visceral yolk sac (Fig. 2, F. Costantini & J. J. Lee, unpublished). Genetic mapping studies have placed this mutation on chromosome 10, 13.5 ± 2.5 cM proximal to the *Steel* locus.

Developmental abnormalities can be observed in Hβ58 homozygous mutant embryos at Day 7.5 p.c.. These embryos display a proportionately smaller and underdeveloped embryonic ectoderm and an unusual folding of the extraembryonic membranes (Radice et al 1991). Histological sections of Hβ58 homozygotes and wild-type embryos at different gestational stages are shown in Fig. 3. Hβ58 homozygotes are grossly abnormal at 9 to 9.5 days p.c., but display a normal neural axis, a beating heart, and extensive development of the allantois, amnion and yolk sac.

Isolation and characterization of mutant ES cell lines

ES cell lines were generated using embryos from the appropriate intercross matings, as described in Robertson (1987). ES cell lines were obtained at a normal frequency. The genotype of the ES cell lines with respect to the two relevant loci was determined by Southern analysis utilizing single copy probes specific for each locus. Of 19 413.d lines examined, three were found to be wild type, five were heterozygous and 11 lines were homozygous with respect to the 413.d locus. Of the 13 Hβ58 lines analysed, four proved to be wild type, five were heterozygous and four were homozygous with respect to the Hβ58 locus.

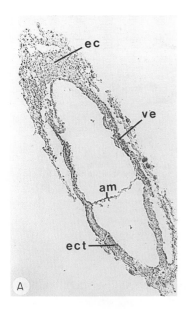 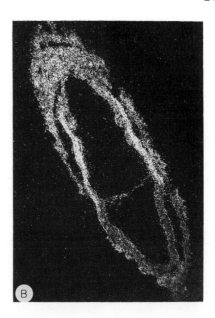

FIG. 2.   Section through a 7.5 day p.c. embryo hybridized with an ^{35}S-labelled probe specific for the Hβ58 mRNA. (A) Bright field photomicrograph, (B) dark field photomicrograph of the same section showing high levels of expression in the visceral yolk sac and ectoplacental cone (ec) and lower but detectable expression in the remaining tissues. am, amnion; ect, ectoderm; ve, visceral endoderm.

None of the homozygous ES cell lines was recognizably different either in morphology or growth characteristics from their wild-type or heterozygous counterparts. Similarly, when both sets of homozygous ES cell lines were subjected to *in vitro* differentiation protocols (Doetschman et al 1985, Robertson 1987), we observed no striking differences in the differentiated cell types formed as compared with those from wild-type cell lines.

To examine a more extensive collection of terminally differentiated cell types, we injected homozygous 413.d and Hβ58 ES cell lines independently into syngeneic 129/Sv and C57BL/6 host animals, respectively, to generate solid tumours. A wide range of mature ectodermal and mesodermal derivatives were detected in the resulting teratocarcinomas. Representative sections from a tumour formed by a 413.d homozygous ES cell line (IMD-16) are shown in Fig. 4.

Developmental potential of mutant ES cell lines

Wild-type ES cell lines efficiently form chimeras after injection into blastocysts (Bradley et al 1984, Robertson & Bradley 1986). As the ES cells homozygous for

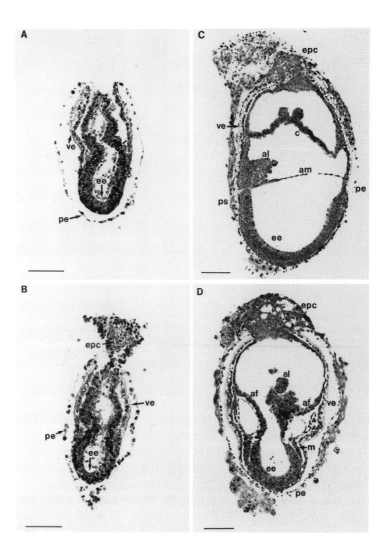

the two mutations were judged to be pluripotent, we wanted to investigate whether these mutations compromised the ability of the ES cell lines to participate in normal embryogenesis.

ES cell lines derived from the 413.d mutant (glucose phosphate isomerase

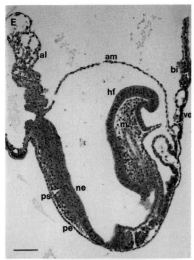

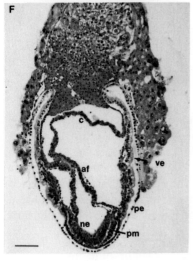

FIG. 3. Histological analysis of early post-implantation stage Hβ58 embryos of confirmed genotype. Embryos were dissected free of decidual tissue and a small portion of extra-embryonic tissue was used to determine genotype (as described in Radice et al 1991). (A) Wild-type, (B) homozygous embryos at Day 6.5 p.c.. (C) Wild-type, (D) homozygous embryos at Day 7.5 p.c.. (E,F) Examples of homozygous embryos at Day 8.5 of development. af, amniotic fold; al, allantois; am, amnion; bi, blood island; c, chorion; ee, extraembryonic ectoderm; epc, ectoplacental cone; hf, head fold; m, mesoderm; ne, neural ectoderm; pe, parietal endoderm; pm, paraxial mesoderm; ps, primitive streak; ve, visceral endoderm.

(GPI) 1^c, agouti coat colour) and from the Hβ58 mutant (GPI 1^b, non-agouti coat colour) were injected into host blastocysts derived from outbred animals of the MF1 strain (GPI 1^a, albino). Homozygous mutant ES cells of each genotype gave rise to overt live-born chimeras with very high efficiency (approximately 80% of the live-born animals). To examine whether mutant ES cells of either genotype are selectively excluded from various adult somatic tissues,

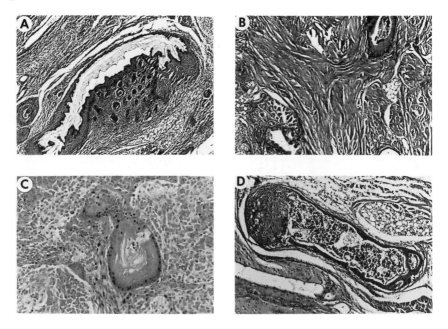

FIG. 4. Examples of mature somatic tissues present in a teratocarcinoma generated by the IMD-16 ES cell line that is homozygous for the 413.d proviral integration site. Numerous well differentiated tissues are formed including (A) keratinizing epithelium, (B) striated muscle, (C) pigmented melanocytes and (D) a small bone. 6 μm sections stained with haematoxylin and eosin.

as has been reported for parthogenetic aggregation chimeras (Fundele et al 1989, Nagy et al 1989), we assessed the quantitative and qualitative distribution of derivatives of mutant cells by GPI analysis. ES cell derivatives were present in all the tissues tested (Conlon et al 1991, J. J. Lee, unpublished data). In some instances, the ES cell derivatives constituted the major population of a given organ. Additionally, breeding studies with chimeric males generated using 413.d mutant ES cells indicated that a number of these animals were germline chimeras.

Analysis of the pattern of colonization of the extraembryonic lineages of chimeric conceptuses

One explanation for the observed failure of the 413.d and Hβ58 mutant embryos is that the affected genes may be required for the establishment or normal functioning of the extraembryonic lineages. *In situ* hybridization analysis showed that expression of the Hβ58 mRNA is highest in the visceral yolk sac and ectoplacental cone (Fig. 2). It was therefore important to examine the distribution of mutant ES cells in the extraembryonic lineages of the developing embryo.

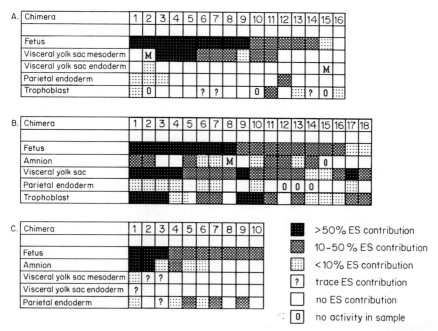

FIG. 5.   Distribution of wild-type and mutant ES cell derivatives in chimeric conceptuses analysed at mid-gestation. The cell lines and host blastocyst combinations are as follows. (A) Contribution by wild-type ES cells of 129/Sv genotype after injection of 10–15 cells into outbred MF1 strain blastocysts. (B) Contribution by IMD-16 ES cells (homozygous for the 413.d mutation) of 129/Sv genotype after injection of 10–15 cells into C57BL/6 host blastocysts. (C) Contribution by ES cell lines homozygous for the Hβ58 insertional mutation (C57BL/6 genotype) after injection of 10–15 cells into MF1 strain host blastocysts. Chimeras 1, 3, 4 and 7 were obtained using the LJ2 cell line: chimeras 2, 5, 6 and 8 were obtained using the LJ3 cell line. The relative contributions were quantified by comparing the intensity of the appropriate bands after electrophoretic separation of isozymal variants (McLaren & Buehr 1981). M, missing.

Tissue samples were recovered from 10.5 day p.c. chimeras and analysed as described previously (Beddington & Robertson 1989). Fig. 5 provides a comparison of the patterns of distribution and relative contributions of wild-type 129/Sv cells, 413.d homozygous cells and Hβ58 homozygous cells.

Results from wild-type cells (Beddington & Robertson 1989) indicate that ES cell derivatives are found predominantly in the fetus, amnion and visceral yolk sac mesoderm. However, there were clear instances in which ES cells made a substantial contribution to derivatives of both the primitive endoderm and the trophectoderm (Fig. 5A). 423.d mutant ES cells showed a similar pattern of colonization, with one notable exception: in the majority of the chimeric conceptuses mutant ES cells made a significant contribution to the trophoblast giant cell populations. This cannot be due to contamination as the GPI 1^c

isozymal marker is unique to the ES cells. In summary, 413.d mutant cells can colonize all three primary lineages (Fig. 5B). Unfortunately, these experiments are uninformative with respect to the ability of the 413.d mutant cells to contribute to the visceral yolk sac endoderm.

Analysis of 10 chimeric conceptuses generated using two independent Hβ58 homozygous cell lines showed a more restricted pattern of distribution (Fig. 5C). As predicted from the adult chimera studies, a high level of chimerism was detected in the fetus proper. However, only sporadic and low level contributions were found in the extraembryonic tissues, even in chimeras where mutant cells constituted the predominant cell population in the embryo proper. A few examples were found where the mutant cells colonized the parietal yolk sac. Thus, these cell lines can contribute to the derivatives of the primitive endoderm. Strikingly, there were few instances of colonization of the visceral yolk sac mesoderm. As this analysis was performed at Day 10.5 p.c., one possibility is that mutant cells are present at earlier stages, but do not persist until mid-gestation. It remains to be determined whether Hβ58 cells can contribute to the trophectodermal lineage.

Discussion

Genes expressed in the embryo can be considered as belonging to three very broad categories. Firstly, it seems likely that there are many ubiquitously expressed proteins with functions associated with cell metabolism and division. A second set of gene products will be confined to a specific cell lineage or expressed at different levels within distinct cell lineages. Expression of these genes may relate to the gradual acquisition of a specialized function during development. Finally, there are genes whose expression is required for imparting or interpreting developmental cues.

Mutational analysis can lead to the identification of novel genes that regulate embryonic development. This unbiased approach offers two advantages: it defines gene products for which there is an absolute requirement for completion of normal embryonic processes. Secondly, the phenotypic characteristics displayed by the mutant embryo may give immediate information about the temporal and spatial requirements for gene expression.

Initially, we used the ability to isolate ES cells from the 413.d and Hβ58 strains to address whether developmental failure might be the result of a cell lethal effect. Previous studies have clearly established that ES cell lines can be derived from embryos homozygous for the t^{w5} haplotype and t^{w18} haplotype (both recessive embryonic lethal alleles) but not from homozygous embryos of the t^0 haplotype (Magnuson et al 1982, Martin et al 1987). This approach has also been used to examine mutations mapping to the albino deletion complex (Niswander et al 1988). For the 413.d and Hβ58 mutations, the availability of cloned single copy genomic probes allowed us to determine that homozygous ES cell lines can readily be established.

The spectrum of cell types formed when the mutant ES cells were induced to differentiate in isolation is indistinguishable from that formed by wild-type ES cell lines; these mutant ES cells clearly have the potential to form primitive endoderm-like cells and terminally differentiated cell types of both mesodermal and ectodermal origin. In solid tumours, a wide variety of mature somatic tissues are formed including epithelia, hair follicles and small bones, suggesting that neither mutation compromises the cellular interactions involved in the formation of at least a subset of normal tissues.

One potentially informative approach for studying embryonic lethal mutations is to determine the fate of mutant cells in chimeric embryos. In the case of two early acting lethal mutations, lethal yellow (A^y) and the t^{w5} haplotype of the t-complex, homozygous cells could not be rescued in aggregation chimeras (Papaioannou & Gardner 1979, Barsh et al 1990, Magnuson et al 1983). In contrast, similar studies have demonstrated the developmental capacities of androgenetic and parthenogenetic embryonic cells (Surani et al 1975, Stevens et al 1977, Nagy et al 1989, Clarke et al 1988, Thompson & Solter 1988, Fundele et al 1989).

Here we assessed the developmental potential of 413.d and Hβ58 mutant ES cell lines. Descendants of homozygous 413.d ES cells contributed to derivatives of all three primary tissue lineages in the mid-gestation chimeras. Thus, 413.d mutant cells behave very similarly to wild-type ES cells, except that they show an increased propensity to contribute to the trophectoderm lineage. Overall our studies suggest that ES cells derived from homozygous 413.d embryos are indistinguishable from wild-type ES cells. Indeed, chimeras in which mutant cells constitute the major population are viable. One possible interpretation of these findings is that integration of the provirus interferes with expression of a gene encoding a secreted or diffusible molecule. This would lead to a non-cell-autonomous effect of the mutation. Insight into the mutant phenotype awaits the molecular characterization of the affected gene product.

The locus disrupted by the Hβ58 mutation has been more extensively characterized (Radice et al 1991, J. J. Lee and F. Costantini, unpublished data). A 2.7 kb mRNA known to be affected by the insertion is widely expressed in adult tissues. *In situ* hybridization analysis shows the mRNA to be expressed in all embryonic lineages at all stages examined. However, the level of expression differs between tissues. At 7.5 d p.c., when phenotypic abnormalities first become apparent in the homozygotes, the highest level of expression is observed in the developing visceral yolk sac and ectoplacental cone. Consistent with this, studies analysing the developmental fate of homozygous ES cells provide little evidence for colonization of the extraembryonic lineages at 10.5 days of development. It will be interesting to learn whether the Hβ58 gene product is required for the correct functioning of these tissues.

The developmental role played by any particular gene product can be inferred from its spatial and temporal pattern of expression. However, to determine whether the gene has a controlling influence on development, it is important to test whether inappropriate expression of the candidate gene perturbs normal developmental processes. The ability to isolate and characterize ES cell lines that carry loss-of-function mutations, or to use ES cells that have been manipulated in tissue culture to generate gain-of-function mutations, will contribute to establishing the role(s) of specific genes. An important future component of this system will be 'marked' ES cell lines to trace the fate of mutant cells in the developing embryo.

Acknowledgements

This work was supported by grants to E.J.R. and F.C. from the NIH and March of Dimes Foundation. E.J.R. holds a Fellowship from the David and Lucile Packard Foundation.

References

Barsh GS, Lovett M, Epstein CJ 1990 Effects of lethal yellow (A^y) mutation in mouse aggregation chimeras. Development 109:683–690

Beddington RSP, Robertson EJ 1989 An assessment of the development potential of embryonic stem cells in the midgestation mouse embryo. Development 105:733–737

Bradley A, Evans M, Kaufman MH, Robertson E 1984 Formation of germ-line chimaeras from embryo-derived teratocarcinoma cell lines. Nature (Lond) 309:255–256

Clarke HJ, Varmuza S, Prideaux VR, Rossant J 1988 The developmental potential of parthogenetically derived cells in chimeric mouse embryos: implications for action of imprinted genes. Development 104:175–182

Conlon FL, Barth KS, Robertson EJ 1991 A novel retrovirally induced embryonic lethal mutation in the mouse: assessment of the developmental fate of embryonic stem cells homozygous for the 413.d proviral integration. Development 111:969–981

Costantini F, Radice G, Lee JJ, Chada KK, Perry W, Son HJ 1989 Insertional mutations in transgenic mice. Prog Nucleic Acids Res Mol Biol 36:159–169

Doetschman TC, Eistetter H, Katz M, Schmidt W, Kemler R 1985 The *in vitro* development of blastocyst-derived embryonic stem cell lines: formation of visceral yolk sac, blood islands and myocardium. J Embryol Exp Morphol 87:27–45

Fundele R, Norris ML, Barton SC, Reik W, Surani MA 1989 Systematic elimination of parthenogenetic cells in mouse chimeras. Development 106:29–35

Gossler A, Joyner AL, Rossant J, Skarnes WC 1989 Mouse embryonic stem cells and reporter constructs to detect developmentally regulated genes. Science (Wash DC) 244:463–465

Gridley T, Soriano P, Jaenisch R 1987 Insertional mutagenesis in mice. Trends Genet 3:162–166

Jaenisch R 1988 Transgenic animals. Science (Wash DC) 240:1468–1475

Jaenisch R, Harbers K, Schnieke A et al 1983 Germline integration of Moloney murine leukemia virus at the Mov13 locus leads to recessive mutation and early embryonic death. Cell 32:209–216

Kessel M, Gruss P 1990 Murine developmental control genes. Science (Wash DC) 249:374–379

Lyon MF, Searle AG 1989 Genetic variants and strains of the laboratory mouse, 2nd edn. Oxford University Press, Oxford

Maas RL, Zeller R, Woychik RP, Vogt TF, Leder P 1990 Disruption of formin encoding transcripts in two mutant *limb deformity* alleles. Nature (Lond) 346:853–855

Magnuson T 1986 Mutations and chromosomal abnormalities: how are they useful for studying genetic control of early mammalian development? In: Rossant J, Pedersen RA (eds) Experimental approaches to mammalian embryonic development. Oxford University Press, Oxford p 437–474

Magnuson T, Epstein CJ, Martin GR, Silver LM 1982 Pluripotent embryonic stem cell lines can be derived from t^{w5}/t^{w5} blastocysts. Nature (Lond) 298:750–753

Magnuson T, Martin GR, Silver LM, Epstein CJ 1983 Studies of the viability of t^{w5}/t^{w5} embryonic cells *in vitro* and *in vivo*. In: Silver LM, Martin GR, Strickland S (eds) Teratocarcinoma stem cells. Cold Spring Harbor Press, Cold Spring Harbor, NY p 671–681

Martin GR, Silver LM, Fox HS, Joyner AL 1987 Establishment of embryonic stem cell lines from preimplantation mouse embryos homozygous for lethal mutations in the t-complex. Dev Biol 121:20–28

McLaren A, Buehr M 1981 GPI expression in female germ cells of the mouse. Genet Res 37:305–309

Nagy A, Sass M, Markkula M 1989 Systematic non-uniform distribution of parthenogenetic cells in adult mouse chimeras. Development 106:321–324

Niswander L, Yee D, Rinchik EM, Russell LB, Magnuson T 1988 The albino deletion complex and early post implantation survival in the mouse. Development 102:45–53

Papaioannou V, Gardner RL 1979 Investigation of the lethal yellow A^y/A^y embryo using mouse chimeras. J Embryol Exp Morphol 52:153–163

Radice G, Lee JJ, Costantini F 1991 Hβ58, an insertional mutation affecting early postimplantation development of the mouse embryo. Development 111:801–811

Robertson EJ 1987 Embryo-derived stem cell lines. In: Robertson EJ (ed) Teratocarcinomas and embryonic stem cells: a practical approach. IRL Press, Oxford p 71–112

Robertson EJ, Bradley A 1986 Production of permanent cell lines from embryos and their use in studying developmental problems. In: Rossant J, Pedersen RA (eds) Experimental approaches to mammalian embryonic development. Oxford University Press, Oxford p 475–508

Robertson E, Bradley A, Kuehn M, Evans M 1986 Germ-line transmission of genes introduced into cultured pluripotent cells by retroviral vector. Nature (Lond) 323:445–448

Schnieke A, Harbers K, Jaenisch R 1983 Embryonic lethal mutation in mice induced by retroviral insertion in the a1(I) collagen gene. Nature (Lond) 304:315–320

Soriano P, Gridley T, Jaenisch R 1987 Retroviruses and insertional mutagenesis in mice: proviral integration at the Mov34 locus leads to early embryonic death. Genes & Dev 1:366–375

Stevens LC, Varnum DS, Eicher EM 1977 Viable chimeras produced from normal and parthenogenetic mouse embryos. Nature (Lond) 269:515–517

Surani MA, Barton SC, Kaufman MH 1975 Development to term of chimeras between diploid parthenogenetic and fertilized embryos. Nature (Lond) 270:601–603

Thompson JA, Solter D 1988 The developmental fate of androgenetic, parthenogenetic and gynogenetic cells in chimeric gastrulating mouse embryos. Genes & Dev 2:1344–1351

Woychik RP, Maas RL, Zeller R, Vogt TF, Leder P 1990 "Formins": proteins deduced from the alternative transcripts of the *limb deformity* gene. Nature (Lond) 346:850–853

DISCUSSION

Joyner: Your homozygous ES cell lines in the chimera analysis, have you tested two or three lines and got the same results?

Robertson: Yes, we tested independent lines.

Joyner: When I was with Gail Martin, we made t^{w18} and t^{w5} homozygous cell lines (Martin et al 1987). Those cell lines, like yours, usually differentiate *in vitro* although the embryos don't *in vivo*. Do you think this is going to be a general property?

Robertson: One of the problems with using ES cells to do chimera analysis is that their pattern of colonization is restricted. It would be much more informative to use aggregation chimeras, which would make a more balanced mosaic. However, the problem then becomes one of having to determine retrospectively the genotype of the embryos that were used in the aggregation.

I don't know if t^{w5} ES cells were ever used in the blastocyst injection experiments. Terry Magnuson's data suggest that aggregation of a t^{w5} mutant embryo with a wild-type embryo results in embryonic failure (Magnuson et al 1983). It is as if the t^{w5} cells fail to be regulated by the presence of wild-type cells. Do you know whether anyone has looked more carefully at the t^{w5} ES cells?

Joyner: No, I don't think they were ever followed up. It makes you wonder whether a lot of these mutations are in genes which encode proteins that are present in the culture medium, thereby allowing the mutant ES cells to differentiate *in vitro* in an unorganized manner.

Robertson: It's difficult to address that specifically. However, if the mutant cells can be rescued by 'factors' in tissue culture medium, presumably the 'factors' are also available in the maternal environment.

Solter: Have you tried to rescue 413.d embryos with Hβ58 cells?

Robertson: No, we haven't tried that.

Hogan: How close are the two mutations?

Robertson: There are no recombinants in 146 progeny, according to Nancy Jenkins. This places the loci within 0.68 cM. We are just setting up the appropriate genetic crosses to establish more precisely how closely linked the two mutations are.

Lawson: Is there a possibility that your 413.d mutant has an insertion in a parentally imprinted gene?

Robertson: I don't think so because the genetic data give no indication of this.

Hogan: What happens if you inject your mutant ES cells into tetraploid blastocysts, so that the trophoblast will die, making it likely that the whole of the mouse is derived from ES cells?

Robertson: In those experiments all the extraembryonic membranes, including the placenta, are largely derived from the tetraploid cells, i.e. the ES cells would contribute almost exclusively to the embryo proper. We need to be able to do the experiment the other way round.

Hogan: It would be really nice to generate a whole mouse from the mutant cells. If one could do so, it would suggest that in this mutation there is some defect in the extraembryonic cells that prevents normal development of the embryo.

Robertson: We are planning to try such experiments with the mutant ES cells. When we generate chimeras in a C57BL/6 background the live-born animals are almost completely ES cell derived, so the experiment might work.

Bradley: Have you tried the experiment the other way around, putting wild-type cells into the mutant embryos?

Robertson: No, for technical reasons. We would have to do a retrospective analysis because we wouldn't know the genotype of the blastocyst we had injected. It is a good idea; it's just tedious to do. You would have to find the chimeras, grow out fibroblast clones and determine their genotype to discover what particular combination you had generated.

Evans: Has it proved possible to get completely viable mice by ES cell injection into tetraploid embryos?

Joyner: Andras Nagy has tried six different ES cell lines without success (Nagy et al 1990).

Robertson: Has he tried picking small clonal clumps and using those as the donor ES cells? My worry about those experiments is that you only have to include one or two aneuploid cells in the chimera to get a lethal effect.

Joyner: Yes, he has.

Goodfellow: Liz, to go back to your strategy: if there are 100 000 genes, the basic problem is how to choose the genes you want to study. With this technique you are putting your hand in at random, picking out a gene and saying this is the gene I am going to study. What criteria are you using to make that choice? One could argue that you should knock out 50 *Hox* genes instead of wasting your time taking a gene completely at random when you have no idea what it is doing.

Robertson: This method finds a gene for which there is an absolute requirement; of course, it could turn out to be something like carbonic anhydrase.

Hogan: You would have said that p53 was a really important gene. Allan Bradley has now knocked out the p53 gene and got a completely normal mouse (personal communication). By this method, at least you have a mutant phenotype which shows you there is a requirement for this gene in normal development.

Robertson: Right, there is nothing that can substitute for that gene at that particular time.

Evans: The important difference between this method and the random 'hand in the bag' method suggested by Peter Goodfellow is that here the locus is selected by the direct functional test that it is essential for development. This is a completely different approach to that where a conceptual choice is made of whether a known locus is 'interesting'. This approach is a method for identifying a class of recessive embryonic lethal alleles that are systematically missed in normal genetic screens of both mice and humans. Unless we presuppose that all informative developmental mammalian loci are to be found only by our prescience or by homology with other species or gene families, a screen of this type is essential.

Goodfellow: But what is the functional test telling you about the gene? The answer is we don't know. Is the functional test you are using ever going to be valid for choosing the genes you should work on? Up to this moment in time, the only criterion that has been given for interesting genes is interesting tissue-specific expression patterns. *Wnt* and *Vgr-1* have interesting patterns of expression and we say therefore we should study them.

Buckingham: At the moment one can do this slightly better in the zebrafish. Walter Gilbert and his colleagues are following the expression of β-galactosidase at a particular insertion site in the transparent zebrafish embryo. Then you have an '*in vivo*' pattern of expression as a criterion as well as the effect on embryonic development.

Beddington: That's like gene trap.

McMahon: There's some confusion here. Widespread expression doesn't mean widespread responsiveness. There is quite a lot of evidence in *Drosophila* to indicate that a number of genes that seem to be key developmental genes have widespread distributions. *armadillo*, for instance, a segment polarity gene, has the protein distributed over the whole embryo, yet it seems to function in only a few cells in each segment. The *Notch* gene product is also widely distributed.

Goodfellow: All these are descriptions and I am not disagreeing with them, but you get into a circular argument where you say a gene is important so you knock out the gene. When there is no phenotype, the interpretation is not that p53 is not an interesting gene or that *engrailed* is not an interesting gene, the interpretation is that we have not learnt anything from doing this type of experiment.

Evans: But Liz would not have picked up p53 by her screen, so she is obviously picking up something with a different interest.

Goodfellow: A dominant mutant in p53 may have killed the embryo, so Liz may have picked it up.

Jaenisch: One criterion is that at least you can exclude the cell lethals.

Joyner: I think many ES lines carrying homozygous cell lethal mutations will grow in culture because the medium is so rich.

Robertson: I agree, but unfortunately no systems exist that you could test those in. I think the 413.d mutation does result in an interesting phenotype. The histological sections don't really reflect how interesting it is. Unlike the Hβ58 mutation, where affected embryos develop so far and then basically peter out, the 413.d embryos are severely disturbed in the normal patterning of the cells.

Beddington: So what do you do next with it?

Robertson: At this stage we have to wait for the molecular analysis. There are a few descriptive experiments we could do but they would not tell us much more. They might indicate that there are some cell lineages for which this gene product is absolutely required.

Evans: Is it possible that it's not a knock out but overexpression? The retroviral long terminal repeat is sitting there and it could be enhancing the expression of this gene.

Robertson: Yes, it could be a gain-of-function mutation.

McLaren: How extensively have the heterozygotes and the ES cell-derived chimeras been examined for abnormalities in the adult?

Robertson: Not very.

McLaren: Do they have the right number of toes and that sort of thing?

Robertson: Yes.

Lawson: You said that there was no mesoderm formation. I thought there was extraembryonic mesoderm formation—some thickening of the epiblast, indicating mesoderm formation posteriorly.

Robertson: At that stage of development, we can't tell whether the section is of a mutant embryo or not. We have to develop a method for establishing the 413.d genotype of early embryos to verify this. What unequivocal markers are there for establishing whether or not these are mesodermal derivatives present in a presumptive mutant?

Lawson: Vimentin.

Copp: In the 413.d mutant, the trophoblast was very extensive. Is that due to increased cell number?

Robertson: I haven't tested that carefully; on the basis of histology, I think it would be. It appears that the most extensive regions of trophoblast development derive from the secondary giant cells coming from the ectoplacental cone, not the derivatives of the wall of the original blastocyst (the mural trophectoderm).

Kaufman: With regard to this excessive proliferation of giant cells, could there be a maternal effect here? We observe exactly the same effect when we make triploids of various types (Kaufman 1991). I wondered whether the extraembryonic tissues in diandric triploids would have a much greater proliferative activity than comparable tissues in digynic triploids. Unfortunately, I have yet to analyse the histological sections of this material.

Robertson: We have not done embryo transfers into wild-type foster mothers to test whether the phenotype is associated with the fact that the mother is heterozygous.

Beddington: Your 413.d chimeras are all made by blastocyst injections. Weren't you getting a rather high frequency of trophectoderm?

Robertson: I think so; that was one of the interesting things that compared to normal 129 ES cells there were lots of situations where by glucose phosphate isomerase analysis more than half the cell population of secondary trophoblast giant cells were homozygous for the 413.d mutation.

Copp: I wonder if your phenotype is a trivial problem of trophoblast–decidual interaction, just starving the embryo.

McLaren: If anything, it looked the other way round, as though there was more maternal blood there than normal.

Robertson: We really haven't discussed this problem of reabsorption. It is so complicated to interpret histological sections taken through the deciduum. Maybe we should try and culture them.

Beddington: I think it might be less useful. If an embryo is defective already and you put it in culture, you may compound that defect with things that would never happen normally *in vivo*. The culture conditions, I think we all agree, are not optimal.

It is like the *situs inversus* mutants (Brown et al, this volume). There is a low level of abnormal development in culture anyway, and trying to study a mutant abnormality under particularly 'stressful' conditions could prove very misleading.

McLaren: From the sections of the Hβ58 mutant, it looked as though the extraembryonic development was more extensive than normal, although rather abnormal in form. Did the greater extent of extraembryonic proliferation include the visceral endoderm? That was the tissue that failed to be colonized in the chimeras. It would seem paradoxical if heterozygotes showed increased development of a tissue that was not colonized in chimeras.

Robertson: I don't think they are larger than normal; the amnion and the allantois are certainly more abnormal than is the visceral yolk sac. Maybe the mutant cells do colonize these lineages initially but then are either selected against or just fail to keep pace with the wild-type cells and hence are not represented in the tissue by mid-gestation.

McMahon: If this approach is to be broadly applicable to the study of mutants, one has to be able to derive ES cell lines from different strains. What's the current consensus on the ease of derivation of ES cell lines from different mouse strains?

Robertson: Some strains, e.g. 129, C57BL, are very easy, whereas others, or indeed F1 combinations, are more problematical.

Jaenisch: We have tried to derive ES cells from FVB mice, a very vigorous inbred strain, which has big pronuclei, gives large litters and therefore is a good strain for producing transgenics. Unfortunately, it was impossible to derive ES cell lines under conditions where we easily obtained ES cell lines from 129 or C57/BL embryos, even by varying the culture conditions, or by using delayed implantation.

McLaren: It sounds as if you have no idea what the properties are of a strain that would make it good or bad for making ES cell lines.

Joyner: Gail Martin and I had the same problem with CD-1. 129 seems to work well in many different labs.

References

Brown NA, McCarthy A, Seo J 1992 Development of the left–right axis. In: Postimplantation development in the mouse. Wiley, Chichester (Ciba Found Symp 165) p 144–161

Kaufman MH 1991 New insights into triploidy and tetraploidy, from an analysis of model systems for these conditions. Hum Reprod 6:8–16

Magnuson T, Martin GR, Silver LM, Epstein CJ 1983 Studies on the viability of t^{w5}/t^{w5} embryonic cells in vitro and in vivo. In: Silver LM, Martin GR, Strickland S (eds) Teratocarcinoma stem cells. Cold Spring Harbor Laboratory Press, Cold Spring Harbor, NY p 671–681

Martin GR, Silver LM, Fox HS, Joyner AL 1987 Establishment of embryonic stem cell lines from preimplantation mouse embryos homozygous for lethal mutations in the t-complex. Dev Biol 121:20–28

Nagy A, Gócza E, Diaz EM et al 1990 Embryonic stem cells alone are able to support fetal development in the mouse. Development 110:815–821

Genetic manipulation of the mouse via gene targeting in embryonic stem cells

Allan Bradley, Ramiro Ramírez-Solis, Hui Zheng, Paul Hasty and Ann Davis

Institute for Molecular Genetics, Baylor College of Medicine, Houston, TX 77030, USA

Abstract. Gene targeting applied to totipotent embryonic stem (ES) cells is a very powerful means of creating highly specific mutations of genes in the mouse. The successful application of this technology is however constrained by both the types of mutations that can be generated at a target locus and the ability to reconstruct a germline chimera from the manipulated cells. We have developed two cell lines that can be routinely transmitted through the germline of chimeras after cloning and prolonged selection in tissue culture. We have also established a variety of methods for generating non-selected mutations at the X-linked *hprt* locus in ES cells. Our observations at this locus have enabled us to generate successfully a subtle mutation at the non-selectable *Hox-2.6* locus.

1992 Postimplantation development in the mouse, Wiley, Chichester (Ciba Foundation Symposium 165) p 256–276

The correlation of genotype and phenotype is a central theme in the analysis of development. Indeed, the observational pathway has often been from phenotype to genotype and molecular analysis in organisms such as *Caenorhabditis elegans* and *Drosophila*. In the mouse this type of reverse genetics is constrained not only by the size of the murine genome, but also by the limited availability of mutations that can readily be identified and maintained (Green 1989). Embryonic lethal mutations not closely linked to a visible phenotype are difficult to observe and keep. There are just a few examples of mutations in mice that correlate with cloned genes (Balling et al 1988, Chabot et al 1988, Li et al 1990, Yoshida et al 1990). While mapping and attempting to identify mutations in candidate genes is valuable, its applicability is clearly limited.

Another approach to link phenotype to genotype in the mouse is derived from transgenic insertions that have mutated endogenous genes at the integration site. The transgene is a molecular flag which can serve as a marker to clone the disrupted gene. A number of retroviral and microinjection transgenic insertions have led to the cloning of novel genes important to the developing organism (Schnieke et al 1983, Maas et al 1990).

While each approach is uniquely important, one of the values of the genetic approach is the ability to generate an allelic series of mutations that functionally dissect a genetic locus. These are available for a limited set of genes in mice (such as *Steel* and *W*) (Chabot et al 1988, Copeland et al 1990), but for most cloned murine genes a mutant mouse does not exist.

The demonstration that embryonic stem (ES) cell lines can repopulate the germline (Bradley et al 1984) after prolonged *in vitro* culture, and that they possess the appropriate homologous recombination machinery (Thomas & Capecchi 1987) will undoubtedly facilitate a genetic analysis of function of many cloned genes. Moreover, unlike in *Drosophila* and *C. elegans*, the genetic dissection of the mouse need not be constrained by the generation of random mutations in functional areas of a genetic target.

The ability to manipulate the mouse genome is currently limited by both the biology of the ES cells and the efficiencies with which gene targeting occurs in mammalian cells. Our studies on vector–chromosome homologous recombination have enabled us to devise efficient strategies that facilitate the construction of site-directed mutations in non-selectable genes (Davis et al 1992, Hasty et al 1991a). The application of these methodologies to recently isolated ES cell lines (McMahon & Bradley 1990) has led to a much higher frequency of chimera formation and germline transmission than was previously possible.

Clonal heterogeneity of ES cells and germline transmission

The ultimate assay for a normal ES cell is its ability to participate in embryogenesis after injection into a host blastocyst. The fate and germline transmission of descendants of the introduced cells are determined by a variety of factors (Bradley 1990), which include the differentiation state of the cell, the chromosomal state (aneuploid or euploid), the sex of both the host and ES cells, and the genotype of both the host and ES cell components.

The first *in vivo* differentiation experiments performed by injecting ES cells into the 3.5 day blastocyst used ES cells that had been expanded from isolated secondary clones from the primary explant of a single embryo (Bradley et al 1984). With the particular host–ES cell combination used at this time, germline transmission was achieved. The significance of a host genotype was not initially realized (Bradley et al 1984, Robertson et al 1986), although studies of transmission from aggregation chimeras had previously detected specific biases with certain embryo combinations (Mullen & Whitten 1977). The first systematic studies of the effect of the host embryo genotype in combination with microinjected 129-derived ES cells revealed a similar dependence on genotype (Schwartzberg et al 1989). In addition, it rapidly became apparent that the successful incorporation of ES cells into a host embryo was very dependent on the specific clone of cells chosen for analysis. The pattern of colonization of the clonal isolates of the original ES cell lines in chimeras does not necessarily reflect the pattern of the parental cell line (Table 1).

Table 1 demonstrates that the assay of chimera formation divides clonal isolates from the CCE ES cell line (Robertson et al 1986) into two distinct subpopulations, one of which has properties analogous to the original population, while the others behave more like embryonal carcinoma (EC) cells (Papaioannou & Rossant 1983). The pattern of colonization of all of the clones derived from AB1 (McMahon & Bradley 1990) and AB2.1 (Soriano et al 1991) cell lines resembles that of the parental population. This observation is particularly significant in the context of gene targeting experiments that may generate very few targeting events. The difference between the CCE and AB cell lines may be a property of the cell lines or related to the culture environment, specifically, the feeder layer and serum batches used for culture.

Gene targeting: generating non-selectable mutations

The introduction of a vector containing homology to a target gene into ES cells can result in either a legitimate (targeted) or illegitimate (random) recombination

TABLE 1   Construction of chimeras using embryonic stem cell populations and clones

Populations

Cell line	Percent chimeric	Extent of ES contribution to chimeras	Frequency of germline transmission (per animal)
CCE	70	90%	9/10
AB1 & 2	88	90%	9/11

Clones

Cell line	Type of clones	Percent chimeric	Extent of ES contribution to chimeras	Frequency of germline chimeras (per clone examined)
CCE	Normal	84	90%	3/3
CCE	Abnormal	38	10%	0/11
AB1	Normal	98	95%	7/8
AB2.1	Normal	91	85%	3/5

CCE, AB1 and AB2.1 cells were isolated and maintained as previously described (Robertson et al 1986, McMahon & Bradley 1990, Soriano et al 1991). The populations and clonal derivations were performed at comparable passage numbers. Populations were injected at passage numbers of 10–15. Clones were isolated in this period. All the results presented here are the result of injection of 10–15 ES cells into C57BL/6 host blastocysts. The percentage chimeric represents the proportion of animals that survived to 10 days that were scored as chimeric. The extent of contribution to the chimera is an estimate based on the amount of agouti in the coat of the mice.

event. The absolute frequency of each event depends on a number of factors, such as the genetic target, the length of homology (Thomas & Capecchi 1987, Hasty et al 1991b) and the type of vector used (insertion or replacement vectors) (Hasty et al 1991c). For many purposes, the relative frequency of these events can also be manipulated by aspects of the vector. For instance, vectors can be designed without a promoter so that the cell can aquire G418 resistance only following a targeted integration event. The most important proviso for this selection is that the target gene is expressed in ES cells (Schwartzberg et al 1989, 1990, Stanton et al 1990). It is also possible to reduce the proportion of illegitimate recombinants by requiring that the targeted recombination event proceeds via double reciprocal recombination. This removes the external negatively selectable genes from the targeting vector (Mansour et al 1988). The majority of random intergration events include these elements and the cells may be selectively killed.

These techniques can be utilized separately or in combination as powerful ways of selecting for rare targeting events. However, their applicability is clearly limited to a subset of genetic changes—insertional inactivation or deletion of coding sequences to generate a null allele. They are not appropriate for directing very specific changes to target genes, which will ultimately be desirable for many genes. Analysis of mutations at a fine level, by definition, requires the exclusion of the selectable marker from the target locus. Targeting an expression cassette into a locus will introduce enhancers, promoters and potential splicing artifacts, which may have a direct impact on the pattern of expression of the target gene. If selection cassettes are to be excluded from the target locus, this introduces another complexity into attempts to generate subtle mutations, since the transfection events cannot be selected for directly. Given that the electroporation efficiency is of the order of 10^{-3}, and considering a gene with a moderate targeting frequency of 10^{-3} targeted/illegitimate integration events, without positive selection for transfection events the ratio would be 10^{-6} targeted events per cell transfected. The generation of a non-selectable mutation would thus become a very rare event. We describe here three strategies to introduce non-selectable mutations: co-electroporation (Davis et al 1992), 'Hit and Run' (Hasty et al 1991a), and 'Double Hit' gene targeting.

One of the basic rationales to which all gene targeting concepts are subject is the belief that sequence modifications made in the homologous sequences in a vector will be faithfully transferred to the chromosomal target. While the fidelity may not be a significant concern when generating a null allele, it is important in attempts to generate precisely defined changes in the genome. Although studies in a variety of gene targeting systems have observed a high frequency of error (Thomas & Capecchi 1986, Brinster et al 1989), our recent survey of approximately 80 kb of junction fragments representing 44 targeted events revealed a very low error rate. We detected just two mutations

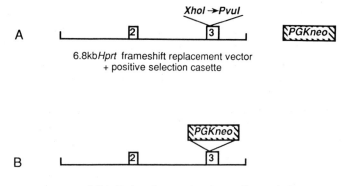

FIG. 1. (A) Co-electroporation vectors. RV6.8XP contains 6.8 kb of the mouse *hprt* gene including exons 2 and 3. Exon 3 has been modified to include a 4 bp frameshift insertion that destroys an *Xho*I site and creates a diagnostic *Pvu*I site. This vector is co-transfected with the unlinked expression cassette *PGKneo*. (B) Control vector for co-electroporation. *PGKneo* inserted into the *Xho*I site in exon 3 of the 6.8 kb *hprt* vector.

in 80 kb: a deletion of a T and a T to G transversion (Zheng et al 1991). Thus, we are confident that minor sequence changes may be faithfully established in the genome by homologous recombination.

Co-electroporation

Targeted recombination events may typically be generated at absolute frequencies of 10^{-5} to 10^{-6} per cell transfected (Thomas & Capecchi 1987, Davis et al 1992, Soriano et al 1991, Hasty et al 1991a,b,c). Since only one in 10^3 cells actually integrates the transfected DNA, this ratio can be enhanced by positive selection for the transfected cells. The relative frequency of targeted events in this subpopulation will be in the range of 10^{-2} to 10^{-3}, thus, positive selection facilitates the direct isolation of targeted events. Positive selection of all transfection events including legitimate as well as illegitimate recombinants should be possible with an unlinked transfection marker (co-transfection), provided that the treated cells can take up and integrate more than one DNA molecule. For co-transfection to work with gene targeting, the additional proviso is that both a random and a targeted integration event must occur in the same cell.

To test if multiple site integration is possible in ES cells, we transfected DNA at various concentrations into these cells. We routinely observed multiple insertions under a variety of conditions and that at least 15% of the G418[r] clones integrated the introduced DNA at two different chromosomal locations.

To test if we could obtain both a targeted and a non-targeted event in ES cells by co-electroporation, we chose to target the X-linked *hprt* (hypoxanthine phosphoribosyl transferase) gene (single copy in XY ES cells). Targeted recombinants generated in this locus can be designed to lose all enzyme activity; these clones can be directly selected in 6-thioguanine (6TG), regardless of the fidelity of the integration event.

We constructed a targeting vector (RV6.8XP) which contains a 6.8 kb fragment of the mouse *hprt* gene, including exons 2 and 3 (Fig. 1). This fragment had been modified to include a 4 bp frameshift insertion in exon 3 that destroyed a native *Xho*I site and created a diagnostic *Pvu*I site. We co-transfected this vector at various concentrations with a cassette carrying the *neomycin* gene (*PGK neo pA*) into CCE cells. We obtained 6TG^r clones at a ratio of approximately 1/500 G418^r clones, very similar to with a control plasmid where the *neo* cassette was cloned in the *Xho*I site in exon 3.

Detailed molecular analysis of the integration structures in the 6TG^r clones revealed that all the clones were targeted, although only a subset could be identified as having the desired integration pattern. Further analysis of these clones using the polymerase chain reaction verified the conversion of the *Xho*I site to the *Pvu*I site. The frequency of *bona fide* clones was 8% of the 6TG^r clones analysed. Although this figure is low, it is not unexpected since the predicted double crossover/gene conversion integration event has occurred in only 10% of the 6TG^r clones examined after transfection with the control vector (Hasty et al 1991c). In the co-transfection experiments, we are also requiring a second random integration event to generate a G418^r clone.

Co-transfection enables one to generate specific mutations with the selectable marker located at a chromosomal site other than the target gene (Davis et al 1991). This makes it possible to target genes which suppress *neo* expression when integrated into the locus. The mutations generated by co-electroporation occur in a single step. This is important for many applications, since ES cells might lose their totipotent state during the prolonged culture that other targeting schemes require.

Hit and Run

The recombination mechanism for this type of approach is indicated in Fig. 2. The vector is designed to undergo a single reciprocal integration reaction utilizing the double-stranded break in the region of homology. Although this type of vector offers no selection step to enrich for targeted events, our observations suggest that the targeting frequency of vectors linearized within the region of homology can be significantly higher than that of vectors that require a double reciprocal/gene conversion exchange (Table 2, Hasty et al 1991c). The single reciprocal exchange is initiated at the homologous ends that constitute the

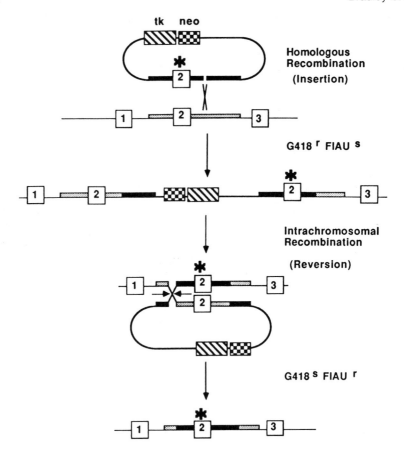

FIG. 2. Two-step recombination procedure used to introduce a small mutation into a non-selectable gene. The initial insertional recombination event involves single reciprocal recombination between the gene target and the insertion vector.The vector contains a small mutation in the homologous DNA with *neo* and *tk* (tyrosine kinase) genes located outside the region of homology. The vector is linearized within the region of homology at a unique restriction site. The entire vector integrates at the target gene and creates a duplication of genomic sequences. One, both or neither of the duplicates may contain the desired mutation. The targeted cell becomes G418^r and FIAU^s. G418^r colonies are screened to find targeted events that included the mutation. Intrachromosomal recombination will subsequently occur within the duplication. This results in the loss of the plasmid, *neo* and *tk* sequences and the duplication. The reverted cell becomes G418^s and FIAU^r: these cells are screened for the presence of the mutation. Thick lines represent homologous DNA: black is derived from the vector and stippled from the genome. Thin lines represent non-homologous genomic DNA. Lines of intermediate thickness are plasmid. Exons are boxed and numbered. * is the small mutation. Horizontal arrows represent the plane of resolution within the Holliday structure. Reproduced with permission from Nature 350:243–246. Copyright © Macmillan Magazines Ltd.

TABLE 2 Insertion and replacement vectors that target the *hprt* locus

Vector	Type	No of electroporation	Total G418^r colonies	Total 6TG^r colonies	6TG^r/G418^r
IV1.3	Insertion	14	854	2	1/427
RV1.3	Replacement	33	4919	0	—
IV6.8	Insertion	46	23884	610	1/39
RV6.8	Replacement	12	10264	43	1/239

IV1.3 is an insertion vector that contains 1.3 kb of sequences homolgous to the *hprt* gene. The *neo* is cloned into the plasmid backbone. RV1.3 is the analogous replacement vector with the same 1.3 kb of homologous sequences except the *neo* is cloned into the homologous sequences. IV6.8 and RV6.8 are constructed in the same way as IV1.3 and RV1.3, except they have 6.8 kb of homologous sequences. For each construct, DNA (25 µg) was linearized at a unique restriction enzyme site; the insertion vectors were cut within the *hprt* region of homology at the *Xho*I site in exon 3 and the replacement vectors were cut at the *Sac*I site that joins the plasmid polylinker to the *hprt* sequences. For a single electroporation the DNA was transfected into AB1 ES cells at 1×10^7 cells/ml phosphate-buffered saline using a BioRad gene pulser at 230 V and 500 µF. After electroporation, each batch of cells was plated onto G418^r STO feeder layers in 90 mm plates and selected in 180 µg/ml G418 active ingredient. After five days 1×10^{-5} M 6-thioguanine was added and both selections were maintained for another 14 days (for each construct tested one plate out of ten was selected in only G418 to obtain the number of G418^r colonies). Expression of MC1neopA (Thomas & Copecchi 1987) is subject to positional effects such that a sevenfold increase in G418^r colonies was seen in constructs with 6.8 kb of *hprt* compared with the 1.3 kb region of *hprt*.

linearization site. The details of the recombination event have some similarity with the double-strand break repair model (Orr-Weaver et al 1981). The resultant recombinant structure is a direct repeat of the homologous sequence separated by the plasmid, *neo* and herpes simplex virus tyrosine kinase (*HSVtk*) genes. Because the Holliday junctions (Holliday 1964) formed during the initial integration event can migrate, the region of the homologous sequence that differs between the vector and chromosomal target may form a region of heteroduplex. The differential repair of heteroduplexed DNA and the point of resolution of the Holliday junction will determine whether the introduced mutation ends up in the 5′ or 3′ regions, both or neither of the duplicates.

Our analysis of the insertions into the *hprt* and *Hox-2.6* genes revealed all these patterns (Hasty et al 1991a). The insertion events generate a direct repeat which may be resolved by intrachromosomal recombination between duplicates on the same chromosome or between sister chromatids at the time of cell division. At the *hprt* locus the reversion events are directly selectable because the reverted cells will contain the wild-type *hprt* gene and be resistant to HAT (hypoxanthine aminopterin thymidine). We have determined the frequency of these events at the *hprt* locus (Fig. 3). A 1.3 kb duplication reverted at a frequency of 10^{-6} per cell generation; a 6.8 kb duplication (a fivefold increase in homology) reverted at a frequency of 10^{-5} per cell generation. Introducing a small

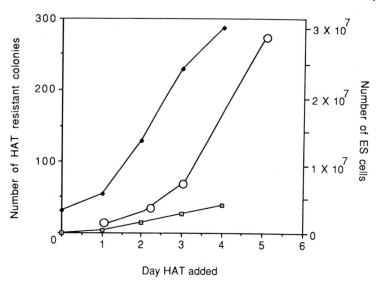

FIG. 3. Reversion of duplicted homology at the *hprt* locus. ▣ , IV1.3 is a 1.3 kb duplication; ♦, IV6.8 is a 6.8 kb duplication. ○, cell growth without selection. Each duplication is separated by the *neo* and plasmid sequences, totalling 4 kb. 6-Thioguanine-resistant (6TG[r]) clones, shown to be targeted by Southern analysis, were grown in 6TG until they were plated to calculate the reversion rate. This was done to avoid over-representation of revertant clones representing daughter cells that arose early during clonal expansion. Cells were plated at 2.5×10^6 cells/90 mm plate on duplicate plates and 6TG selection was removed. HAT was added to a final concentration of 1×10^{-4} M hypoxanthine, 4×10^{-7} M aminopterin and 1.6×10^{-5} M thymidine to one plate every 24 hours starting at Day 0 for five days. HAT[r] colonies were counted on Day 10. Cell generation time was determined by plating 2.5×10^6 ES cells/90 mm plate and counting the number of cells on Days 1, 2, 3 and 5. Each total was adjusted to account for the feeder cells.

oligonucleotide mismatch into the element of the duplication resulted in a 10-fold decrease in frequency. At the non-selectable *Hox-2.6* locus, the reversion events can be selected for because they excise the *HSVtk* gene and the cells become resistant to the nucleoside analogue FIAU (1-(2-deoxy-2-fluoro-β-D-arabinofuranosyl)-5-iodouracil). Interestingly, the *Hox-2.6* locus displays a much higher reversion frequency than the *hprt* locus: with a 3.1 kb duplication and a single mutation, revertants could be selected at a frequency of 10^{-3} per cell division (Hasty et al 1991a). We examined the reversion events from six primary clones and were able to detect excision of the inserted vector in revertants from only two. Approximately 15% of the revertant clones that excised the vector retained the oligonucleotide mutation. These revertant clones were derived from a primary recombinant which had the mutation in the 5' duplicate. In these cases either the crossover occurred in just 0.5 kb or it occurred elsewhere in association with gene conversion. This agrees with studies on intrachromosomal recombination in other mammalian cell types (Bollag & Liskay 1988).

The application of the Hit and Run technique at the *Hox-2.6* locus has enabled us to generate targeted clones which have a premature termination codon in the third helix of the homeodomain. Several clones carrying this mutation contribute extensively to chimeric mice.

Double Hit gene replacement

The basis of this procedure is the introduction of a negatively selectable marker into the target site of interest. Clones of cells carrying this mutated allele become

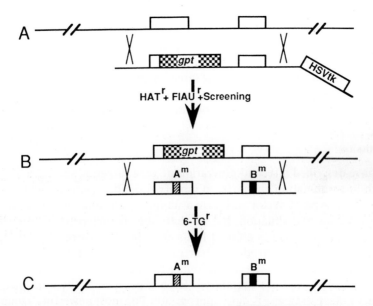

FIG. 4.   Double Hit gene replacement scheme. (A) A hypothetical gene replacement event replacing sequence elements in a target gene by double reciprocal recombination. Positive selection for *gpt* (guanine phosphoribosyl transferase) is in HAT in *hprt⁻* cells and negative selection for the targeting event is accomplished by the herpes simplex virus tyrosine kinase gene. Targeted clones are enriched by HAT + FIAU selection and screened by analysis using Southern blotting and the polymerase chain reaction. (B) The targeted allele resulting from the recombination event depicted in (A) becomes the target for a second recombination event, the 'Double Hit'. The hypothetical vector is drawn with two areas of change, A^m in exon 1 and B^m in exon 2. Targeted events will excise the *gpt* gene and may be selected in 6-thioguanine (6-TG). (C) The desired 'Double Hit' recombinant product where areas A^m and B^m have been exchanged with the target gene. FIAU, 1-(2-deoxy-2-fluoro-β-D-arabinofuranosyl)-5-iodouracil); HAT, hypoxanthine aminopterin thymidine).

the target cells for secondary gene targeting experiments to perform gene replacement events at the allele that has already been targeted. The secondary gene replacement events are directly selectable because the negative marker should be removed by gene replacement/gene conversion (Fig. 4).

Negative selection is a very powerful means of selecting for the loss of gene activity at the target locus. However, for it to be effective, the background number of cells that survive selection must be 10^{-6} to 10^{-7} per selected cell, since secondary recombination events per transfected cell are likely to be generated at this frequency. We have targeted negatively selectable genes to both the *Hox-2.6* and c-*myc* loci and placed these targeted clones under negative selection. In both cases there is a very high background number of cells that survive negative selection. We have determined that the clones that survive have lost the negative selection cassette from the c-*myc* and *Hox-2.6* loci. The loss of the negative marker could be explained by gene conversion between the homologous chromosomes or by chromosome loss (followed by reduplication). The high background frequency is, however, a significant problem for isolating the secondary targeting events. In principle, 'Double Hit' gene replacement is the most flexible and powerful gene targeting technique, because once the allele is set up those cells may be used repeatedly for generating targeted events; in practice, its use has yet to be demonstrated.

Discussion

We have described a second generation of gene targeting vectors and ES cells which in combination enable us to modify the mouse genome in a variety of ways. Each type of vector will have a unique applicability, depending on the exact experimental situation, but all share the characteristic of being able to direct specific changes in a target locus at its resident chromosomal location. These vectors do not introduce other transcription units into a locus and thus constitute a major advance in our ability to interpret the consequences of the genetic changes we have engineered. Each experimental situation will dictate which of these strategies is most appropriate. The most powerful, 'Double Hit' gene replacement, relies on the exchange of a negative marker at the target site with vector sequences. Virtually any genetic change (deletions, insertions, exchange of control regions or exchange of coding sequences) should be able to be made and isolated by selection. However, this particular strategy has yet to be demonstated to work experimentally.

Co-electroporation can mediate the same types of genetic changes, namely gene replacement events. In this experimental situation, the frequency of the desired exchange is relatively modest, 1/6000 at the *hprt* locus, although it does not differ dramatically from that of gene replacement events that do not require co-electroporation in the same region of *hprt*. For genes with a relatively high targeting efficiency, co-electroporation is a viable procedure. It will also be a very valuable technique where selective markers become silenced by the target locus. In situations where ES culture conditions are sub-optimal, co-electroporation has the advantage of generating recombinants in a single step. This can be very important because the probablity of isolating a totipotent clone through the two rounds of cloning required for the Double Hit gene replacement approach may be low.

The Hit and Run procedure utilizes the most efficient pathway for gene targeting in ES cells (Hasty et al 1991c). Our observations on this pathway suggest that this vector–chromosome interaction has some similarities with, as well as important differences from, the double-strand break repair model (Orr-Weaver et al 1981). In experimental scenarios where considerable effort must be dedicated to screen/select rare recombinants from large populations of transfected cells, increasing the relative abundance of the required recombinant is desirable. Our results have shown that insertion vectors will form the predicted recombinant structures about 100-fold more efficiently than replacement vectors can mediate the desired replacement events (Hasty et al 1991c). The high frequency of targeting with insertion vectors was a necessary prerequisite for the success of the Hit and Run strategy. If the primary recombinants were difficult to detect, then the system would be more difficult to work with. Although the reversion events occurred at modest frequency, the ability to kill non-revertant clones selectively negates the relative importance of the frequency of the reversion event. We have used the Hit and Run approach to introduce an oligonucleotide insertion (Hasty et al 1991a), but it may also be used for larger genome modifications.

One of the advantages of insertion vectors is the possibility of generating more than one mutation in a single experiment. The primary recombinant clones may represent null mutations of a gene. Mice can be generated with these clones provided the frequency of somatic reversion is not so high that the phenotype of the resultant mouse becomes attenuated by selective growth of wild-type revertant clones.

It is now possible to modify virtually any gene and establish the mutation in the germline. As developmental biologists attempting to unravel the genetic control of mouse development, our limitations are no longer what we can do, but understanding what we have done!

Acknowledgements

This work was supported by grants from the NIH (A.B.), the Searle Scholars Program (Chicago Community Trust (A.B.), the Cystic Fibrosis Foundation (A.B. and P.H.), the MRC Canada (A.D.), The Centro International de Biologia Molecular, A. C. (R.R.S.) and the March of Dimes Birth Defects Foundation (H.Z.).

References

Balling R, Deutsch U, Gruss P 1988 *Undulated*, a mutation affecting the development of the mouse skeleton, has a point mutation in the paired box of *Pax 1*. Cell 55:531–535

Bollag RJ, Liskay RM 1988 Conservative intrachromosomal recombination between inverted repeats in mouse cells. Association between reciprocal exchange and gene conversion. Genetics 119:161–169

Bradley A 1990 Embryonic stem cells: proliferation and differentiation. Curr Opin Cell Biol 2:1013–1017

Bradley A, Evans M, Kaufman MH, Robertson E 1984 Formation of germ-line chimaeras from embryo-derived teratocarcinoma cell lines. Nature (Lond) 309:255–256

Brinster RL, Braun RE, Lo D, Avarbock MR, Oram F, Palmiter RD 1989 Targeted correction of a major histocompatibility class II α gene by DNA microinjected into mouse eggs. Proc Natl Acad Sci USA 86:7087–7091

Chabot B, Stephenson DA, Chapman VM, Besmer P, Bernstein A 1988 The proto-oncogene c-kit encoding a transmembrane tyrosine kinase receptor maps to the mouse W locus. Nature (Lond) 335:88–89

Copeland NG, Gilbert DJ, Cho BC et al 1990 Mast cell growth factor maps near the steel locus on mouse chromosome 10 and is deleted in a number of steel alleles. Cell 63:175–183

Davis AC, Wims M, Bradley A 1992 Investigation of co-electroporation as a method for introducing non-selectable mutations into embryonic stem cells. submitted

Green MC 1989 Catalog of mutant genes and polymorphic loci. In: Lyon MF, Searle AG (eds) Genetic variants and strains of the laboratory mouse, 2nd edn. Oxford University Press, London

Hasty P, Ramirez-Solis R, Krumlauf R, Bradley A 1991a Introduction of a subtle mutation into the Hox 2.6 locus in embryonic stem cells. Nature (Lond) 350:243–246

Hasty P, Rivera-Perez J, Bradley A 1991b The length of homology required for gene targeting in embryonic stem cells. Mol Cell Biol 11:5586–5591

Hasty P, Rivera-Perez J, Chang C, Bradley A 1991c Targeting frequency and integration pattern for insertion and replacement vectors in ES cells. Mol Cell Biol 11:4509–4517

Holliday R 1964 A mechanism for gene conversion in fungi. Genet Res 5:282–304

Li S, Crenshaw EB, Rawson EJ, Simmons DM, Swanson LW, Rosenfeld MC 1990 Dwarf locus mutants lacking three pituitary cell types result from mutations in the POU-domain gene pit-1. Nature (Lond) 347:528–533

Maas RL, Zeller R, Woychik RP, Vogt TF, Leder P 1990 Disruption of formin-encoding transcripts in two mutant limb deformity alleles. Nature (Lond) 346:853–855

Mansour SL, Thomas KR, Capecchi MR 1988 Disruption of the protooncogene int-2 in mouse embryo derived stem cells: a general strategy for targeting mutations to non-selectable genes. Nature (Lond) 336:348–352

McMahon A, Bradley A 1990 The wnt-1 (int-1) proto-oncogene is required for development of a large region of the mouse brain. Cell 62:1073–1085

Mullen RJ, Whitten WK 1977 Relationship of genotype and degree of coat color to sex ratios and gametogenesis in chimaeric mice. J Exp Zool 9:111–129

Orr-Weaver TL, Szostak JW, Rothstein RJ 1981 Yeast transformation, a model system to study recombination. Proc Natl Acad Sci USA 78:6354–6358

Papaioannou VE, Rossant J 1983 Effects of the embryonic environment on the proliferation and differentiation of embryonal carcinoma cells. Cancer Surv 2:165–183

Robertson EJ, Bradley A, Kuehn M, Evans M 1986 Germ line transmission of genes introduced into cultured pluripotential cells by retroviral vector. Nature (Lond) 323:445–448

Schnieke A, Harbers K, Jaenisch R 1983 Embryonic lethal mutation in mice induced by retrovirus insertion into the α1(I) collagen gene. Nature (Lond) 304:315–320

Schwartzberg PL, Robertson EJ, Goff SP 1990 Targeted disruption of the endogenous c-abl locus by homologous recombination with DNA encoding a selectable fusion protein. Pro Natl Acad Sci USA 87:3210–3214

Schwartzberg PL, Goff SP, Robertson EJ 1989 Germ-Line transmission of a c-*abl* mutation produced by targeted gene disruption in ES cells. Science (Wash DC) 246:799–803

Soriano P, Montgomery C, Geske R, Bradley A 1991 Targeted disruption of the c-src proto-oncogene leads to osteopetrosis in mice. Cell 64:693–702

Stanton B, Reid SW, Parada LF 1990 Germ line transmission of an inactive N-myc allele generated by homologous recombination in mouse embryonic stem cells. Mol Cell Biol 10:6755–6758

Thomas KR, Capecchi MR 1986 Introduction of homologous DNA sequences into mammalian cells induces mutations in the cognate gene. Nature (Lond) 324:34–38

Thomas KR, Capecchi MR 1987 Site directed mutagenesis by gene targeting in mouse embryo-derived stem cells. Cell 51:503–512

Yoshida H, Hayashi S-I, Kunisada T et al 1990 The murine mutation osteopetrosis is in the coding region of the macrophage colony stimulating factor gene. Nature (Lond) 345:442–444

Zheng H, Hasty P, Brenneman M et al 1991 Fidelity of targeted recombination in human fibroblasts and murine embryonic stem cells. Proc Natl Acad Sci USA 88:8067–8071

DISCUSSION

Jaenisch: Have you seen a difference in targeting frequency depending on where you linearize your plasmid, either in the replacement or in the insertion vectors?

Bradley: If we linearize a replacement vector in the region that is homologous to the target locus, we see a higher targeting frequency. Incidentally, this linearization would result in the integration of the vector in the manner of an insertion vector that undergoes a single reciprocal exchange. If we linearize the same vector at the edge of the region of homology, the vector is predicted to integrate by double reciprocal recombination, but this occurs at a much lower frequency.

Jaenisch: But in a replacement vector you can linearize on one side or the other side without affecting the internal homology; does this make a difference?

Bradley: No, it doesn't. Whether you linearize a replacement vector on one side or both sides doesn't make any difference to the frequency. If you linearize in the region of homology, the frequency jumps by at least an order of magnitude.

McLaren: Is there an optimal length of homology?

Bradley: We find that with less than 2 kb the frequency is quite low. We see a fairly linear relationship between the length of homology and the frequency working with the *hprt* locus: we don't have data for any other locus at this point. Above about 6 kb we don't see a dramatic improvement in frequency with added homology. I would argue that using more homology beyond 6 kb is not going to be helpful; not because it is not going to increase the frequency—which it might do to some extent—but mainly because the integration structures are much more difficult to analyse. If duplications are occurring at the locus, you are going to be looking at 20 kb fragments and it's going to be very difficult to understand the integration structure.

Joyner: Are those numbers just for insertion vectors or for replacement vectors as well?

Bradley: Insertion vectors have a clean integration structure, so there probably isn't a problem with adding more homology. With insertion vectors we see a similar, fairly linear increase in frequency with length.

Robertson: What was the largest deletion that you made within the targeting vector? How much sequence can you remove and still get an adequate homologous recombination frequency?

Bradley: We have deleted 7 kb in one step.

Robertson: How does removing 2 or 3 kb compare with taking out 7 kb? Does the homologous recombination frequency drop off dramatically?

Bradley: We haven't done a direct comparison. There is a day to day difference in targeting frequency; some days it may decrease or increase by up to a factor of 10. Unless you did these comparisons side by side, on the same day with the same gene and everything, it would be really difficult to know. Our basic feeling is that there isn't a problem with making deletions.

Robertson: It wouldn't jeopardize your entire experiment if you wanted to delete genomic sequences?

Bradley: No, I don't think it would.

Evans: But I take it you can only delete with a replacement vector because you can't introduce much deletion with an insertion vector.

Bradley: You can delete with an insertion vector using the Hit and Run approach we have described, except the mutation would be a deletion rather than an oligonucleotide insertion. One problem with this is that the mechanism of recombination relies on exonuclease activity; if the exonuclease digestion proceeds to the gap, it will be repaired. The important thing if you are using an insertion vector to make a deletion is to have a large distance between the linearization site and the gap.

Jaenisch: You generate the subtle mutation by placing a linker into the vector to introduce a point mutation or a stop codon. How big can this linker be? Does its size affect the targeting frequency?

Bradley: I don't think it does. You can have a point mutation or an insertion of 1 kb or a deletion of 1 kb. I don't think that makes a lot of difference to the frequency.

Jaenisch: M. R. Capecci has claimed that a few point mutations can decrease the targeting frequency measurably. T. Jacks (personal communication) has measured the targeting frequency of the *Rb* gene using a vector containing or not containing a linker: the frequency of targeting differed by 10-fold.

Bradley: We have looked at this only in the context of insertion vectors. If we make a vector and linearize it to produce a single reciprocal cross-over event, as I described previously, and put in a mutation that turns the endogenous *Eco*RI site into a *Bam*H1 site, there is only about a twofold reduction in targeting frequency.

Jaenisch: How long is the region of heterology at this site?

Bradley: It's about 20 bp. If we take the 6.8 kb insertion vector and put the *neo* gene in the homology region and linearize in that region of homology, the targeting frequency is about the same as when the sequences are not interrupted and *neo* is in the backbone of the plasmid.

The only direct observation we have that relates to the effect of the interruption of homologous sequences by heterologous ones on the frequency is in the reversion steps. Reversions from insertion events where there is heterology in one duplicate (an endogenous *Eco*RI site changed to a *Bam*H1 site) occur at a frequency of about 10^{-6} per cell generation. When these duplicates are totally homologous, the frequency is about 10^{-5} per cell generation. So there is a relationship between having absolute homology and not, as far as we can tell, but only in the interchromosomal reversion event.

Goodfellow: In yeast a single recombination fork is severely disrupted when you get more than 20 bp of non-homology. In the test-tube *RecA* can't drive a single strand exchange past more than about 25 bp of non-sequence homology.

Bradley: These are two different things. You are talking about migration of the Holliday junction; I agree these junctions are stalled by about 20 bp of an insertion or deletion. The other question is how does that affect the recombination frequency. Recombination frequency is not, as far as we know, related to migration of Holliday junctions but is related to their formation. These mismatches in homology don't dramatically interfere with the initial strand exchange, which we believe is the rate-limiting step in gene targeting.

Jaenisch: One concern is that if your source of DNA for the targeting vector is not mouse strain 129 like your ES cell, then polymorphisms in the vector could affect your target.

Bradley: Our evidence is limited to this experiment and we haven't seen a dramatic difference. We have made vectors that have two mismatches by linker insertions. Again we see a very similar frequency. The only time linkers affect the targeting frequency is when we put them on the homologous end. They affect the targeting frequency by a factor of 2–3; but those ends are incorporated so this also has an impact on the integration structure.

Goodfellow: I didn't understand what was going on in the co-electroporation experiment. Were you getting homologous recombination plus integration of the *neo* at that site or was the *neo* always targeted to the non-homologous site?

Bradley: Of the *hprt⁻* colonies, 8% have the predicted structure. The other 92%, as far as we can tell, have a major disruption of the *hprt* locus in the region that we have targeted. The *neo* insertion in that case is at the *hprt* locus and there are some concatameric structures in there. It is very complex and we haven't been able to resolve any one of those structures.

In the case where we have the mutations we desired, as far as we can tell the *neo* mutation is elsewhere in the genome, it's not in *hprt*. It is probably in a different chromosome.

Joyner: Have any of those been transmitted into the germline to test whether the targeted mutation and the inserted *neo* sequence can be separated?

Bradley: No; we haven't made mice using those lines to check.

Evans: What environments has the third method, the Double Hit gene replacement, been used with?

Bradley: We have put it into *Hox-2.6* and into c-*myc*. Alex Joyner has put it into *En-2*. We both see the same thing, this high background frequency.

Joyner: We have analysed eight such revertants by Southern blotting and in all cases the selectable gene had gone.

Bradley: The reversion frequency we find varies from clone to clone. We can get the frequency of background down to 10^{-5} in some cases, so it has been possible for us to isolate these second events, but they are still relatively infrequent.

Goodfellow: Have you looked at the structure where the *hprt* is missing?

Bradley: It looks wild-type, so it was probably recombination between the homologous chromosomes.

Joyner: Or a loss of a chromosome? Have you checked whether you still have two copies of the chromosome by karyotyping?

Robertson: I don't think that would happen for the *hprt* locus; don't forget the ES cells are XY!

Bradley: It's also difficult to tell because it could be a loss of a chromosome and then a gain. In many p53 tumours there is loss of heterozygosity by chromosome loss and endoreduplication.

Joyner: This problem presumably also occurs with your Hit and Run technique.

Bradley: I agree. With the Hit and Run method, some clones give a low frequency of pulling out the correct structure. In that case I think the intrachromosomal recombination event occurs at a similar frequency to the reversion event; that's why we couldn't pull those clones out.

McLaren: Do you get unintended Hit and Run occurring with ordinary insertion vectors, so that your lines once you have them aren't stable?

Bradley: The insertion vectors will give an unstable structure. When you make a mouse with them, before you revert them, potentially you get a somatic mosaic. Even when you transmit it through the germline, you will get a somatic mosaic. That could be quite interesting, but it can be a real problem.

McLaren: Has such inherited somatic mosaicism ever been shown?

Bradley: We have never bothered transmitting them, but I think *dilute* is a good example of that. There is the reversion between the two direct repeats in the long terminal repeats, so there is the somatic reversion frequency and the germline reversion frequency.

McLaren: Could it also be the basis of the *pink-eye* unstable mutation?

Bradley: Yes, Brilliant et al (1991) described that recently.

Jaenisch: The *dilute* phenotype is not based on somatic reversion. Expression of the gene in the central nervous system is due to differential splicing which removes the viral sequences.

Bradley: I am talking about the reversion of *dilute*.

Jaenisch: Reversions of *dilute* have been shown to be due to proviral excisions in the germline, not in somatic tissues. Nobody, as far as I know, has demonstrated somatic reversion of a retroviral insertion.

Bradley: I think it can occur. It is a one in a million event, so it's going to be difficult to see.

Joyner: In collaboration with Janet Rossant, we have transmitted into the germline an insertion mutation in N-*myc*. Janet's student, C. Moens, has analysed offspring from heterozygous intercrosses and found by Southern blotting of tail DNA that one quarter of the offspring are homozygous and show no evidence of a wild-type locus. Some of these samples were analysed further by PCR and there was no evidence for a small population of cells carrying a reversion event. Thus, the insertion mutations can be quite stable.

Goodfellow: Did you do this experiment in the *hprt* system by knocking out *hprt*, then putting back *hprt* and then doing the selection again?

Bradley: Yes, we have done this in *hprt*⁻ cells.

Goodfellow: So in that experiment what's the explanation for the loss of the *hprt* sequence? Is the insert in the X chromosome *hprt* gene?

Bradley: This is TG4, a mouse strain made in Martin Evans's lab a long time ago. The only important thing about them is that they are *hprt*⁻.

Evans: There were two strains made, one of which had an insertion in the first intron, the other, TG4, had an insertion in the 5th intron. This is not a clean insertion of the vector used, there is some rearrangement but it seems to be a completely stable *hprt*⁻ mutation. TG3, which is a clean retroviral vector insertion, is unstable and reverts at a frequency of about one in 10^5 per cell division by excising the virus and leaving a single long terminal repeat.

Beddington: I assume the frequency you see here of loss of *hprt* is not assumed to be related to the targeting? If you randomly insert a minigene into the genome, do you see the same frequency of loss?

Bradley: Yes, you do.

McLaren: You have done all this fiddling around and you stressed the large number of subclones, cell generations and so on. You sounded as if you were quite surprised that you were able to make a mouse at the end of it. What is the present experience with aneuploidy in ES cell lines? How soon do chromosome abnormalities arise and can you somehow avoid them?

Bradley: I would describe my response as relieved not surprised. Our feeling is that under optimal culture conditions, aneuploidy is not really a problem up until 15–20 passages. We don't take it to those extremes of course. Whenever you subclone, basically you re-set the clock. When we work within 15–20 passages, which is a lot, we find a very high frequency of somatic mosaicism and germline transmission.

McLaren: You said that you re-set the clock every time you subclone the ES cells. Do you always karyotype when you subclone?

Bradley: No, we never karyotype, we just inject and wait.

McLaren: So you are re-setting the clock but you don't know what you are setting it to.

Bradley: We don't until the mice are born, that's correct. Karyotyping is a reasonably good assay of the quality of a cell but smaller genetic changes go undetected by the resolution of that method. I regard the blastocyst injection as the formal proof.

Jaenisch: Do you think there are intrinsically more stable ES lines and intrinsically less stable ones? The D3 line seems to be a particularly stable line and nobody knows of what passage number the cells are.

Bradley: I don't think there is a great difference; I think the variation comes from the way the cells are cultured and what you are inadvertently selecting for. The apparent stability is a property of the laboratory not of the cell lines.

Robertson: I agree with Allan that the important thing is the culture you start with. There is a long-standing debate about karyotypic stability of ES cells under different culture conditions and I don't think that has been properly tested. There are a few reports that you can take ES cells off a feeder layer and just grow them in LIF-supplemented medium for 15 culture generations and still make a germline chimera. I think all of those experiments were done with polyclonal populations of cells, so that if one euploid cell was included, it might have contributed to the germline. The feeling is that you don't really get enough long-term stability unless you grow the cells on a feeder cell layer.

Beddington: There are not enough extensive data to decide one way or the other.

Bradley: Nobody I know has grown cells in LIF alone, cloned them and looked at the quality of the clones. Until that is done, there is no formal proof that LIF has any value in this system. We have STO cell lines that make LIF and other STO cell lines that don't: there's no correlation between expression of LIF and being a good feeder layer.

Beddington: I am not sure that is necessarily surprising because we haven't characterized what STO cells are providing, so it may be at saturation level and putting more LIF on top of that has no effect at all. Martin Hooper (personal communication) has put selected clones grown in LIF alone through the germline at very high frequency.

Robertson: What we have noticed in long-term culture in LIF alone is that it retards the kinetics of endoderm differentiation. When the cells are grown in suspension, they are much more reluctant to start making endoderm.

Evans: That is true, but they can be retrained. If you then put the same LIF-passaged cell culture, which you might think has progressed away from a differentiation potential, back on feeders for a few passages, it will come back to its previous differentiation ability. I don't know what that means; I suspect the time scales are wrong for it to be a complete reselection. It could be the result of modulation of cell surface LIF receptors.

Hogan: If you make subclones of your STO cells, what kind of variability do you see in their ability to support ES cells?

Bradley: The experiments are difficult because formally you have to grow the cells on these feeder layers, subclone them and make chimeras. So for every cloned STO you have, you really have to look at ten clones of ES cells by injecting them into blastocysts and breeding the resultant chimeras to score germline transmission. By looking at the morphology of ES cells on these feeder cells, we can detect distinct differences between one STO clone and the next. Unfortunately, it is just experience that allows me to say one looks better than another. The criterion is no more discriminating than that. So far it's been alright! The subclone that a number of people use now has worked exceptionally well. In our hands, of ES cells grown on this subclone, about 80–90% of clones go through the germline after 15–20 passages, so it's pretty good.

Buckingham: What would you say at the moment about the desirability of using the STO line as a feeder layer, particularly this good subclone, compared with using primary embryonic fibroblasts, for example?

Bradley: I have never used primary cells. The STO clones we have isolated have worked well and have the advantage that you don't have constantly to remake the primary fibroblasts, which do not have a very constant phenotype.

Jaenisch: We are using primary embryonic fibroblast cultures. One reason was that we didn't want to use permanent lines because we didn't trust them. The results with primary cells are very reproducible, giving a similar frequency of clones going through the germline as with the STO lines.

Hogan: But the primary cells have to come from a *neo*-resistant mouse?

Jaenisch: Yes, but a targeting event very often gives a *neo*-resistant mouse anyway. We are using, for example, the β_2-microglobulin knock out, which is a viable homozygous line and generates perfect *neo*-resistant feeder cells.

Robertson: Do you think it matters what strain?

Jaenisch: We have not seen strain differences.

McLaren: What tissue is the primary culture made from?

Jaenisch: Fifteen-day-old embryos are explanted and the cells, after reaching confluency, are frozen. They can be passaged up to four times. This is not that much more work than using an established feeder line.

Joyner: We were talking earlier about the need to catalogue all the gene expression patterns. Are we rapidly getting to that point with all these new mouse mutants? I am not sure we need to coordinate who does what, because it doesn't necessarily hurt to have more than one mutation made in a gene. It is more a question of where are all these mice going to end up and what kind of access are people going to have to them.

I feel that once they are published they should be made available. Then there's the practical point: does everybody have to house them, which is a major effort, or can The Jackson Lab or somewhere accommodate these mice? That would make them more freely available.

Solter: Basically, they are going to be like any mutant mouse line found by accident during the last fifty years and they should probably be treated in the same way. The only difference is that previously the number of useful mutants increased very slowly, now it is going to increase rapidly.

McLaren: The situation has also changed through the advent of embryo freezing and storage. There are mutants that were described in the old days and, alas, have been lost.

Solter: Should every lab freeze their own insertional mutant embryos? The recovery of those frozen embryos is not trivial.

McLaren: I would have thought there should be a freezing centre; perhaps one in the States and one in Europe. But it would not be productive to freeze embryos unless the mutation had been characterized to some extent.

Jaenisch: The issue is much less severe now where mutant mice are derived from homologous recombinations in ES cells. If you have a clone that goes into the germline well and you have the targeted cells frozen, it will probably take only three months longer to derive a heterozygous mouse from the ES cell line than to derive it from a frozen embryo.

Joyner: Except that not all cell lines go through the germline easily. Also, many labs are not set up for making chimeras.

Hogan: Maybe you could make ES cells from the mutants once you have them.

Bradley: It is easier to pass a vial of cells around than to send mice, particularly to different countries.

Hogan: Rick Woychik of the Oak Ridge National Laboratory is making a computer databank of all the transgenic mice that have been made. He has a contract from NIH to do that. So one could also catalogue all the knock out mice.

Reference

Brilliant MH, Gondon Y, Eicher EM 1991 Direct identification of the mouse *pinkeyed* unstable mutation by genome scanning. Science (Wash DC) 252:566–569

The gene trap approach in embryonic stem cells: the potential for genetic screens in mice

Alexandra L. Joyner, Anna Auerbach and William C. Skarnes

Division of Molecular and Developmental Biology, Samuel Lunenfeld Research Institute, Mount Sinai Hospital, 600 University Avenue, Toronto, Canada M5G 1X5

Abstract. The gene trap approach in embryonic stem cells was developed as a means to screen for genes expressed during early postimplantation development in the mouse. We have validated the approach by showing that *lacZ* from the integrated vector is activated by splicing to endogenous exons and expressed in embryos in patterns that mimic those of the endogenous genes. These insertions can produce developmental defects in homozygous mice. The results indicate that a large screen of gene trap cell lines on the basis of embryonic *lacZ* expression is feasible and should provide a new source of genes, mouse mutants and mouse strains that express *lacZ* in particular domains and lineages. The gene trap approach could be extended to a smaller screen for genes based on mutant phenotypes.

1992 Postimplantation development in the mouse. Wiley, Chichester (Ciba Foundation Symposium 165) p 277–297

The study of the genetic control of postimplantation development in the mouse has been hampered by an inability to carry out large-scale screens for specific types of developmental mutants (reviewed in Rinchik 1991). Nevertheless, over the past century approximately 600 mutant strains have been described and many of these are still available for study (Lyon & Searle 1989). Two problems concerning the study of mouse mutants have limited our ability to utilize these mice effectively. First, it is difficult to determine from the phenotype alone whether the mutated gene is required for cell viability or for directing the normal programme of development. Analysis of the developmental potential of homozygous mutant cells in chimeras can in part address this question. More recently, an alternative approach has been to attempt to establish mouse embryonic stem (ES) cell lines homozygous for a mutation (Magnuson et al 1982, Martin et al 1987, Niswander et al 1988, Conlon et al 1991). The second problem in the analysis of existing mutants is the difficulties encountered in cloning genes responsible for the mutant phenotypes. Only two mutant genes (*Brachyury*, Herrmann et al 1990, and *Sex-reversed*, Gubbay et al 1990) have been cloned by conventional genetic approaches. The majority of pre-existing mutants have been identified by chance, when cloned genes have been found to be allelic with developmental mutants. In addition, a small number of mutant genes have been cloned by fortuitous insertion into the gene of a retrovirus or DNA marker in transgenic mice.

In invertebrate organisms, notably *Drosophila melanogaster* and *Caenorhabditis elegans*, large-scale mutagenesis screens that approach saturation for particular types of developmental phenotypes have been achieved. In *Drosophila* a major effort has been made in identifying the genes that control early pattern formation. In *C. elegans*, mutant screens have concentrated on particular mutant phenotypes recognizable in the adult. Analysis of these mutants has provided a means of assigning a hierarchy of gene activities leading to the proper functioning of particular developmental processes.

The major factors that have made such screens feasible in flies and worms are: the small size of their genomes, the availability of strains carrying large chromosomal deletions, the fact that the embryos develop outside of the mother and thus can be easily screened for defects, and the ability to breed large numbers of animals efficiently using limited laboratory space. In contrast, the mouse genome is approximately twenty times larger than that of fruit flies and nematodes; mouse embryos develop *in utero* thus the mothers have to be sacrificed at different developmental stages to analyse embryonic mutant phenotypes, and it is very costly to breed mice.

A second type of genetic screen recently developed in *Drosophila* is the 'enhancer trap' (O'Kane & Gehring 1987, Bier et al 1989, Bellen et al 1989, Wilson et al 1989). This approach identifies genes on the basis of their expression patterns rather than mutant phenotype. In such a screen, a *lacZ* reporter construct carrying a minimal promoter contained within a transposable element is introduced into the fly genome and then mobilized to generate random insertions. Embryos derived from animals carrying new insertions are analysed for *lacZ* expression that is regulated by *cis*-acting DNA transcription regulatory sequences located near the site of integration.

We have explored the feasibility of setting up a similar type of screen in mice for genes expressed in spatially defined patterns during gastrulation. We have adapted the enhancer trap approach for mice by designing a new vector, the 'gene trap', that requires splicing to exons of a gene for activation of *lacZ* (Gossler et al 1989). Such a vector allows easy cloning of the interrupted endogenous gene expressing *lacZ* and produces a mutation in the gene (reviewed in Skarnes 1990). To screen a large number of gene trap insertions efficiently, we have introduced the gene trap vector into ES cells and then analysed the expression of *lacZ* in ES cell chimeric embryos. In this paper, we describe our experiments that demonstrate the validity of the gene trap approach in ES cells and discuss the design and expected results of a large-scale screen for genes expressed during gastrulation.

Design of the *lacZ* gene trap vector

The rationale for the gene trap vector was that *lacZ* would be activated and expressed as a spliced fusion transcript if it became integrated into the intron of a gene (see Fig. 1). In cases when *lacZ* was in the correct orientation and reading frame to be compatible with endogenous gene sequences, it would be transcribed and translated as part of a functional fusion protein. The vector contained, in addition to *En-2* intron and exon sequences and *lacZ*, the bacterial gene for neomycin resistance (*neo*) which permits selection of cells in which the vector has been integrated. In our preliminary screen with a gene trap vector that did not include an ATG for the start of translation of *lacZ* (Gossler et al 1989) we found that one in sixty *neo*[R] ES colonies expressed *lacZ*. Other types of vectors have been designed that increase the frequency of *lacZ* activation. For example, it has been reported that including an ATG for the start of translation of *lacZ* increases the percentage of cells expressing *lacZ* in fibroblasts (Brenner et al 1989). A second improvement could be to use a retrovirus vector to deliver the gene trap vector into

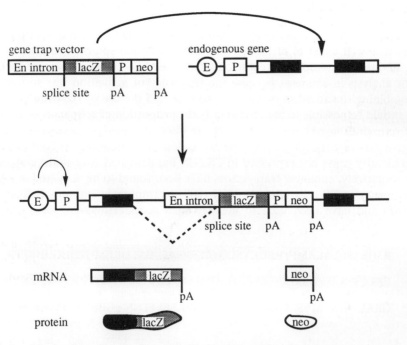

FIG. 1.   Mechanism of activation of *lacZ* after integration of a gene trap vector into an intron. A gene trap vector is shown at upper left and a generic gene at the right. Below is shown insertion of the gene trap vector into the gene and activation of *lacZ*. The exons of the gene are indicated by rectangles and the coding sequences by filled-in regions. The stippled line indicates splicing from the first exon to *lacZ*. E, enhancer; En intron, 1.5 kb of *En-2* intron sequence; P, promoter; pA, poly A addition site.

TABLE 1   Summary of the molecular and mutational analysis of four gene trap ES cell lines

Gene trap cell line	Size of endogenous mRNA upstream of lacZ (endogenous transcript size)	Expression of endogenous gene versus lacZ (expression pattern)	% wild-type transcript in homozygous mutants	Homozygous mutant phenotype in mice
GT10	0.5 (3.5)	Same (restricted)	ND	ND
GT4-1	0.5 (7)	ND	<1	Perinatal death
GT4-2 (zinc finger)	6 (10)	Same (nervous system)	<1	Deficient postnatal growth
GT2	ND	ND	ND	Viable

ND, not determined.

ES cells such that a majority of infected cells would carry multiple proviral insertions (Brenner et al 1989, P. Soriano, personal communication).

In our earlier studies we chose only those *neo*[R] ES clones that expressed *lacZ* for analysis in chimeras because the frequency of activation was so low. By combining the addition of an ATG to *lacZ* and the use of retroviral vectors, it should be possible to increase greatly the proportion of cells expressing *lacZ*. If retroviral-based gene trap vectors can integrate into non-expressed loci and accurately report endogenous gene expression, then these vectors could be used to identify genes not expressed in ES cells but activated later in development. Alternatively, enhancer trap vectors have been found to be activated at a high frequency in ES cells (Gossler et al 1989) or transgenic embryos (Allen et al 1988) and have been used to identify such genes. However, there are two

RSHERSHLALAMFTREDKYSCQYCSFVSAFRHNLDRHMQTHHGHHKPFR

CKLCSFKSSYNSRLFTHILKAHAGEHAYKCSWCSFSTMTISQLKEHSLKVH

GKALTLPRPRIVSLLSSHAHPSSQKATPAEEVEDSN ↑GPRSRKPKKKNP
                                      splice

FIG. 2.   *lacZ* fusion transcripts are properly spliced and contain an open reading frame upstream of *lacZ*. The amino acid sequences spanning the splice site upstream of *lacZ* in the cell line GT4-2 are shown. The unique sequences encoded by the upstream endogenous exon are shown in regular type and the *engrailed* gene sequences downstream of the splice site are shown in bold. *lacZ* begins 50 amino acids downstream of the splice site. The zinc fingers are indicated by shading.

drawbacks to the use of enhancer trap vectors: it can be difficult to identify the endogenous gene that activates *lacZ* expression and the integrations do not necessarily disrupt the transcription unit.

Before we embarked on a large-scale screen with the gene trap vector, it was necessary to determine whether *lacZ* is activated by splicing to endogenous gene sequences and expressed in a pattern similar to the endogenous gene. In addition, it was of interest to test whether the insertions cause mutant phenotypes when transmitted into mice. In order to address these questions we carried out a molecular genetic analysis of three gene trap insertions and transmitted two of these and one additional insertion into the germline for analysis of the phenotype (Table 1).

Molecular genetic analysis of gene trap insertions

The first evidence that *lacZ* was activated by a splicing event that created *lacZ* fusion proteins with the products of endogenous genes came from the subcellular localization of β-galactosidase activity in different gene trap cell lines. β-galactosidase was observed in different cellular compartments such as the nucleus and a dot in the cytoplasm (Gossler et al 1989). Northern blot analysis of total RNA extracted from these lines with a *lacZ* probe gave further support for a spliced fusion transcript. Each line had a unique *lacZ*-containing transcript that was larger than the 3.7 kb *lacZ* transcript (Skarnes & Joyner 1992).

In order to confirm that the splice acceptor upstream of *lacZ* was being used properly, we cloned cDNAs spanning the splice junction (Skarnes & Joyner 1992) using the RACE (rapid amplification of cDNA ends) protocol (Frohman et al 1988) to amplify sequences 5′ to *lacZ*. At least three independent clones that contained approximately 350 bp of DNA upstream of the splice acceptor from each cell line were sequenced. In all cases the splice acceptor was used properly. Furthermore, there was a potential open reading frame upstream and in frame with *lacZ*, whereas the other two reading frames contained stop codons (for example see Fig. 2). Sequences obtained from the amplified cDNAs were used to search the Genbank nucleotide database and all were found to be novel. A search of the EMBL protein database revealed that the GT4-2 transcript encodes a putative novel zinc finger-containing protein (Fig. 2). Thus, our results prove that the gene trap vector is activated by splicing to upstream endogenous gene exons to generate fusion proteins.

The normal endogenous mRNAs transcribed from each gene were studied by Northern blot and RNase protection analyses using the cloned probes. GT4-1, GT4-2 and GT10 recognized transcripts of 7 kb, 10 kb and 3.5 kb, respectively, in RNA from ES cells (Skarnes & Joyner 1992). In addition, each probe detected the expected fusion transcript in RNA from the corresponding gene trap ES cell line or transgenic embryo.

To investigate whether splicing around the *lacZ* insertion occurred to generate a normal endogenous transcript, we did an RNase protection analysis on RNA

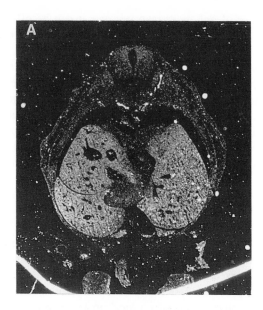

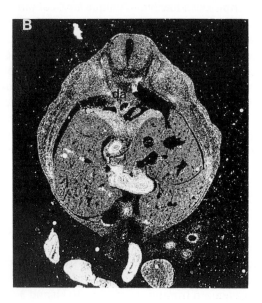

FIG. 3. *lacZ* mimics endogenous embryonic gene expression. Bright field photo-micrographs are shown of cross sections of a 12.5 day embryo hybridized with (A) sense and (B) antisense probes from the GT10 cloned endogenous exon sequences. Highest levels of expression are seen in the developing intestine (i), dorsal aorta (da) and mesenchyme and absent in the liver (l) and spinal cord.

extracted from heterozygous and homozygous animals carrying the GT4-1 and GT4-2 insertions (see below). A probe that spanned the *lacZ* splice acceptor site was used for each gene. For both gene trap insertions the RNA from homozygous animals showed the larger protected band representing the *lacZ* fusion transcript and negligible amounts of the smaller band representing the normal spliced endogenous transcript. Thus, insertion of the gene trap vector appears to inhibit effectively normal splicing of the endogenous transcript to the exon downstream of the insertion.

Analysis of gene trap insertions in mice

For the gene trap vector to be of value in a screen for genes expressed during gastrulation, *lacZ* must be expressed in the same cells as the endogenous gene. To address this question, we used the cloned exon sequences from GT10 and GT4-2 in an *in situ* hybridization analysis of 9.5 and 12.5 day embryos. The expression patterns detected were then compared to the distribution of β-galactosidase activity in chimeras generated from the GT10 cell line and transgenic animals carrying the GT4-2 insert. In all cases there was a good correlation between the *lacZ* and endogenous gene expression patterns (Skarnes et al 1992). Fig. 3 shows an example of the analysis of the GT10 endogenous gene. We previously described the *lacZ* expression at this stage as being high in the dorsal aorta, intestine and lateral mesenchyme (Gossler et al 1989). The bright field photomicrograph shown in Fig. 3 demonstrates that the endogenous gene is also expressed at highest levels in these tissues. GT4-2 was found to be expressed in all tissues at 9.5 days, although highest in the neural tube. At 12.5 days GT4-2 was expressed primarily in the nervous system. The high correlation between *lacZ* and endogenous gene expression demonstrates that meaningful results can be obtained from analysing gene trap-activated *lacZ* expression patterns in embryos, although there are likely to be some cases where the gene trap vector interferes with regulatory elements.

Our molecular data suggest that each gene trap insertion that results in transcription of *lacZ* should cause a mutation in the endogenous gene. To test this prediction, we transmitted three gene trap insertions, GT4-1, GT4-2 and GT2 into the germline by chimera production. Homozygous animals were obtained for each *lacZ* insertion after outcrossing heterozygous animals and then intercrossing heterozygotes. The GT4-1 homozygotes were found to have open eyes at birth and to die within one week for no obvious reason. The GT4-2 homozygotes displayed a postnatal growth defect which was recognizable by three weeks of age. The GT2 homozygotes were viable and had no apparent defect. Thus, the gene trap insertions can cause developmental defects, however, as with many mutations in developmental genes (for example Joyner et al 1991), there was not a good correlation between the phenotype observed and the gene expression pattern.

Large-scale screens in ES cells with the gene trap

Fig. 4 summarizes one design of a large-scale screen for genes on the basis of their expression patterns during gastrulation. The first part of a screen would involve generating and freezing away thousands of cell lines carrying independent gene trap insertions. One person working full-time could likely freeze 2000 cell lines in a year. The next part of the screen, involving analysis of *lacZ* expression patterns in chimeric embryos, is much more labour intensive. If two different embryonic stages (8.5 and 12.5 days) were examined, we estimate that two people could analyse approximately 500 lines in a year. This number could be increased by looking at only one stage, however, examination of eye pigment in 12.5 day

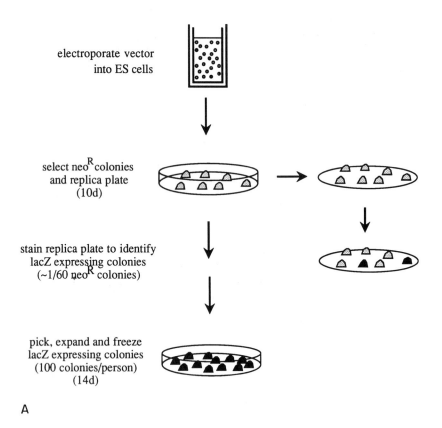

electroporate vector
into ES cells

select neo^R colonies
and replica plate
(10d)

stain replica plate to identify
lacZ expressing colonies
(~1/60 neo^R colonies)

pick, expand and freeze
lacZ expressing colonies
(100 colonies/person)
(14d)

A

FIG. 4.   Strategy for screening gene trap ES cell clones for *lacZ* expression during gastrulation. (A) The steps taken to isolate and freeze ES cell clones expressing *lacZ*.

embryos serves in part as an indicator of the proportion of chimeras obtained in each experiment. In addition, in our earlier experiment (Gossler et al 1989) and more recent work, we found that whereas only approximately 20% of the lines showed restricted *lacZ* expression at 8.5 days, the majority showed restricted patterns later in development. GT4-2 is a good example of this.

We have begun a large screen using the gene trap vector in which we intend to analyse 500 gene trap insertions for expression during gastrulation and at 12.5 days of embryogenesis (S.-L. Ang, A. Auerbach, S. Gasca, F. Guillemont,

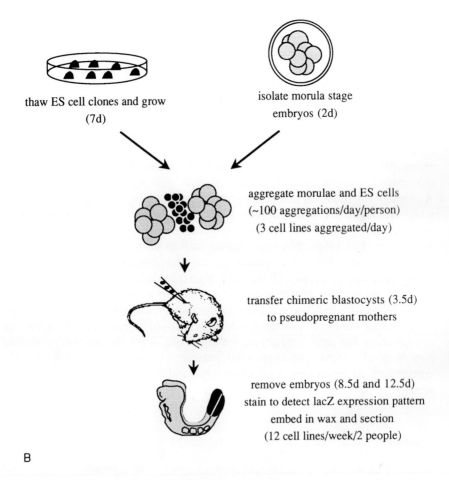

thaw ES cell clones and grow
(7d)

isolate morula stage
embryos (2d)

aggregate morulae and ES cells
(~100 aggregations/day/person)
(3 cell lines aggregated/day)

transfer chimeric blastocysts (3.5d)
to pseudopregnant mothers

remove embryos (8.5d and 12.5d)
stain to detect lacZ expression pattern
embed in wax and section
(12 cell lines/week/2 people)

B

FIG. 4.   (B) Analysis of *lacZ* expression patterns in chimeric embryos. (x) indicates number of days for each step.

D. Hill, A. Joyner, V. Prideaux, J. Rossant, W. Wurst). One question is whether expression data alone will provide new insights into our understanding of developmental processes. Our screening efforts and those by other groups, together with embryonic expression data available for cloned genes, should provide material for a catalogue of gene expression patterns during embryogenesis. If a large enough database is assembled, we might expect to find recurring themes such as expression in particular domains or lineages that will give clues to the genetic control of pattern formation.

Perhaps the more compelling reason for carrying out these screens is that the cell lines would provide important starting material for future experiments. First, a bank of gene trap cell lines would contain a source of genes expressed in different regions of the embryo for further cloning, sequencing and studies of mutant phenotypes. Second, mice produced from gene trap cell lines that express *lacZ* in all cells or in particular lineages would provide a source of marked cells for studies involving grafting or cell mixing experiments. Third, mice expressing *lacZ* in particular cell types could be used as marker strains for following cell lineages. These strains could be particularly useful for the analysis of embryonic phenotypes in breeding experiments with developmental mutants.

Our results demonstrating that the splice acceptor in the gene trap acts as a potent mutagen indicate that this approach can be used to screen efficiently for genes on the basis of mutant phenotypes. Any integration in the correct orientation into a transcription unit should disrupt the gene. A strategy for inducing and cloning germline mutations has been proposed in which ES cells are infected with a retrovirus and the proviral insertions subsequently transmitted into the germline for analysis (Robertson 1986, Robertson et al 1986). The rate-limiting step in such a screen is the analysis in mice of each proviral insertion for the five percent that show a visible phenotype (Jaenisch 1988). The use of a gene trap vector for such a screen could decrease the number of cell lines that need to be tested in mice by analysing only cell lines that express *lacZ* and thus contain a mutation. Alternatively, since the gene trap vector is an effective mutagen, it could be used in a screen without pre-selection for *lacZ* expression. Furthermore, once an interesting mutant had been identified, the expression pattern of the gene could be readily determined by examining *lacZ* expression and the interrupted gene rapidly cloned.

Summary

We have cloned the upstream endogenous genes from a number of gene trap cell lines and demonstrated that the *lacZ* reporter is activated during embryogenesis in a manner that mimics that of the endogenous gene. We did not detect splicing around the *lacZ* insertions to generate wild-type transcripts and two out of three insertions resulted in developmental defects in mice homozygous for the mutations. We therefore conclude that the gene trap vector offers a versatile tool to screen for developmental genes in the mouse on the basis of gene expression patterns and/or mutant phenotypes.

Acknowledgements

We would like to thank all the members of our lab and Janet Rossant's for their work mentioned in this paper and many helpful discussions on the gene trap approach. We also thank Phil Soriano for discussing his data before publication. This work was supported from grants to A. J. from the NIH, and the NCI and MRC of Canada. A.J. is a scholar of the MRC and W.S. is supported by an N.C.I. studentship.

References

Allen ND, Cran DG, Barton SC, Hettle S, Reik W, Surani MA 1988 Transgenes as probes for active chromosomal domains in mouse development. Nature (Lond) 333:852–855
Bellen HJ, O'Kane CJ, Wilson C, Grossniklaus U, Pearson RK, Gehring WJ 1989 P-element-mediated enhancer detection: a versatile method to study development in *Drosophila*. Genes & Dev 3:1288–1300
Bier E, Vaessin H, Shepherd S et al 1989 Searching for pattern and mutation in the *Drosophila* genome with a P-*lacZ* vector. Genes & Dev 3:1273–1287
Brenner DG, Lin-Chao S, Cohen SN 1989 Analysis of mammalian cell genetic regulation *in situ* by using retrovirus-derived "portable exons" carrying the *Escherichia coli lacZ* gene. Proc Natl Acad Sci USA 86:5517–5521
Conlon FL, Barth KS, Robertson EJ 1991 A novel retrovirally induced embryonic lethal mutation in the mouse: assessment of the developmental fate of embryonic stem cells homozygous for the 413.d proviral integration. Development 111:969–981
Frohman MA, Dush MK, Martin GR 1988 Rapid production of full-length cDNAs from rare transcripts: amplification using a single gene-specific oligonucleotide primer. Proc Natl Acad Sci USA 85:8998–9002
Gossler A, Joyner AL, Rossant J, Skarnes WJ 1989 Mouse embryonic stem cells and reporter constructs to detect developmentally regulated genes. Science (Wash DC) 244:463–465
Gubbay J, Collingnon J, Koopman P et al 1990 A member of a novel family of embryonically expressed genes mapping to the sex-determining region of the Y chromosome. Nature (Lond) 346:245–250
Herrmann BG, Labeit S, Poustka A, King TR, Lehrach H 1990 Cloning of the T gene required in mesoderm formation in the mouse. Nature (Lond) 343:617–622
Jaenisch R 1988 Transgenic animals. Science (Wash DC) 240:1468–1473
Joyner AL, Herrup K, Auerbach BA, Davis CA, Rossant J 1991 Subtle cerebellar phenotype in mice homozygous for a targeted deletion of the *En-2* homeobox. Science (Wash DC) 251:1239–1243
Lyon MF, Searle AG 1989 Genetic variants and strains of the laboratory mouse. Oxford University Press, Oxford
Magnuson T, Epstein CJ, Silver LM, Martin GR 1982 Pluripotent embryonic stem cell lines can be derived from t^{w5}/t^{w5} blastocysts. Nature (Lond) 298:750–753

Martin G, Silver LM, Fox HS, Joyner AL 1987 Establishment of embryonic stem cell lines from preimplantation mouse embryos homozygous for lethal mutations in the *t*-complex. Dev Biol 121:20–28

Niswander L, Yee D, Rinchik EM, Russell LB, Magnuson T 1988 The albino deletion complex and early post implantation survival in the mouse. Development 102:45–53

O'Kane CJ, Gehring WG 1987 Detection *in situ* of genomic regulatory elements in *Drosophila*. Proc Natl Acad Sci USA 84:9123–9127

Rinchik EM 1991 Chemical mutagenesis and fine-structure functional analysis of the mouse genome. Trends Genet 7:15–21

Robertson EJ 1986 Pluripotential stem cells as a route into the mouse germ line. Trends Genet 2:9–13

Robertson EJ, Bradley A, Kuehn M, Evans M 1986 Germ-line transmission of genes introduced into culture pluripotent cells by retroviral vectors. Nature (Lond) 323:445–447

Skarnes WC 1990 Entrapment vectors: a new tool for mammalian genetics. Bio/Technology 8:827–831

Skarnes WC, Joyner AL 1992 A gene trap approach in mouse embryonic stem cells: I. The *lacZ* reporter is activated via splicing. Submitted

Skarnes WC, Auerbach A, Jenkins N et al 1992 A gene trap approach in mouse embryonic stem cells: II. The *lacZ* reporter reflects endogenous gene expression and is mutagenic in mice. Submitted

Wilson C, Pearson RK, Bellen HJ, O'Kane CJ, Grossniklaus U, Gehring WJ 1989 P-element-mediated enhancer detection: an efficient method for isolating and characterizing developmentally regulated genes in *Drosophila*. Genes & Dev 3:1301–1313

DISCUSSION

McLaren: Alex, one of the questions relating to this work is whether the value of the resource is greater than the effort. One has to ask whose effort and to whom is the value. I think it's tremendous that you and Janet Rossant have embarked on this project. From a long-term point of view, the cumulative value of the resource would be well worth the effort, but the effort would be all yours, whereas the value would be much more widely spread.

Joyner: There are two other questions we need to consider: how will the results of different screens be coordinated and cell lines shared? Also, when is it time to stop screening for more genes?

Hogan: Are you going to publish a catalogue of your patterns, like a Sears catalogue!?

Joyner: We have to: this information has to be disseminated. We are taking pictures of every line; that is how we are doing the initial cataloguing.

McMahon: You could put single frames on video.

Joyner: Could you scan them in colour? That would be the best.

Jaenisch: When you generate a pattern, would you offer the cells to people or after you have put the insertion through the germline?

Joyner: Cells!

McLaren: You mentioned aggregation chimeras made with ES cells (Fig. 4), have you also tried making injection chimeras? Do you know which technique is better?

Joyner: We are using both techniques now when we look at *lacZ* expression in embryonic chimeras. We haven't been using aggregation chimeras extensively to make germline chimeras.

Jaenisch: Is there evidence for rearrangements of the DNA after DNA electroporation, which would make it difficult to analyse the gene? Presumably, when a retrovirus vector is used that does not happen.

Joyner: My student, Bill Skarnes, has analysed the integration sites in three gene trap cell lines by Southern blotting. In one case we can clearly demonstrate that no large rearrangements or deletions have occurred.

Bradley: The mutated gene can be cloned by rapid amplification of cDNA ends (RACE), which will ensure that you get the cDNA. Provided that you can ultimately clone the gene, it doesn't really matter what the exact structure of the mutated gene is, as long as you have a good mutation and can prove it is a null allele.

Jaenisch: Unless there is a large deletion or several genes have been deleted.

Robertson: How random is the insertion of the retrovirus or of the DNA?

Jaenisch: DNA transfection is clearly not random. The same genes can be mutated by independent insertion events. Concerning retroviruses, the only thing known is that they integrate preferentially at hypersensitive sites, which may be close to actively expressed genes. So you would probably get only a subset of the possible sites.

Joyner: However you do it, you get a subset. Electroporation may differ from microinjection in the way the DNA integrates, in terms of whether it causes gross rearrangements.

Krumlauf: Phil Soriano has a fusion between *lacZ* and *neo*: in that case, supposedly only a few molecules of *neo* are needed for resistance (Friedrich & Soriano 1991). Can he get things that are *neo* resistant that aren't expressing β-galactosidase?

Joyner: The vector is a *lacZ/neo* fusion, so anything that's *neo* resistant is expressing *lacZ* as a fusion protein (Friedrich & Soriano 1991).

Bradley: He does have clones in that category which look white as stem cells but then there was expression of *lacZ* in the embryo.

Joyner: That's nice, so it's picking up genes expressed at very low levels.

Bradley: It is presumed that expression is low.

Robertson: Our lab has found using promoterless gene targeting vectors that you need only an incredibly low level of *neo* transcripts to get a G418-resistant phenotype. One of my concerns would be whether the use of the *lacZ* as a reporter gene is completely neutral. When you ultimately get these insertion sites in the germline and find that they result in a recessive lethal mutation, how can you rule out that the lethality results from particularly high levels of expression of the β-galactosidase?

Joyner: That's certainly an important question. We haven't characterized the *lacZ* expression pattern of the one of our mice that is viable. The obvious alternative to *lacZ* is analysis of RNA or protein using whole mounts; however, that would definitely be a lot more work.

Beddington: If *lacZ* has a general deleterious effect, you might expect to start seeing a lot of phenotypes that were all the same because of *lacZ* toxicity.

Joyner: You mean some cell types might be more sensitive to *lacZ* than others?

Jaenisch: So this could cause apparent mutations. You may find a mutant phenotype which has nothing to do with the gene into which *lacZ* has integrated.

Hogan: An important feature of this approach is the pre-screening, to choose interesting patterns. For that, it is important to have some preconceived idea of what's an interesting pattern.

You look at several chimeras; if there was a low amount of chimerism, would you miss an interesting pattern, for example, if cells were colonizing only one part of the 8.5 day embryo and not another? In your experience, how many chimeras do you have to look at in order to obtain a reproducible pattern?

Joyner: We find that the patterns are very similar from chimera to chimera, even with weak chimeras.

Hogan: You seem to have drawn a distinction between a tissue-restricted pattern of expression and a ubiquitous pattern of expression. There was a nice paper by Bonnerot et al (1990) reporting a very diverse set of patterns of *lacZ* staining in transgenic embryos after microinjection of fertilized eggs.

Joyner: Was that using an enhancer trap?

Hogan: Yes, in effect, because they had an *hprt* promoter, acting as a very weak promoter in front of the *E. coli lacZ* gene, so that all the positives were due to chromosomal integration site effects. The point is that only a few of the embryos had *lacZ* staining restricted to one cell type or one position. In most of the embryos, different cell types in different spatial domains expressed *lacZ*, generating complex patterns of *lacZ* expression in multiple tissues. How would you decide which pattern was interesting? How many different ones are you getting? Are you getting recurrent themes?

Joyner: It is hard to judge from a screen of 23. I don't think we should be worrying at this point about what is interesting and what is not interesting. Everybody is going to have a completely different opinion on what is their favourite tissue, developmental stage, etc. Each person should choose a cell line based on that.

Bradley: What is the correlation between the expression pattern as shown by β-gal and the phenotype as determined genetically looking at homozygotes? This comes back to what you define as an interesting phenotype. Just because a gene is ubiquitously expressed, it doesn't mean that it's not required in a very defined subset of cells, which will determine the phenotype. What are we screening for here, and would it be better to make these mutations into mice?

Joyner: And choose everything by mutant phenotype?

Bradley: Both. The expression patterns are important, but they don't necessarily tell you about function.

Joyner: The problem with that is the numbers. If you want to look at the mutant phenotypes, you are going to reduce your numbers probably by 10-fold. If you can't look at a large number, it's no longer a screen.

Bradley: But a screen is only valuable if the information is there. Thinking back to a lot of the homologous recombinant mutants that people have generated, the expression pattern does not necessarily correlate with the mutant phenotype.

Hogan: Genes like c-*kit* and *W*, for example: they are both expressed extensively in very interesting patterns in the brain, but the homozygous deletion mutants don't seem to have any neurological defects.

Joyner: But again it could be redundancy, we don't know yet.

Balling: Would it be worthwhile determining the chromosomal integration sites?

Joyner: Yes, but that requires adding cloning and mapping into the screen.

Balling: You could do it simply by fluorescent *in situ* hybridization using the *lacZ* probe.

Joyner: That would be great!

Goodfellow: Can you use cell biology techniques directly without worrying too much about what the gene is? Can you start looking for suppressors or enhancers of the original pattern? If you have expression of β-gal in the cell, you have a reasonable way of selecting for mutants which affect the expression of β-gal in that cell. Then you could go on directly and ask whether that expression has perturbed the developmental pattern. Is that a way to get to enhancers and suppressors of genes?

Herrmann: But then you are looking for dominant mutations.

Joyner: Would you do this by mutagenizing the cells?

Goodfellow: You could either mutagenize or do a second transfection with a second marker colour. There must be something like β-gal which gives cells of a different colour.

Joyner: This would likely be a problem, since we are seeing a loss of inserted vectors at a frequency of one in a thousand.

Evans: Are you going to look in all three translation reading frames? You are doing a broad screen to try to catch everything you can of interest but you have restricted yourself to one third of those things.

Goodfellow: It is not a broad screen: at this stage of development many thousands of different transcripts are being expressed. That means you are just putting your hand in and taking out the few at random: why worry about how you put your hand in? The fact that you missed by one base pair and didn't pick up one gene rather than another is neither here nor there. If you ever hit the same gene twice, you should worry.

Joyner: Because we are looking at expression by X-gal staining we are restricted to one reading frame, but you could envisage doing it just by looking for the RNA transcripts using whole-mount RNA *in situ* analysis. Also, if a gene contains introns inserted in different reading frames, one of them is likely to be compatible with our vector.

Bradley: You are also pre-screening for expression.

Goodfellow: What can you say from the numbers you have at the moment? Can you estimate the number of genes which are expressed, the number of genes switched on, switched off? We know the total genome size, so we must know something about the target size. We have estimates of the numbers of genes there are and how much of the genome they take up.

Joyner: The problem comes back to what Rudi Jaenisch and I were talking about—whether or not the DNA inserts anywhere or only in certain regions.

Goodfellow: But you can put limits on that. If I told you there was random integration, what could you tell me about the number of genes that are expressed and the proportion that are switched on and switched off?

Joyner: I don't think we have enough data yet to make meaningful conclusions.

Evans: She will be able to answer your question properly when she starts hitting the same gene over and over again.

Hogan: No, because there are large genes, like muscular dystrophy, which you are more likely to hit twice than you are to make one hit in small genes like β-globin.

Jaenisch: Muscular dystrophy would not appear in the screen because it is not expressed in ES cells. Isn't it also worth considering a vector where the *neo* gene has its own promoter, so that one can obtain insertions into non-expressed genes? One could ask whether a given cell would generate a pattern after injection into a chimera. With the present approach you are missing the whole set of genes that are not expressed until after the inner cell mass stage. That is a severe restriction in your screen.

Joyner: That's where the enhancer trap has an advantage and it has been used, primarily by Achim Gossler, to identify such genes. The gene trap could potentially be used in a retroviral vector.

Beddington: In Janet Rossant's gene trap insertion that produces *lacZ* expression specifically in the notochord, floor plate and roof of the gut, if you look earlier, is *lacZ* expressed and then turned off everywhere but in the notochord or are you actually picking up artifactual expression in the ES cell that you don't see in the blastocyst or egg cylinder?

Joyner: For the gene D3-628 that was identified using the gene trap vector (Gossler et al 1989), Achim Gossler has done a careful analysis. It is expressed in ES cells and preimplantation embryos; it turns off soon after implantation and then on again at 8.5 days.

Beddington: Is the endogenous gene affected in that gene trap clone expressed in the blastocyst or are you looking at a hang over of 'tissue culture-specific' expression in ES cells?

Joyner: In the mice carrying the insertion there is some *lacZ* expression in the blastocyst, then it turns off and then back on.

Robertson: Francoise Poirier has screened an ES cell cDNA library; she found little correlation between the expression of genes in ES cells and their normal pattern of expression in the developing embryo. Two of the three genes she studied in detail turned out to be the reciprocal from what would have been predicted: they were very highly expressed in ES cells but not expressed in the early embryonic ecto-derm or inner cell mass but they are expressed in the trophectoderm.

Smith: Take my remark as one from an ignorant experimental embryologist who works on a tetraploid organism which has a generation time of 18 months! I need to be convinced a little more that the value of the resource is worth more than the effort. From what we have heard at this meeting you have in the mouse an awful lot of genes with quite well defined and interesting expression patterns. What you don't know is how the expression of these genes is controlled, or what they are doing in development. Shouldn't you sort these out before you start looking at another 500 genes?

Hogan: Particularly given that we don't have unlimited resources. Would money be better spent following up more *Wnt* genes or more tyrosine kinases? Or looking at existing mutants? There is a wealth of mutants in the Jackson lab, like *disorganization*, that have not yet been studied.

Jaenisch: You have to get to the gene of those existing mutations, which is often not easy.

McMahon: Alex and her colleagues hope to generate 500 new mutants or survey 500 ES cell lines in one year. It is unlikely that one can clone a gene that one knows very little about in that time.

Goodfellow: But Alex does not put them into the germline, so you haven't got the mutant phenotype of 500 genes in one year.

Hogan: Alex's screen is being done by nine people part time. You might say that resource could be better spent using the polymerase chain reaction to generate, from single short primers of arbitrary sequence, products which co-segregate with known mutations. The products could then be isolated and mapped in RI strains or interspecific backcross panels. This technique is being used by Joe Nadeau at The Jackson Lab.

Jaenisch: That will leave probably still megabases between the gene and the identified markers.

Goodfellow: This is trivial. The technology is coming on so fast that what it took two years to do a year ago now takes a few months. The cloning is not difficult, identifying genes is difficult.

Herrmann: I agree, the limiting factor is not the cloning but the genetics. Even if you construct a complete linkage map for the mouse genome at good resolution, you still have to do a cross with each individual mutation to localize the mutation on the chromosome map.

The other point is that by random integrations people are making transgenic mice and generating homozygotes and quite a few of the previously known mutants have been identified already. I think this is because a lot of these mutants are visible mutations. So if we continue this work, many of the genetically defined mutations will be identified within the next 10 years.

Also, it is important that everybody maps the genes they are finding by these insertion events. This will improve the map of the mouse genome a lot and will help correlating cloned genes with known mutations.

Balling: But if I want to isolate, for example the *tail kinks* mutation, because I am interested in the posterior half of the sclerotome, I can't wait until somebody gets an insertion mutation in that locus by chance. I have to go after it.

Bradley: I perceive part of the purpose of this work being to link some of these genes into genetic pathways and biochemical pathways. (This also comes back to the discussion about making some sort of coherent map of *in situ* hybridization patterns; see p 286.) I think that is one of the most important values of Alex's kind of effort. Whether 500 insertions will be sufficient, I don't know.

Joyner: Yes, you can hope to find genes that have similar expression patterns and then go after them. I would question the value of a mutant phenotype; I have very little confidence in that.

Buckingham: It seems to me that these are different things. Alex Joyner is talking about accessing or obtaining knowledge of genes which one would not have suspected otherwise, whereas what we do at the moment is look more closely at genes or phenotypes that we do suspect or know about. In the long term it is very important to access genes we don't have an idea about at present.

Joyner: We should use both approaches.

Goodfellow: There is another brute strength approach which is to do basically what Davor Solter has done: either screen one library against another library or just take sections from four or five stages in development and do *in situ* hybridization with random cDNA clones. You won't get the mutant, I accept that, but you could get a much larger sample of the number of possible gene expression patterns that may exist. In the time Alex and her colleagues will do 500, maybe they could do 5000.

Joyner: Can you make subtractive libraries so that you will not pick up the same gene several times?

Goodfellow: You don't do subtractive libraries, you do it by brute strength. You normalize your libraries and then screen every clone from one library against the other library. The technology enables you to do that now.

Joyner: Are those libraries really normalized?

Balling: There seems to be a lot of worry about the interpretation of the phenotype, whether it really represents a null allele or a complicated mutation and so on. What does anyone think about the necessity of making overlapping deletions in the genome? Maybe taking those insertions and treating them with deletion mutagens?

Jaenisch: I think this is very important, particularly with viral insertions, and I assume also for DNA insertions. The mutant we have studied in detail is the collagen mutation in Mov13 mice. We know the virus is expressed in certain lineages but not in others. The lethal phenotype is due to collagen not being expressed in the early embryos. In other lineages the virus is spliced out, permitting collagen gene expression. I think any proviral insertion could be similarly suppressed. *dilute* appears to be another example where some lineages are affected by the viral insert and some are not.

Two lessons can be learned from this; first, deletions are the preferable form of mutation to generate. Secondly, when you have a phenotype due to an insertion, the experiment is only complete once you have rescued the phenotype, preferably with a cDNA. Using the collagen mutations as an example, even after 10 years we have not been able to rescue the phenotype completely. There are many reasons, but one possibility is that the virus has affected another gene which is within the collagen gene but transcribed in the brain from the other strand. We can rescue the collagen mutation only up to birth by introducing a genomic clone into the mice, then they die. We have not been able to figure out why the mutants are only partially rescued.

Joyner: What's coming out is that nothing is perfect—homologous recombination, deletion, insertion. We have to take that into account and use more than one approach for each problem.

Jaenisch: It's most important that the molecular biology is as simple as possible, so I would stay away from DNA injection, then at least you don't have to worry about large deletions or genomic alterations, although if there are genes within genes, you can have a problem!

Evans: In mammals, there is a recurrent problem of a specific inability to screen for completely recessive embryonic lethals unless you do it by a technique such as Alex is describing or one like that described earlier by Liz Robertson (this volume) or unless you use the technique that Brigid Hogan is espousing of exploiting cross-homology with other species. If there are useful mutations to be found that would present as an embryonic recessive lethal but which have not been localized by cross-homology, then a screen for function or expression pattern is the only way they will be found.

McMahon: One other benefit of this approach is to provide marked populations of cells that might be useful in a variety of experiments on development. Would anyone like to elaborate on the usefulness of having marked cell lineages?

Joyner: One is for grafting experiments. The other is for analysing mutant phenotypes, so you can follow certain lineages and determine whether they are affected in the mutants.

Beddington: I agree. The only caveat is that, depending on what you are doing with it, you have to know whether the expression is heritable from cell to cell or whether it is a transient, tissue-specific phase of expression, which would mean you could not trace cells using gene trap expression as a marker. You may waste a certain amount of time chasing changes in expression patterns that do not reflect cell deployment.

Joyner: For grafting I meant that it would be useful to have a mouse that expressed *lacZ* in all cells.

Goodfellow: Would it be worthwhile trying to put more effort into designing dominant negative mutation trap systems (to put a lot of trendy words in one sentence!)? There are two approaches. One—I never understood why it doesn't work—is using retrovirus enhancers to switch genes on at random and get dominant mutations that way. If it works in tumours, why can't you use it in other systems?

Jaenisch: The ectopic expression of oncogenes gives a flurry of interesting or uninteresting phenotypes which are often not very informative. I would be very afraid of effects that you cannot interpret at all.

Goodfellow: The other approach is specifically trying to alter a biochemical phenotype in the cell and then see what the correlation of that is. We know there are proteins that bind to DNA and regulate gene expression. Can we design something which would perturb that regulation? If the answer is yes, then you have a handle on whatever is being perturbed. There are dimerization domains that we know are common to classes of transcription factors. They are potential targets and perhaps one could generalize that approach to different biochemical areas.

Buckingham: Peter, your strategy is for looking at the function of a gene which is already identified.

Goodfellow: No. We know enough about families of growth factors, for instance. Say we don't know if there is a gene working through the EGF receptor in a particular system. If we knock out the EGF receptor, we may pick up that gene. It is a more directed approach than the global approach. Do we know enough about biochemistry to take those types of approaches?

Beddington: If you try to design something that's generic, you may disrupt so many genes or parallel pathways that you end up with an uninterpretable developmental mess. I agree with Rudi Jaenisch that you will just add to the repertoire of completely uninterpretable phenomenology.

References

Bonnerot C, Grimber G, Briand P, Nicolas V-F 1990 Patterns of expression of position-dependent integrated transgenes in mouse embryos. Proc Natl Acad Sci USA 87:6331–6335

Friedrich G, Soriano P 1991 Promoter traps in embryonic stem cells: a genetic screen to identify and mutate developmental genes in mice. Genes & Dev 5: 1513–1523

Gossler A, Joyner AL, Rossant J, Skarnes WJ 1989 Mouse embryonic stem cells and reporter constructs to detect developmentally regulated genes. Science (Wash DC) 244:463–465

Robertson EJ, Conlon FL, Barth KS, Costantini F, Lee JJ 1991 Use of embryonic stem cells to study mutations affecting postimplantation development in the mouse. In: Postimplantation development in the mouse. Wiley, Chichester (Ciba Found Symp 165) p 237–255

Final discussion

Functional redundancy

McLaren: I would like to come back to the question of functional redundancy. People have said that if a gene is redundant it is unnecessary, so why has it been conserved in evolution. But surely, redundancy in this context is being used as in information science rather than as in sociology. In information transfer, redundancy is not unnecessary but crucial. As Shannon (1949) first realized, '. . . by sending the information in a redundant form the probability of errors can be reduced'. In other words, if you don't have redundancy, you are less likely to get correct transmission of your messages.

I presume that if two or three gene products are doing the same job biochemically, it is in a way comparable to the double assurance of classical experimental embryology. The more value each embryo has, the less you can afford in evolutionary terms for something to go wrong without having a back-up. A species that produces a very large number of embryos need bother less about redundancy. Flies produce many embryos and appear to have rather little genetic redundancy. Mammals produce fewer embryos and can't afford for many of them to die. If you looked at elephants and whales, where each embryo is of enormous value to the species, you would probably find even more redundancy. Does that make sense?

Goodfellow: There has to be an advantage to the individual.

McLaren: There has to be an advantage for the parents, so their germ plasm gets successfully transmitted. One needs only a minute selective advantage for something to get built in during the course of evolution.

Solter: If an embryo without redundancy is going to die, then very soon all those without redundancy will be gone. If there are embryos with one particular character and nine out of ten of those survive, and another group of embryos without the same character where eight out of ten survive, in time, the second group will die out.

McLaren: Redundant does not mean totally unnecessary; it means that 99% or 99.9% of the time the gene product will have no role, but when one copy falters, the other one is still there to carry you through. That is how redundancy is defined in information science.

Goodfellow: Why don't you see a high frequency of mutations in the other gene? The selection pressure against mutation in that gene would be very small, so you would be generating mutations all the time and you would expect to find quite a high frequency of mutations in, say, *engrailed-2* (*En-2*).

Hogan: But how do you know which is the right one? Where there are slightly overlapping patterns, one may be selected for one function and the other would be selected for the other.

Goodfellow: I am just making a straight prediction. If Anne's model is correct, you would expect to find a higher frequency of mutation in the population in those genes.

McLaren: Yes, I agree.

Jaenisch: Interleukin 2 is clearly a very important growth factor, but homozygous mice in which this factor has been knocked out are completely normal. So it looks like as if the haemopoietic system uses a network of these factors and they may substitute for each other. I am not sure whether the lack of phenotype is due to redundancy or whether it is compensation by some other pathway.

Krumlauf: If you put those homozygous mice into the wild, would they survive?

McMahon: There's no case I know of where two genes in the mouse have been shown to have wholly overlapping patterns of expression throughout development. They might appear to be redundant in one localized area but they have unique sites as well. It is the unique sites that are selected for.

For instance, there is good evidence from the patterns of expression that *Wnt-1* is the only member of the *Wnt* family to be expressed in the midbrain area, which is affected in the *Wnt-1* mutants, but its expression overlaps with that of another member of the family in the hindbrain and spinal cord. So there might be redundancy, as we define it, in those regions. The other family member has a unique site of expression somewhere else, and its function there is the one that is being selected for.

Copp: I don't think that you can explain the maintenance of redundancy on that basis.

McMahon: I am not explaining the maintenance of redundancy; I am explaining the evolution of new functions.

Copp: You might select for a gene on the basis of specific expression but surely over evolutionary time the expression in the overlapping site would be lost.

McMahon: Yes. That expression is not lost because it is useful, it is not actually redundant: it has a marginal role which is important.

Joyner: I think there are two things. There must be selection in the long run for the redundant expression pattern and then in the short run for the unique expression of the gene itself.

Hogan: If you have overlapping expression and unique expression, how do you get rid of the common expression? You can't select against it.

Joyner: En-2 could lose the enhancers for expression in the mid hindbrain but conserve the enhancers for the cerebellar expression. But it doesn't, fish have the same pattern of expression as mice, so the regulatory regions have been conserved.

Evans: Evolution of the genome in many cases occurs by reduplications and slippages. So we would expect to find numbers of copies of genes probably altered and partly evolved. Could we be seeing evolution in progress?

Joyner: But for *En-1* and *En-2*, fish, birds, amphibians and mice all express them in the mid hindbrain region.

When does an embryo become a fetus?

Kaufman: I have a straightforward question but I think the answer may be more difficult to interpret. With mice, when do you stop using the term embryo and start using the term fetus? In the human, the cut-off point was established as roughly eight weeks, when you first see ossification in the humerus. In the mouse that may be at about Day 14.5–15. There may be other cut-off points that people would wish to use. One possibility is when the total number of somites is achieved.

Would there be strong objections if we dropped the term fetus in the mouse? Could we just use the term embryo right the way through to term? I don't see a great advantage in saying that after Day 14 or 15 we should call it a fetus. I am trying to produce an atlas that people will find at least reasonably acceptable and this is one of the most contentious points of all.

McLaren: I certainly think that the transition between embryo and fetus in the postimplantation period is a continuous process, so little would be lost by using the same term throughout. More dubious is the use of the term embryo from the fertilized egg through to the postgastrulation period, since here a real discontinuity in meaning is involved. Initially, one is using embryo to mean the whole conceptus, including the 'extraembryonic' parts; then, from about the primitive streak stage onwards, one refers to the embryo as that subset of the conceptus that is going to develop directly into the fetus.

Saxén: Would you extend your suggestion to human embryos and fetuses as well?

Kaufman: The human work is now so well established that I don't think anyone would try to argue against what Carnegie scholars have been advocating for so many years.

Buckingham: For people who work on muscle, the tendency is to use 'embryonic' up to the period when innervation occurs and muscle masses begin to acquire distinct phenotypes; for later stages people use 'fetal'.

Lawson: Another, I thought generally accepted, transition from embryo to fetus terminology is the moment when all the organ primordia are evident, the moment when everything has been allocated. That is pretty good except for one or two exceptions, like the parotid gland which doesn't emerge until 14 days in the mouse, a few days after everything else.

McLaren: My own feeling about terminology is that it doesn't matter what you call things providing you make it abundantly clear at the outset what you mean by the term that you are using. Matt, I don't think that anybody is going to have any objection to your using the term embryo all the way through your atlas. We will all just be extraordinarily glad to see your atlas in print!

Reference

Shannon CE 1949 The mathematical theory of communication. University of Illinois, Urbana

Summary

I have enjoyed this symposium immensely and I have learnt a lot. Kirstie Lawson started with some very beautiful descriptive classical embryology but using the powerful modern techniques of cell labelling. Some of us, at least, came to understand the principles of cell lineage better. I finally realized that the cells at the tip of the egg cylinder at the beginning of gastrulation don't remain there but are shifted anteriorly by differential growth. The discussion also confirmed my suspicion that the mouse equivalent of the dorsal lip of the blastopore is not the primitive streak, but rather the structure that we used to call the archenteron, equivalent to Hensen's node in the chick, and that we resolved henceforth to refer to simply as the node.

Patrick Tam then introduced the idea that the germ layers, once induced, acquire their anteroposterior positional information by passing through the primitive streak. We had an interesting discussion on the comparison between mouse and *Xenopus*. Rosa's paper seemed to me to form the bridge between the descriptive embryology and the molecular approach. She discussed her *Hox-1.1* studies, looking at the difficult question of cell and tissue determination. Both she and then Bernard Herrmann described the beautiful *Brachyury* system. I suppose *Brachyury* must qualify as the first classical mouse mutant to have been isolated by reverse genetics.

The biggest single advance in developmental biology in the last ten years has been the isolation and analysis of numerous developmental regulatory genes. Many of them, like the homeobox genes and the paired box genes, were discovered in *Drosophila* and then their homologues isolated in mice. More recently, traffic has been going in the other direction, with the *Oct* genes and some of the *Wnt* genes first discovered in mice and then isolated in *Drosophila*. It seems to me that as work on transcription factors in mice gets more extensive and the ES gene traps get perfected, more regulatory genes are going to be isolated directly in mice rather than via other species. I may be wrong about that. Obviously we have only uncovered the tip of the iceberg so far in these regulatory families.

Then we were treated to elegant molecular analyses: of the development of muscle by Margaret Buckingham, and the skeletal system by Rudi Balling and David Wilkinson on the central nervous system. Mostly, these systems involved regulatory genes coding for DNA-binding proteins, probably transcription factors. It looked as though much of the future research effort in those areas was likely to go into establishing the upstream and downstream connections of those cell-autonomous genes, finding out how the regulatory genes are themselves regulated and with what DNA sequences their gene products interact. These are the hard tasks at present, but perhaps new technical approaches will be found before too long.

However sophisticated the understanding of DNA–protein and protein-protein interactions, chromatin conformation and so on finally becomes, development can only get a certain distance on cell-autonomous gene expression alone. Sooner or later, probably sooner, cell–cell interactions are going to become of overriding importance. So we moved on to talk about signalling molecules. We heard about signalling molecules for big T; we don't quite know what they are. We heard talk of mesoderm-inducing factor but it's not yet isolated in the mouse. The systems studied by Lauri Saxén and Irma Thesleff and also by Nigel Brown require signalling molecules, but again we don't know what they are. Brigid Hogan and Andy McMahon told us about putative signalling molecules: we don't yet know what signal they are transmitting and to whom. It seems to me that the developing embryo must be making use of all sorts of signals, including cell surface molecules, paracrine factors, morphogens including perhaps retinoic acid, and even rather vague environmental clues such as pH, ionic concentrations, perhaps electric fields. Of course, for every signalling molecule or signalling system there should be a receptor or receptors waiting to be cloned, so there's a lot of work still to be done.

Then finally we heard from Liz Robertson, Allan Bradley and Alex Joyner about three very different ES cell systems that allow genes to be analysed or modified, added or trapped. Those techniques are not easy but they have truly revolutionized and will continue to revolutionize the genetic analysis of development.

In conclusion, what of the future? I suppose eventually the plasticine models that we played with during the symposium will be replaced by 3D computer representations of the developing embryo from the fertilized egg on, stage by stage, with the facility to look at sections in any plane and to call up the database to find what genes are or are not expressed in any tissue or organ or domain. I was very happy that at this meeting we set in motion some activities towards that end. I am sure that within the next decade it will actually come about.

So what about the next Ciba Symposium in this area? First, because the field is growing so fast and advancing so rapidly, the next meeting will have to be in five years, not ten. Then I predict that it won't be just mice or even just mammals, it will embrace chick and frog and zebrafish too, because for some developmental problems those other systems offer much better and easier approaches than does the mouse. I doubt that it will address postimplantation development in general or genetic control of vertebrate development or molecular aspects of developmental genetics, or however you put those global terms together. It might focus on a particular process, such as gastrulation, or on a particular system such as the nervous system, or on the role of signalling molecules or transcription factors in development. Best of all, it might celebrate a hitherto unsolved problem such as axis formation or even the establishment of the germ cell lineage: that would really be exciting!

Index of contributors

Non-participating co-authors are indicated by asterisks. Entries in bold type indicate papers; other entries refer to discussion contributions.

Indexes compiled by Liza Weinkove

304

Subject index